Handbook of Optical Communications

Handbook of Optical Communications

Editor: Josephine Winston

NY RESEARCH
P R E S S

New York

Published by NY Research Press
118-35 Queens Blvd., Suite 400,
Forest Hills, NY 11375, USA
www.nyresearchpress.com

Handbook of Optical Communications
Edited by Josephine Winston

© 2019 NY Research Press

International Standard Book Number: 978-1-63238-679-3 (Hardback)

Cataloging-in-Publication Data

Handbook of optical communications / edited by Josephine Winston.
p. cm.
Includes bibliographical references and index.
ISBN 978-1-63238-679-3
1. Optical communications. 2. Telecommunication. I. Winston, Josephine.
TK5103.59.H36 2019
621.382 7--dc23

Contents

Preface

Optical communication is the process of using light as a medium of communication. The communication can be achieved by employing visual techniques or by using an electronic device. An optical communication system comprises of a transmitter, a channel and a receiver. The transmitter encodes a message into an optical signal, which is then carried through a channel and received by a receiver. The most commonly used channel for such communication is optical fiber. It commonly transmits infrared light through either laser diodes or light-emitting diodes. This book traces the progress of optical communication technology and highlights some of its key aspects and applications. It strives to provide a fair idea about this discipline and to help develop a better understanding of the latest advances within this field. Those in search of information to further their knowledge will be greatly assisted by this book.

This book is a result of research of several months to collate the most relevant data in the field.

When I was approached with the idea of this book and the proposal to edit it, I was overwhelmed. It gave me an opportunity to reach out to all those who share a common interest with me in this field. I had 3 main parameters for editing this text:

1. Accuracy – The data and information provided in this book should be up-to-date and valuable to the readers.

2. Structure – The data must be presented in a structured format for easy understanding and better grasping of the readers.

3. Universal Approach – This book not only targets students but also experts and innovators in the field, thus my aim was to present topics which are of use to all.

Thus, it took me a couple of months to finish the editing of this book.

I would like to make a special mention of my publisher who considered me worthy of this opportunity and also supported me throughout the editing process. I would also like to thank the editing team at the back-end who extended their help whenever required.

Editor

1

Challenges and Opportunities of Optical Wireless Communication Technologies

Isiaka Alimi, Ali Shahpari, Artur Sousa,
Ricardo Ferreira, Paulo Monteiro and
António Teixeira

Additional information is available at the end of the chapter

Abstract

In this chapter, we present various opportunities of using optical wireless communication (OWC) technologies in each sector of optical communication networks. Moreover, challenges of optical wireless network implementations are investigated. We characterized the optical wireless communication channel through the channel measurements and present different models for the OWC link performance evaluations. In addition, we present some technologies for the OWC performance enhancement in order to address the last-mile transmission bottleneck of the system efficiently. The technologies can be of great help in alleviating the stringent requirement by the cloud radio access network (C-RAN) backhaul/fronthaul as well as in the evolution toward an efficient backhaul/fronthaul for the 5G network. Furthermore, we present a proof-of-concept experiment in order to demonstrate and evaluate high capacity/flexible coherent PON and OWC links for different network configurations in the terrestrial links. To achieve this, we employ advanced modulation format and digital signal processing (DSP) techniques in the offline and real-time mode of the operation. The proposed configuration has the capability to support different applications, services, and multiple operators over a shared optical fiber infrastructure.

Keywords: atmospheric turbulence, bit error rate (BER) performance, channel characterization, ergodic capacity, free-space optical (FSO) communication, relay-assisted transmission, RF/FSO technology, scintillation index

1. Introduction

The Internet is experiencing high growth with varieties of bandwidth-intensive mobile applications on an unprecedented scale. One of the potential reasons for the growth is the Internet of Things (IoT) technologies that have brought exceptional revolutions into the number of devices in the network. Conceptually, IoT entails ubiquitous existence of a variety of *things* such as mobile phones, sensors, actuators, and radio-frequency identification (RFID) tags. These entities are capable of interacting with each other as well as cooperating with their neighbors in order to accomplish common goals via unique addressing scheme [1]. It is envisaged that by the year 2020, billions of devices with an average of six to seven devices per person will be connected to the Internet [2]. The fifth generation (5G) wireless communication systems in which millimeter-wave (mm-wave) and massive multiple-input multiple-output (M-MIMO) antenna technologies are expected to be integrated are the promising solutions for supporting the huge amount of anticipated devices. However, the radio-frequency (RF)-based wireless mobile technologies transmission speeds are limited by the available RF spectrum in the regulated RF spectrum. This is due to various innovative wireless technologies and standards like WiMAX (IEEE 802.16), UWB (IEEE 802.15), Wi-Fi (IEEE 802.11), iBurst (IEEE 802.20), the cellular-based 3G and 4G [3]. Moreover, because of various advanced technologies being employed in the optical communications, there have been considerable advancements in the optical system capacity, network reach, and number of supported users. For example, the optical-fiber-based broadband network architectures like fiber to the home (FTTH) and fiber to the building (FTTB) present commercial solutions to the communication barriers by progressively rendering services closer to the customers via the passive optical network (PON) technologies such as gigabit PON (GPON), 10Gbps PON (XG-PON), and Ethernet PON (EPON). Currently, one of the major challenges is the capability to support various service requirements so as to achieve elastic and ubiquitous connections [4]. Consequently, convergence of wireless and optical networks is highly essential for cost-effective and pervasive network penetration for the next-generation network (NGN). The convergence will help in exploiting the mobility benefit offered by the wireless connectivity and the inherent bandwidth provided by the optical systems. This will help in achieving the anticipated capacity and energy-efficiency objectives of the NGNs [3]. Furthermore, optical wireless communication (OWC) system is one of attractive broadband access technologies that offer high speed as well as improved capacity. Consequently, the OWC can attend to the bandwidth requirements of different services and applications of the NGNs at relatively low cost [3, 5].

The OWC can be an alternative and/or complementary technology for the current wireless RF solutions. For instance, OWC operating at 350–1550 nm wavelength band can offer high data rate of about 30 Gb/s data rate. This advantage makes it an attractive solution for addressing the prevailing "last mile" and "last-leg" problems in the access network. Furthermore, in mobile communication, resources re-use is an important requirement in order to enhance the network coverage and capacity. OWC technology is able to meet this requirement with the aid of spatial diversity [5, 6]. OWC link can be of different configurations such as

1. Directed line of sight (LOS)

2. Nondirected LOS

3. Diffuse

4. Quasidiffuse

5. Multispot LOS

Out of these configurations, the LOS links have the highest data rates, lowest bit error rate (BER) performance, and less complex protocol. These features make LOS link the extensively employed configuration in the outdoor applications. Nevertheless, the major deficiencies of the LOS link are lack of mobility and susceptibility to blockage. The diffuse and nondirected LOS configurations, on the other hand, give better mobility advantages and are less susceptible to shading. However, noise, path loss, and multipath-induced dispersion relatively hinder their achievable data rate for high-speed links. Intensity modulation/direct detection (IM/DD) is the most widely used scheme in OWC systems. Furthermore, coherent scheme can also be employed to enhance channel usage. Implementation of coherent scheme relatively improves system performance at the expense of increased system complexity. This can be attributed to the fact that, precise wave-front matching between the incoming signal and the local oscillator (LO) is required to guarantee efficient coherent reception. Furthermore, DD application is uncomplicated as just low-cost transceiver devices are required without the necessity for the intricate high-frequency circuit designs relative to coherent systems [5, 6].

1.1. Block diagram of OWC system

A terrestrial OWC system consists of the transmitter, channel, and receiver. **Figure 1** illustrates a schematic of a terrestrial OWC system. The source at the transmitter generates information waveforms which are modulated onto an optical carrier. The optical field produced is then radiated over the atmospheric channel to the destination. The optically collected field at the receiver is then transformed to an electrical current. The detected electrical current is processed in order to recover the original transmitted information [8]. However, the received information may not be an exact replica of the original transmitted information because of the transmission loss experienced over the channel by the signal. This factor significantly limits the performance of wireless communications systems.

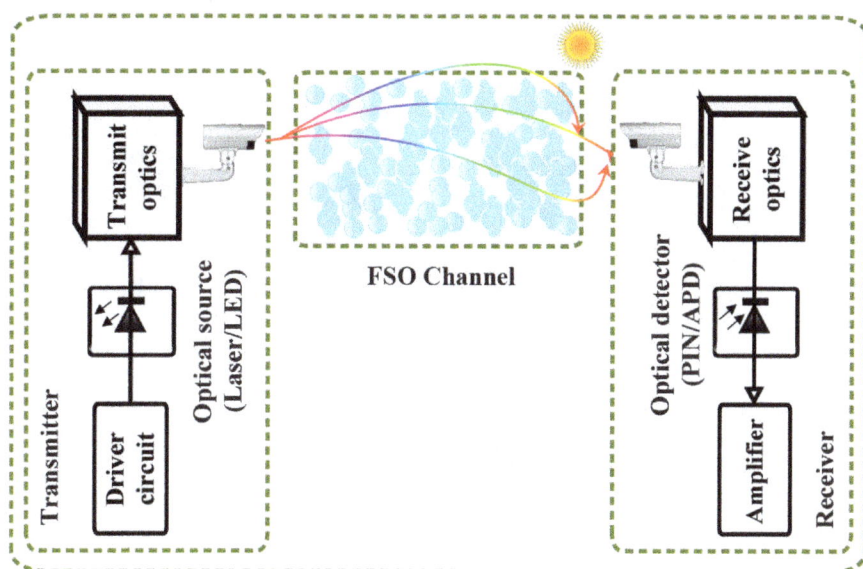

Figure 1. Block diagram of a terrestrial OWC system.

The transmission loss is mainly due to the resultant effects of scattering and absorption which are being introduced by the molecular constituents and aerosols along the transmission path. Therefore, scattering and wavelength-dependent absorption are the key components of atmospheric attenuation. Since absorption is a function of wavelength and wavelength selective, there are a range of wavelength windows that experience comparatively minimal absorptions. These transmittance windows in the absorption spectra of the atmospheric molecules are as shown in **Figure 2**.

In general, the wavelength ranges of 780–850 nm and 1520–1600 nm commonly used in the current OWC equipment are located in the atmospheric transmission windows where molecular absorption is negligible. This helps in mitigating the atmospheric absorption losses. Furthermore, certain wavelength windows that are located in the region of four specific wavelengths such as 850, 1060, 1250, and 1550 nm normally experience an attenuation of less than 0.2 dB/km. It is worth noting that the 850- and 1550-nm transmission windows coincide with the standard transmission windows of fiber communication systems. For this reason, majority of commercial OWC systems operate at these two windows in order to encourage the use of the available off-the-shelf components. Also, wavelengths like 10 μm and ultraviolet (UV) have also been considered for OWC systems. The 10-μm wavelength has better fog transmission characteristics, whereas the UV wavelength is more robust against impairment such as pointing errors and beam blockage. Also, the UV wavelength is less susceptible to solar and other background interferences [8].

Furthermore, it is worth noting that 1520–1600-nm wavelengths are compatible with erbium-doped fiber amplifier (EDFA) technology. This is highly essential in order to achieve high-power and high-data rate systems. Moreover, 1520–1600-nm wavelengths enable transmission of about 50–65 times more average output power than can be transmitted at 780–850 nm for a

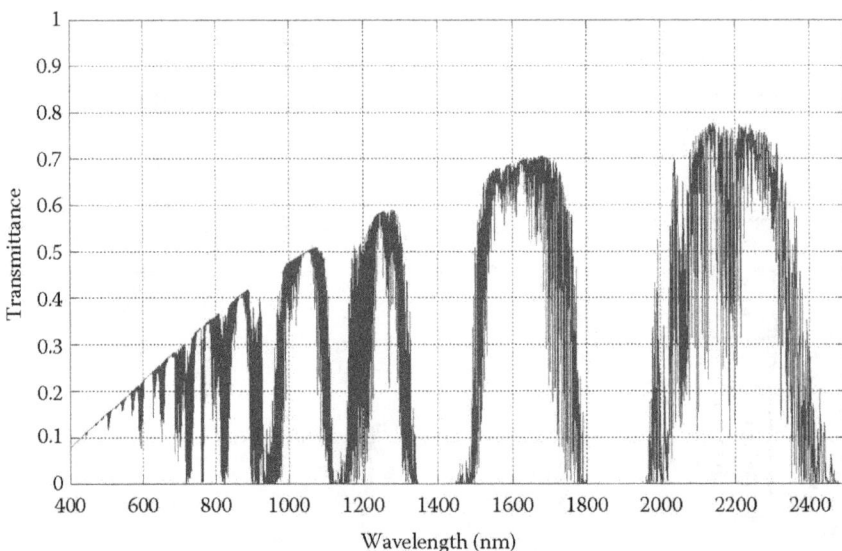

Figure 2. Atmospheric transmittance window with absorption contribution (adapted from Ref. [7]).

specified eye safety classification. This can be attributed to the low transmission of the human eye at these wavelengths [9].

1.2. Safety and regulations

One of the key factors for laser transmitter design is the safety issue. The infrared (IR) light sources can be likely safety threats to human if they are operated inappropriately. Also, exposure to certain optical beams may injure human skin and eye. However, the likely harm to the eye is comparatively more severe due to the eyes' ability to focus and concentrate optical energy. For instance, the eye can focus wavelength range of 0.4–1.4 µm on the retina with enough intensity to damage it; however, other wavelengths can to be absorbed by the front part of the eye before being focused on the retina. It should be noted that laser that is deemed to be "eye-safe" is also "skin-safe" [5, 9].

Moreover, it has been shown that the absorption coefficient at the front part of the eye is considerably greater for longer wavelengths (>1400 nm). Consequently, the permissible average transmission power for lasers operating at 1550 nm is relatively higher. Therefore, they are usually employed for longer transmission range [5, 9].

Figure 3 depicts the absorption of the eye at different wavelengths. At 700–1000-nm spectral range, the cost of optical sources and detectors are relatively low; however, the eye safety regulations are mainly strict. The maximum permissible exposure (MPE) at 900 nm wavelength is ~143 mW/sr. On the other hand, at longer wavelengths (≥1500 nm), the eye safety regulations are relatively less stringent; however, devices operating in these wavelengths are comparatively expensive. The guidelines on the safety of optical beams have been specified by several international standard organizations such as [5, 9]

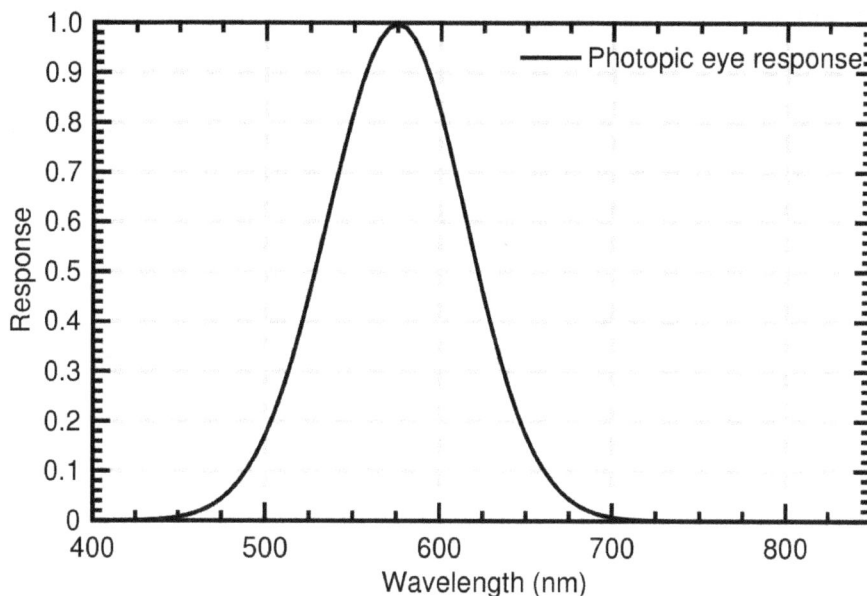

Figure 3. Response of the human eye at different wavelengths.

1.   International Electrotechnical Commission (IEC)

2.   Center for Devices and Radiological Health (CDRH)

3.   European Committee for Electrotechnical Standardization (CENELEC)

4.   American National Standards Institute (ANSI)

5.   Laser Institute of America (LIA)

The aforementioned organizations have established mechanisms for categorizing lasers in accordance with their type and power. Generally, the classification is based on four groups which are Class 1 through Class 4. Comparatively, Class 1 is the least powerful whereas Class 4 is the most powerful. Also, each of the classes is specified by the accessible emission limits (AELs) metric. The AEL is determined by the optical source wavelength, the emitter geometry, and the source intensity [5, 9]. Consequently, the AEL varies from one OWC category to another. In the subsequent section, we present the major OWC categories.

2. OWC system classification

There have been growing research interests in the OWC system as a viable solution to attend to the NGN requirements in cost-effective ways. The two generic groups of OWC are indoor and outdoor optical wireless communications. The unlimited bandwidth offered by the OWC can be attributed to different bands such as IR, visible (VL), and UV being employed for communication purposes. **Figure 4** shows the electromagnetic spectrum for different applications. Furthermore, the spectrum illustrates the frequency and wavelength ranges being occupied by the bands in OWC. The indoor OWC employs IR or VL light for an in-building wireless solution. It is of high importance especially in scenarios in which the probability of offering network connectivity through physical wired connections is challenging. Moreover, the indoor OWC systems can be categorized into four broad configurations such as tracked, diffused, nondirected LOS, and directed line of sight (LOS). Furthermore, the outdoor OWC employs optical carrier for transporting information from one point to another over an unguided channel that could be an atmosphere or a free space. So, this OWC technology is also known as a free-space optical (FSO) communication system. The FSO communication systems operate

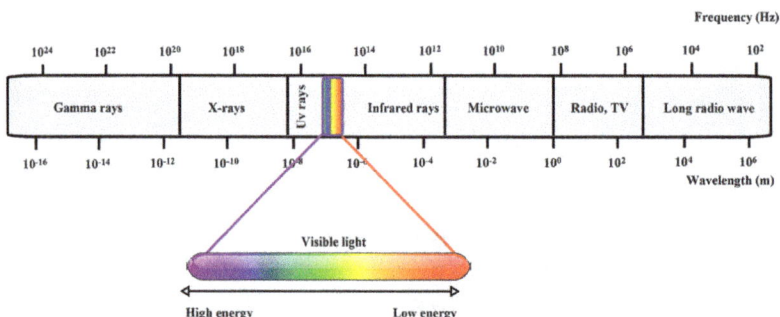

Figure 4. Electromagnetic spectrum.

at the near IR frequencies and are classified into terrestrial and space optical links. These consist of building-to-building, satellite-to-ground, ground-to-satellite, satellite-to-satellite, and satellite-to-airborne platforms (unmanned aerial vehicles [UAVs] or balloons) [10]. The tree diagram in **Figure 5** shows the OWC system classification.

2.1. Underwater optical wireless communications (UOWCs)

The underwater wireless communications are a process of transmitting data in unguided water environments via wireless carriers such as acoustic wave, RF wave, and optical wave. Compared to the RF or acoustic alternatives, UOWC offers higher data rate and transmission bandwidth. Basically, the UOWC uses optical wave as wireless carrier for an unguided data transmission. The UOWC systems are applicable in disaster precaution, offshore exploration, environmental monitoring, as well as military operations. Nevertheless, UOWC systems are susceptible to absorption and scattering which are normally created by the underwater channels. These conditions lead to severe attenuation of optical wave and eventually hindered the system performance. Different viable techniques have been presented in the literature to attend to the associated technical challenges of a UOWC. One of such is an underwater wireless sensor network (UWSN). **Figure 6** depicts a UWSN with aerospace and terrestrial communications.

The major entities in the UWSN are distributed nodes such as relay buoys, seabed sensors, autonomous underwater vehicles (AUVs), and remotely operated underwater vehicles (ROVs). The network entities are capable of performing tasks like processing, sensing, and communication in order to sustain collaborative monitoring of the underwater environment. The acquired data by the sensors that are located at the seabed are transmitted through acoustic/optical links to the AUVs and ROVs which in turn relay the signal to the ships, communication buoys, and other underwater vehicles. Furthermore, the onshore data center that is above the sea surface then processes the data and communicates with the satellite and the ships via RF/FSO links [11].

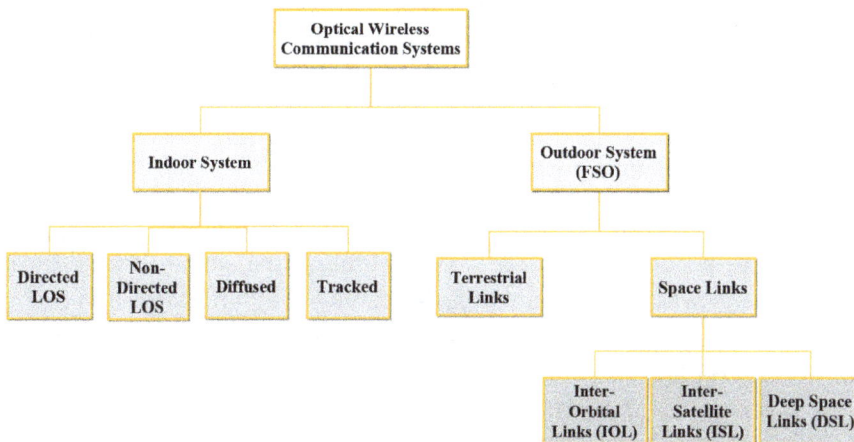

Figure 5. Optical wireless communication system classification (adapted from Ref. [10]).

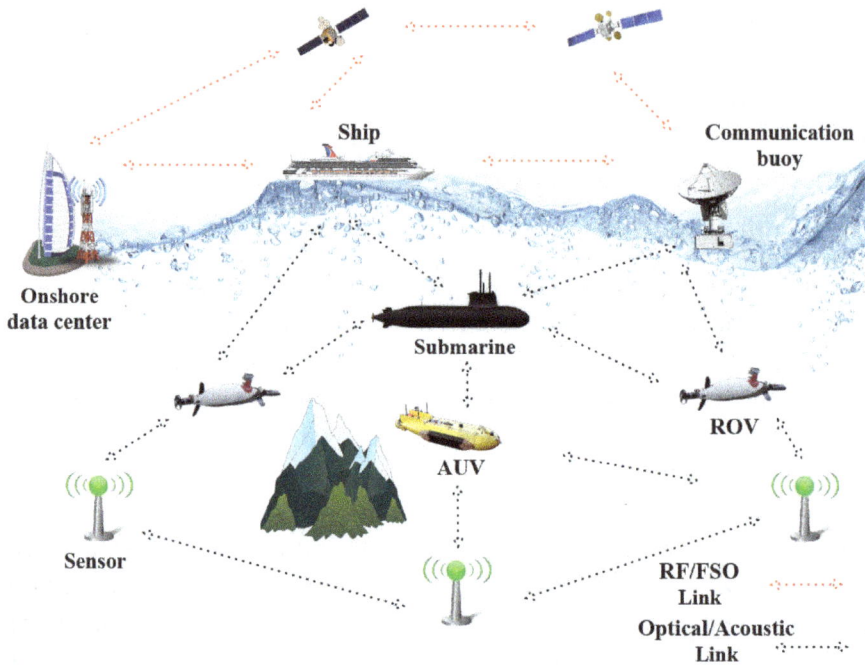

Figure 6. UWSN with aerospace and terrestrial communication.

2.1.1. UOWC link configuration

The four categories of UOWC based on the link configurations of the UWSN's nodes are as follows:

1. **Point-to-point line-of-sight (PTP-LOS) configuration**

   The PTP-LOS arrangement is the widely adopted link configuration in the UOWC. In this configuration, the receiver detects the light beam in the direct path to the transmitter. Due to the fact that light sources like lasers are generally used in the PTP-LOS-based systems, precise pointing between the transceiver is essential. Hence, the requirement limits the performance of UOWC systems in turbulent water environments. Also, this can even be more stringent when the transmitter and the receiver of the underwater vehicles are not stationary [11].

2. **Diffused LOS (D-LOS) configuration**

   Unlike the PTP-LOS scheme that employs light sources with narrow divergence angle, the D-LOS is a point-to-multipoint (PTMP) configuration that uses diffused light. The light sources like high-power light-emitting diodes (LEDs) that have large divergence angle are used for the UOWC transmission from one node to multiple nodes. The broadcast nature of this configuration helps in relaxing the requirement for precise pointing. Nonetheless, the scheme is prone to aquatic attenuation which is as a result of the large interaction area with the water. Consequently, this limits D-LOS application to comparatively short distances and lower data rate communications [11].

3. Retroreflector-based LOS (R-LOS) configuration

The R-LOS configuration is a special form of PTP-LOS scheme. The R-LOS scheme is appropriate for duplex UOWC systems in which the sensor nodes are expected to have low form factors and consume less power. Due to the fact that there is no laser or other light sources in the retroreflector end, its power consumption, weight, and volume are significantly reduced. However, the R-LOS performance is limited by the backscatter of the transmitted optical signal that may interfere with the reflected signal [11].

4. Non-line-of-sight (NLOS) configuration

Another category of UOWC that does not require alignment restriction of LOS is NLOS configuration. The concept of this configuration is that the transmitter launches the beam of light to the sea surface in such a way that the angle of incidence is greater than the critical angle. This ensures that the light beam experiences a total internal reflection. In this scheme, the receiver is directed toward the sea surface in a direction approximately parallel to the reflected light for appropriate signal reception. However, signal dispersion may occur due to the random sea surface slopes (i.e., induced by wind or other turbulence sources) that may reflect the light back to the transmitter [11].

2.1.2. Advantages and challenges of UOWC

The three means of realizing an underwater wireless communication for UWSNs are acoustics, RF, and optics [11]. We compare all schemes so as to point out some salient features and advantages of UOWC:

1. Acoustic communication

Acoustic communication is the widely employed method in the underwater wireless communication due to the fact that it is the most viable means of realizing the longest underwater link range. However, the frequency range of operation of acoustic communication is limited (between tens of hertz and hundreds of kilohertz). So, comparatively, this results in low transmission rate in the order of Kbps. Furthermore, the speed of sound wave in the fresh water is about 481 m/s at 2°C and in the salt water it is about 1500 m/s (1500.235 m/s at 1 MPa, 10°C and 3% salinity). These comparative slow propagation speeds make acoustic links prone to serious communication delay. Consequently, acoustic communication cannot be employed in applications that require large volume of data transmission in real time. Additionally, acoustic technology not only impacts marine life but also involves the use of costly, bulky, and energy-consuming transceivers [11].

2. RF communication

There are two main advantages of underwater RF communication. In relation to other UWC methods, the RF wave can achieve a smooth transition through the air/water interface. The feature enables a cross-boundary communication that unifies both terrestrial and underwater RF communication systems. It is remarkable that underwater RF communication is the method that is most tolerant to the water turbulence and turbidity. Nevertheless, the method is limited by the associated short link range. Likewise, the underwater RF methods are cost-ineffective due to the required high energy-consuming transceivers, costly and huge transmission antenna [11].

3. Optical communication

The outstanding technical merits of UOWC are the lowest link delay, highest communication security, highest transmission rate, and lowest implementation costs compared to other methods. The comparative high-speed benefit of UOWC makes it a promising candidate for real-time applications like underwater video transmission. In UOWC systems, LOS configuration is normally employed in the transmission. This helps in preventing eavesdroppers, and hence improves communication security. Besides, UOWC is the most cost-effective and energy efficient means of underwater wireless communication. This can be attributed to the comparatively small, low-energy consuming, and low-cost optical transceivers which are normally employed. This aids in system scalability and large-scale commercialization of UOWC [11]. However, UOWC also has certain associated challenges, which makes its realization highly demanding. The key challenges of UOWC are enumerated as follows:

a. Poor system performance

When light propagates in water, the photon interacts with the water molecules as well as other particulate matters inside the water. The interaction results in absorption and scattering which attenuate the transmitted optical signal and eventually bring about multipath fading. Consequently, in turbid water environment, UOWC experiences poor BER performance over just a few hundred meters link distance. Furthermore, the presence of matters like chlorophyll and other colored dissolved organic material (CDOM) leads to an increase in the turbidity of the water. So, they reduce the propagation distance of the transmitted optical signal. Besides, since the concentration of CDOM varies with ocean depth variations, the associated light attenuation coefficients also change. This anomaly results in an increase in the system complexity [11, 12].

b. Intermittent connection

In practice, owing to the narrow divergence feature of blue/green lasers or LEDs, they are usually employed as the light sources in UOWC systems. Therefore, the narrow feature demands a precise alignment between the optical transceivers. However, there is high tendency for intermittent misalignment of optical transceivers because of the turbulence in the underwater environment. Hence, sporadic movements of the sea surface will bring about severe connectivity loss [11, 12].

c. Energy constraint and system reliability

Factors such as the pressure, temperature, flow, and salinity of the seawater have high impact on the performance and lifetime of UOWC devices. Furthermore, UOWC devices are susceptible to failures due to corrosion and fouling. So, reliable underwater devices are required for effective communications. Also, due to the fact that exploitation of solar energy is extremely difficult and UOWC devices are expected to experience extended operation time, then, reliability of the batteries and power consumption efficiency of device are essential in a UOWC environment [11, 12].

2.2. Visible light communication (VLC) systems

The current enhancement of LED chip design with swift nanosecond-switching times and extensive deployment of LEDs for energy efficiency paves the way for visible light communication (VLC) system [13, 14]. So, the VLC system has become an attractive technology for addressing challenges such as energy efficiency, bandwidth limitation, electromagnetic radiation, and safety in wireless communications [15, 16]. The VLC system operates in the wavelength range of ~390–750 nm. **Table 1** shows the seven colors with specific frequency and wavelength for visible spectrum. **Figure 7** illustrates VLC system implementation. The concurrent support for communication and illumination by the VLC offers the following advantages over the RF communications.

Color	Wavelength (nm)	Frequency (THz)
Violet	440–405	680–740
Indigo	455–440	660–680
Blue	475–455	630–660
Green	560–475	535–630
Yellow	585–560	515–535
Orange	605–585	495–515
Red	685–605	435–495

Adapted from Ref. [17].

Table 1. Visible spectrum colors.

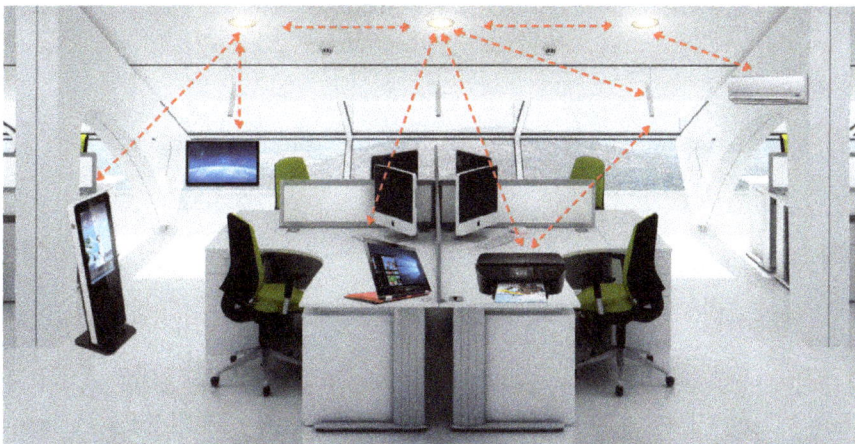

Figure 7. Visible light communication system.

2.2.1. Huge bandwidth

It exhibit almost unlimited and unlicensed bandwidth which approximately ranges from 380 to 780 nm. Therefore, VLC has 350 THz that can support multi-gigabit-per-second data rates with LED arrays in a multiple-input multiple-output (MIMO) configuration [14]. This makes VLC a good alternative to the indoor IR that operates at 780–950 nm for the access technologies [6].

2.2.2. Low power consumption

VLC provides both communication and lighting, giving Gbps data rates with only unsophisticated LEDs and photodetectors (PDs) that consume low power compared to costly RF alternatives that demand high power consumption for sampling, processing, and transmitting Gbps data [14].

2.2.3. Low cost

The required optical components such as LEDs and photodetectors are inexpensive, compact, lightweight, amenable to dense integration, and have very long lifespan [13]. Moreover, with large unlicensed optical spectrum as well as much lower power-per-bit cost compared to the RF communications, VLC is relatively cheaper.

2.2.4. No health concerns

VLC does not generate radiation that leads to public health concern. Besides, it lowers the carbon dioxide emission owing to the little extra power consumption for communication purposes [18].

2.2.5. Ubiquitous computing

Due to the fact that there are various luminous devices like traffic signs, commercial displays, indoor/outdoor lamps, TVs, car headlights/taillights, and so on being used everywhere, VLC can be employed for a wide range of network connectivity [19].

2.2.6. Inherent security

VLC offers comparatively higher security due to the fact that it is highly intricate for a network intruder that is outside to pick up the signal [14].

2.2.7. Indoor localization

The existing RF-based global positioning system (GPS) gives inadequate or no network coverage in the indoor and underground (e.g., tunnel) environment. This is as a result of high attenuation, multipath, and the safety regulation. These factors lead to an accuracy of only up to a few meters for the RF-based GPS. The VLC-based indoor positioning can be employed to attend to the issues in the enclosed environments. So, the VLC-based indoor navigation services offer very high accuracy to within a few centimeters. In essence, VLC offers good indoor localization system using the white LEDs. Furthermore, LEDs give better light source that is more than 400 lux. This is sufficient for high-speed data transmission compared to the incandescent and fluorescent sources. Moreover, LEDs have longer lifespan that results in ecological and financial benefits [6].

In addition, VLC is an alternative technology in sensitive or hazardous environments like airplanes, hospitals, and industrial gas production plants where the employment of RF technology is not permitted. The worldwide research community through bodies like the Wireless World Research Forum, the IEEE standardization body, the VLC consortium, the UK research council, and the European OMEGA project has embraced the indoor short-range VLC as a promising scheme because of the associated excellent attributes [6].

However, simultaneous employment of light sources for the data communication as well as illumination causes certain challenges that require consideration for VLC system implementations to be viable. Flicker mitigation and dimming support are the two major challenges of visible light spectrum [14, 20]. In the contemporary lighting systems, the light sources are equipped with dimming control functionality that enables the users to control the average brightness of light sources to their preferred level. Flicker is the variation in the brightness of light perceived by the human naked eye. Flicker is as a result of continuous switching on and off of the light source during data transmission. It is essential to mitigate any possible flicker because it can instigate negative/harmful physiological changes in humans. Flicker can be prevented by making the changes in brightness be within the maximum flickering time period (MFTP). The MFTP is the maximum time period within which the light intensity can be changed without any perception by the human eye [14]. Also, various modulation formats have been recommended for VLC in view of dimming control and flicker mitigation. For instance, the IEEE 802.15.7 standard proposes variable pulse position modulation (VPPM) for VLC system because of its notable ability to control dimming. It integrates the pulse position modulation (PPM) and pulse width modulation (PWM) in order to support communication with dimming control [14, 20]. Furthermore, other challenges are high path losses, multipath-induced intersymbol interference (ISI), artificial light-induced interference, and blocking. Moreover, LED electro-optic response nonlinearity has to be taken into consideration [6].

2.3. Wireless body area network (WBAN)

Wireless body area network (WBAN) is a system that comprises a set of miniaturized low-power, lightweight sensor nodes, which form wireless sensor networks (WSNs). **Figure 8** shows a WBAN system for medical monitoring. With the help of the sensor nodes, WBANs have emerged as an attractive alternative to the conventional wired medical network. Also, there has been noteworthy increase in the WBAN systems because of the IEEE 802.15.6 standard that regulates their commercial applications. Furthermore, there is an obligation on 2.36–2.40-GHz frequency band as a medical-only WBAN band by the Federal Communications Commission (FCC). The restriction is purposely for service provisions for the indoor health-care facilities as well as for supporting the patients' health-care information and management [21, 22]. Furthermore, the constraint is also to guarantee high quality of service for the health information transmission.

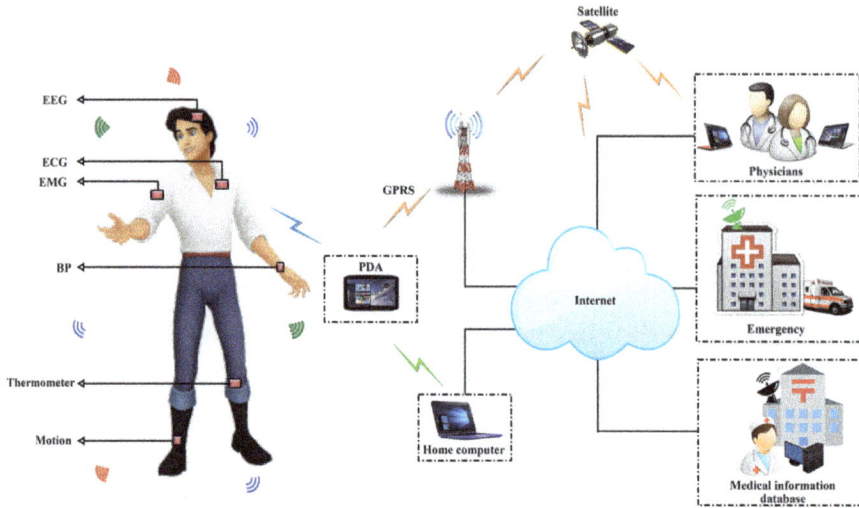

Figure 8. Wireless body area network.

The distributed sensors in the WBAN can be implanted in or on the human body in order to monitor physiological parameters in real time. The implanted sensors that are wirelessly connected to the outside network through a central unit collect different vital health information. The monitored physiological data include electromyogram (EMG), electroencephalogram (EEG), electrocardiogram (ECG), temperature, heart rate, blood pressure (BP), and glucose level [22]. The ZigBee, Bluetooth, as well as the current Bluetooth Low Energy (BLE) are contending for market share of wireless health devices. Their major appealing advantages are low power consumption and the added mobility [23]. Health devices that employ these technologies are operating in the industrial, scientific, and medical (ISM) radio bands. High emissions from these devices can create electromagnetic interference (EMI) and eventually disrupt communication. Similarly, there are security issues concerning data transmission for patient monitoring which are susceptible to hacking [23].

Furthermore, it should be noted that the existing WBANs that use ultra-wide band (UWB) transmissions are RF based. However, their implementation in the hospitals and medical facilities where RF-based system deployment is restricted or prohibited can be challenging. This is due to the potential effects of EMI from various RF transceivers on medical devices. The EMI effects can lead to medical equipment malfunction. Also, RF wave propagation on and/or in the human body is highly complex to examine. Consequently, to address these challenges OWC can be employed as an alternative solution [24].

The ECG signal and patient information can be transmitted concurrently with the help of VLC technology. Moreover, certain medical equipment like the one for the cardiac stress test (or cardiac diagnostic test) can be improved on by incorporating LEDs on the sensor units. This implementation will help in minimizing the large amount of cables (e.g., electrodes) that are normally required. Besides, VLC employment is greener (green communication and networking), safer, more secure in RF-restricted/prohibited hospitals and medical facilities [23]. Furthermore, it is worth noting that the current advancements in the organic LED (OLED) technology enable the integration of VLC transceivers into wearable gadgets and clothing [24].

2.4. Optical space communications

An effective communication links between satellites enable better flexibility, extended coverage, and improved connectivity to be achieved in satellite systems. This is applicable in interorbit links between satellites in low earth orbit (LEO) and geostationary orbit (GEO). Furthermore, intersatellite links between satellites in the same GEO or LEO orbit are other scenarios for application. Also, the satellite system connectivity can be enhanced by exploiting the free-space links between satellites. This will result in an improved capacity for telecommunication systems. Moreover, the capability to relay data from the earth observation satellite to the ground through a GEO relay satellite enables real-time data flow and minimizes the number of ground stations required for service delivery in the system [25].

Generally, the space link implementation can be realized at microwave, millimetric, and optical frequencies. Also, all technologies require the communication beam that emanates from the transmitting terminal to be pointed toward the receiving terminal with sufficient accuracy. This is necessary in order to meet the required link power budget. **Table 2** presents typical antenna gains and beamwidths for intersatellite link technologies [25].

It should be noted that out of the technologies, optical frequency offers an exceptionally high antenna gain for comparatively small antenna size. Therefore, optical links provide significant advantages such as low mass, low power consumption, and reduced size. As demonstrated in [26], the optical antenna diameter may be relatively smaller than that of RF by a factor of 13. Also, its mass and power can be half of that for RF systems. **Table 3** compares different communication technologies in terms of mass, power, and antenna diameter. An architecture that depicts ground-to-satellite optical links that are connected to satellite network and then satellite-to-ground optical links is illustrated in **Figure 9**. The salient features of optical links enable small terminals that can be easily accommodated on the satellite [25]. Also, optical space-based communication offers high-data rate, large capacity, minimized interference risk with other communication systems, and efficient utilization of frequency resources [10, 26].

Nevertheless, the space FSO communications are susceptible to atmospheric effects which make the channel a random function of space and time. Consequently, the uplink/downlink transmissions suffer from losses such as atmospheric scintillation, beam divergence, absorption, scattering, misalignment, cloud blockage, background noise and angle-of-arrival fluctuations. It should be noted that the losses confronted by the beam in the FSO uplink transmission

Band	Parameters			
	Frequency	Antenna diameter (m)	Antenna gain (dBi)	3 dB beamwidth degrees
S-band	2 GHz	2	30	5.25
Ka-band	26 GHz	2	52	0.40
Millimeter	60 GHz	1	53	0.35
Optical	0.36 PHz	0.07	108	6.7×10^{-4}

Adapted from Ref. [25].

Table 2. Intersatellite link technologies comparison.

Link	Frequency band								
	Optical			Ka			Millimeter		
	Antenna diameter	Mass	Power	Antenna diameter	Mass	Power	Antenna diameter	Mass	Power
GEO-LEO	10.2 cm (1.0)	65.3 kg (1.0)	93.8 W (1.0)	2.2 m (21.6)	152.8 kg (2.3)	213.9 W (2.3)	1.9 m (18.6)	131.9 kg (2.0)	184.7 W (2.0)
GEO-GEO	13.5 cm (1.0)	86.4 kg (1.0)	124.2 W (1.0)	2.1 m (15.6)	145.8 kg (1.7)	204.2 W (1.6)	1.8 m (13.3)	125.0 kg (1.4)	175.0 W (1.4)
LEO-LEO	3.6 cm (1.0)	23.0 kg (1.0)	33.1 W (1.0)	0.8 m (22.2)	55.6 kg (2.4)	77.8 W (2.3)	0.7 m (19.4)	48.6 kg (2.1)	68.1 W (2.1)

Adapted from Ref. [26].

Table 3. Comparison of different technologies with transmit power of 10, 50, and 20 W for optical-, Ka-, and millimeter-band systems, respectively.

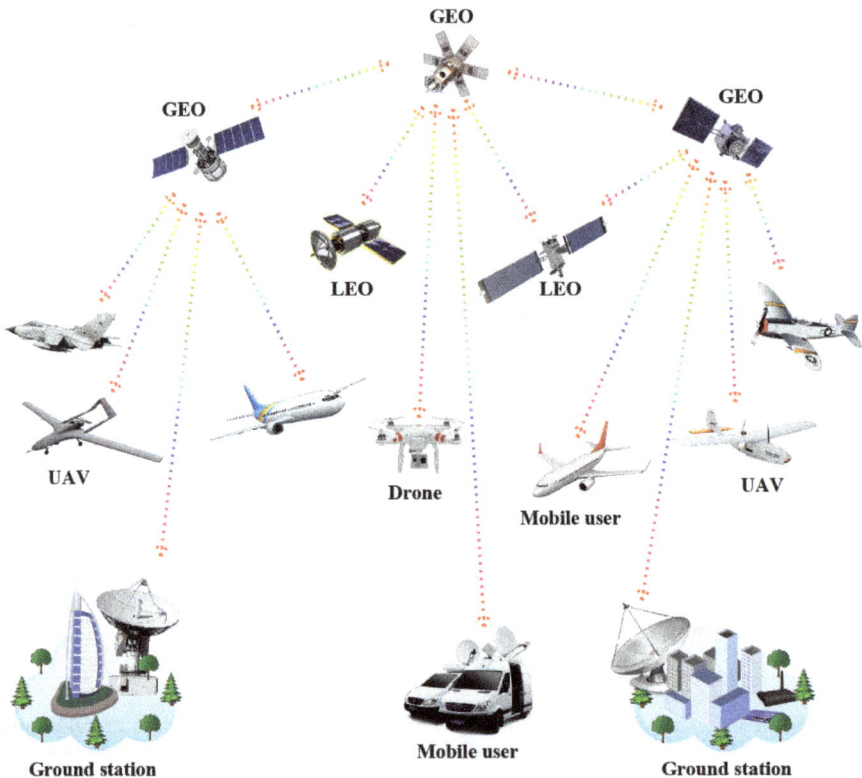

Figure 9. Space optical communication.

are very large compared to downlink transmission. Furthermore, the intersatellite FSO links are insusceptible to weather conditions due to the fact that the satellite orbits are at a considerable distance over the atmosphere [10]. However, the associated narrow optical beamwidths bring about stringent pointing, acquisition, and tracking (PAT) requirements for optical systems. This can also be attributed to different relative velocity of the satellite. So, a compact low-mass PAT system is required to address PAT requirements [26].

2.5. Terrestrial free-space optical (FSO) communications

There have been much more research effort in terrestrial FSO partly because of some successful field trials and commercial deployments [27–31]. **Figure 10** shows a scenario for FSO system deployment as a universal platform for a nippy as well as an efficient ubiquitous wireless service provision for the future broadband access networks. The significant attentions being attracted by the FSO systems are primarily due to their inherent advantages such as cost-effectiveness, lower power consumption (high-energy-efficiency-green communication), ease of deployment, higher bandwidth/capacity, more compact/low-mass equipment, reduced time-to-market, immunity to EMI, high degree of security against eavesdropping, license-free operation, as well as better protection against interferences, compared with the traditional RF communication systems [32, 33]. These salient features make FSO communication systems very appealing for a variety of applications in disaster recovery, radio astronomy, remote-

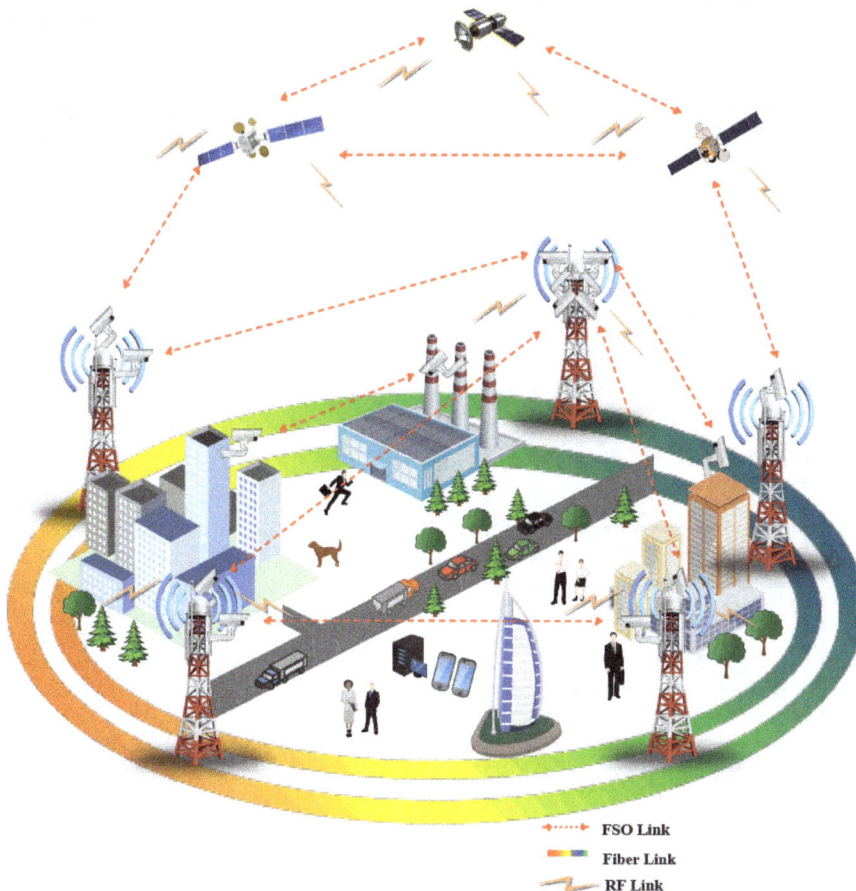

Figure 10. A scenario for OWC system deployment for access networks.

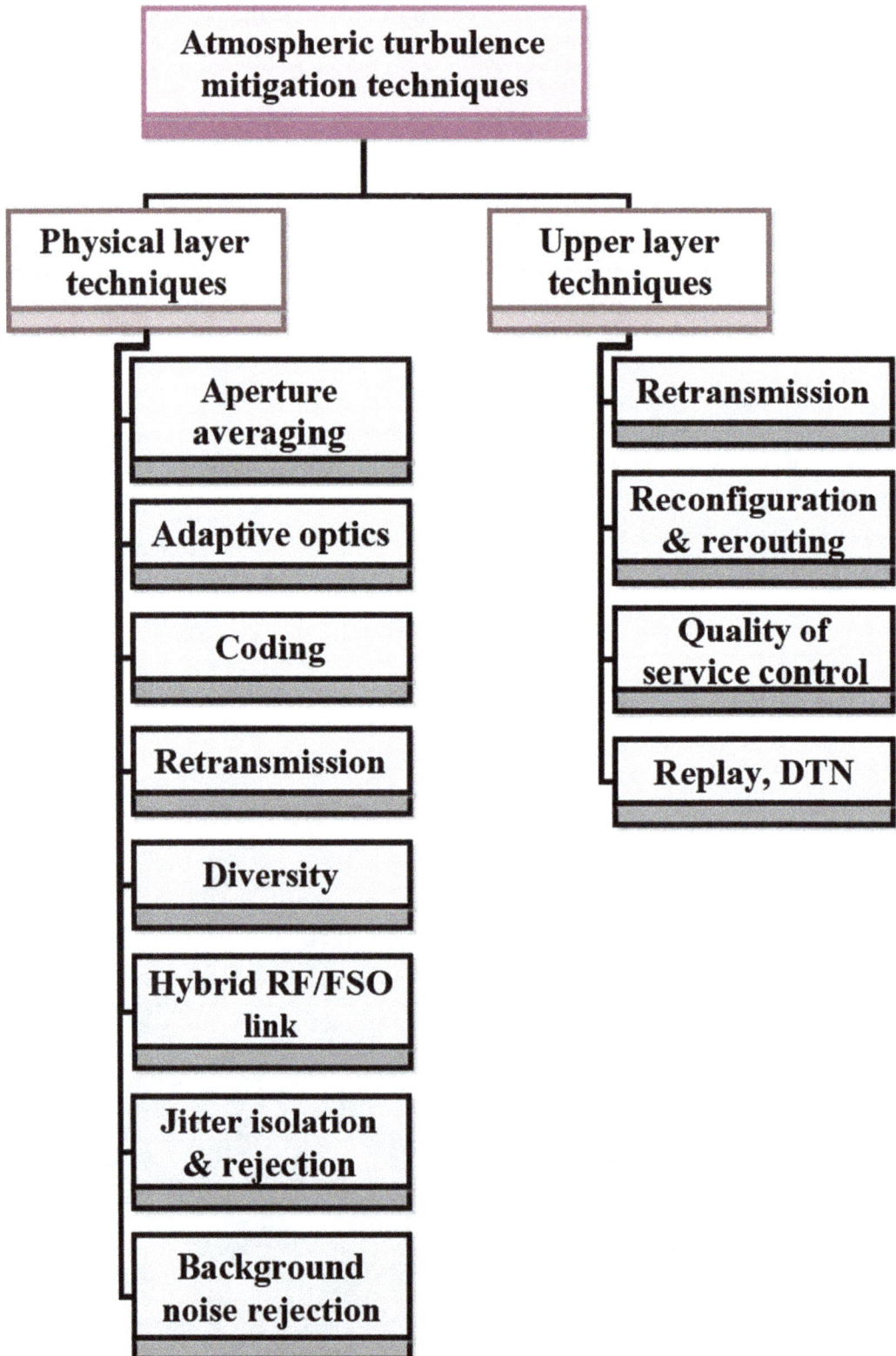

Figure 11. Atmospheric turbulence mitigation techniques (adapted from Ref. [10]).

sensing/surveillance/monitoring, metropolitan area network extension, high-definition TV transmission, sharing of medical imaging in real time, and fronthaul and backhaul for wireless cellular networks [3, 8]. Moreover, apart from being used for establishing terrestrial links, they are applicable for launching high-speed interplanetary space links such as intersatellite/deep space and ground-to-satellite/satellite-to-ground links [10].

In spite of the advantages of FSO communication and diverse application, its extensive use is hindered by some challenges in real-life scenarios. For instance, the FSO links are susceptible to scattering caused by adverse weather conditions like snow, rain, and fog [8, 33]. Moreover, building sway caused by factors such as thermal expansion, wind loads, and weak earthquakes also impairs the FSO link performance [33]. Also, atmospheric turbulence-induced fading has been recognized as the main contributor of the FSO link impairment [33, 34]. Consequently, the system performance is impeded as a result of atmospheric effects which cause loss of spatial coherence, beam spreading, and temporal irradiance fluctuation known as scintillation or fading [8, 35]. Scintillation manifests as temporal and spatial variation in the light intensity along the transmission path. This is due to the random changes in the refractive index which is as a result of the inhomogeneities in the air temperature and pressure [8, 32]. For these reasons, the dispersive nature of the link needs considerable attention when modeling an FSO system. This will help in supporting the stringent requirements of various bandwidth-intensive mobile applications of the NGNs [3]. The FSO link impairment modelings are discussed in Section 3. The tree diagram in **Figure 11** presents different atmospheric turbulence mitigation techniques.

3. Optical system and channel model

Assuming a practical FSO link with intensity-modulation/direct-detection (IM/DD) using OOK modulation, the data are modulated onto the instantaneous intensity of an optical beam at the transmitter. The optical power emanated from the transmit aperture into the free space is assumed to be affected by factors such as misalignment fading (pointing errors), atmospheric turbulence-induced fading, and background noise or ambient noise before reaching the receive aperture. These factors lead to the signal intensity fluctuation. Consequently, the subsequent received electrical signal, r, at the receive aperture can be modeled as [3, 35–37]

$$r = \eta_e h x + n,$$ (1)

where $x \in \{0,1\}$ denotes the transmitted information bit, η_e is the effective photoelectric conversion ratio of the receiver, n is additive white Gaussian noise (AWGN) with zero mean and variance, $\sigma_n^2 = N_0/2$, N_0 is a one-sided noise power spectral density in watts/Hz, and $h = h_\ell h_a h_p$ represents the irradiance that influences the channel state. The channel irradiance in Eq. (1) is a product of the deterministic path loss, h_ℓ: the random attenuation (i.e., atmospheric turbulence-induced fading), h_a, as well as the random attenuation (due to geometric spread and pointing errors), h_p. The h_a and h_p are random variables with probability density functions (pdfs) $f_{h_a}(h_a)$ and $f_{h_p}(h_p)$, respectively.

3.1. Atmospheric attenuation

When an optical beam passes through the atmosphere, it experiences atmospheric loss. The attenuation suffered by the signal power according to the exponential Beers-Lambert law is given by [36–38]

$$h_\ell(\lambda, z) = \frac{P(\lambda, z)}{P(\lambda, 0)} = \exp\left(-\sigma(\lambda)z\right) \tag{2}$$

where $h_\ell(\lambda, z)$ denotes the loss that is a function of propagation path of length z at wavelength λ, $P(\lambda, z)$ and $P(\lambda, 0)$ are the signal power and the emitted power at distance z, respectively. The attenuation coefficient or the total extinction coefficient, $\sigma(\lambda)$ per unit of length, is given by [36]

$$\sigma(\lambda) = \alpha_m(\lambda) + \alpha_a(\lambda) + \beta_m(\lambda) + \beta_a(\lambda), \tag{3}$$

where $\alpha_{m,a}$ represent molecular and aerosol absorption coefficients, respectively, and $\beta_{m,a}$ denote molecular and aerosol scattering coefficients, respectively.

The attenuation h_ℓ is assumed to be a constant scaling factor over a long time period, so, there is no haphazard behavior. Moreover, it is subject to the distribution and size of the scattering particles as well as the wavelength employed. It can be expressed in terms of visibility, which can be measured directly from the atmosphere. Empirically, attenuation can be defined in terms of visibility as [36]

$$\sigma(\lambda) = \frac{3.912}{V}\left(\frac{\lambda}{550}\right)^{-q}, \tag{4}$$

where V is the visibility (in kilometers) and q is a parameter that depends on the particle size distribution in the atmosphere expressed by the Kruse model as [36]

$$q = \begin{cases} 1.6 & V > 50\,\text{km} \\ 1.3 & 6\,\text{km} < V < 50\,\text{km} \\ 0.585 V^{1/3} & V < 6\,\text{km} \end{cases} \tag{5}$$

Furthermore, Kim presents an extended model in order to achieve a better accuracy at lower visibility scenarios. The Kim model is expressed as [36]

$$q = \begin{cases} 1.6 & V > 50\,\text{km} \\ 1.3 & 6\,\text{km} < V < 50\,\text{km} \\ 0.16V + 0.34 & 1\,\text{km} < V < 6\,\text{km} \\ V - 0.5 & 0.5\,\text{km} < V < 1\,\text{km} \\ 0 & V < 0.5\,\text{km} \end{cases} \tag{6}$$

3.2. Pointing error or misalignment fading

An FSO link is an LOS communication with narrow optical beamwidth that brings stringent pointing accuracy requirements for efficient performance and reliability of optical systems.

Pointing errors and signal fading normally occur at the receiver because of the wind loads and thermal expansions that lead to random building sways. Assuming a Gaussian spatial intensity profile of beam waist w_z at the receiver plane, located at distance z away from the transmitter. Also, suppose a circular aperture of radius r, the collected power fraction owing to the geometric spread with radial displacement a, from the detector origin can be approximated as the Gaussian form [37, 38]

$$h_{\mathrm{p}}(a) \approx A_0 \exp\left(-\frac{2a^2}{w_{z_{eq}}^2}\right) \tag{7}$$

where $w_{z_{eq}}^2 = w_z^2 \sqrt{\pi}\,\mathrm{erf}(v)/2v \exp\left(-v^2\right)$, $v = \sqrt{\pi}r/\sqrt{2}w_z$, $A_0 = [\mathrm{erf}(v)]^2$, $w_{z_{eq}}$ is the equivalent beamwidth, and $erf(\cdot)$ denotes the error function given by [39]

$$\mathrm{erf}(x) = (2/\sqrt{\pi})\int_0^x e^{-u^2}\,du \tag{8}$$

If we assume that the elevation and the horizontal sway are independent and identical Gaussian distributions, then the radial displacement, a, follows a Rayleigh distribution. Then, the $f_{h_{\mathrm{p}}}(h_{\mathrm{p}})$ can be defined as [37, 38]

$$f_{h_{\mathrm{p}}}(h_{\mathrm{p}}) = \frac{\gamma^2}{A_0^{\gamma^2}} h_{\mathrm{p}}^{\gamma^2-1}, \quad 0 \le h_{\mathrm{p}} \le A_0 \tag{9}$$

where $\gamma = w_{z_{eq}}/2\sigma_s$ represents the ratio between the equivalent beam radius and the standard deviation (jitter) of the pointing error displacement at the receiver and σ_s^2 is the jitter variance at the receiver.

3.3. Atmospheric turbulence

The intensity fluctuation over the FSO channel has been defined by several statistical models in the literature for different turbulence regimes. One of such model is the log-normal (LN) distribution that has been extensively employed because of its significant match with the experimental measurements. This is discussed further in Section 6. Some of other widely employed models are gamma-gamma ($\Gamma\Gamma$), negative exponential, K distribution, and I-K distribution. However, in this work, we focus on the LN, $\Gamma\Gamma$, and generic Málaga (M)-distribution models.

3.3.1. Log-normal distribution (LN)

In general, the LN model is only suitable for weak turbulence conditions and for link range that is less than 100 m [40]. So, the intensity fluctuation pdf for the weak turbulence modeled by the LN distribution is given by [3, 35]

$$f_{h_a}(h_a) = \frac{1}{2h_a\sqrt{2\pi\sigma_x^2}} \exp\left(-\frac{(\ln(h_a) + 2\sigma_x)^2}{8\sigma_x^2}\right), \tag{10}$$

where $\sigma_x^2 = \sigma_l^2/4$ represents the log-amplitude variance defined for plane wave and spherical waves, respectively as [3, 35]

$$\sigma_x^2|_{\text{plane}} = 0.307C_n^2 k^{7/6} L^{11/6}, \tag{11}$$

$$\sigma_x^2|_{\text{spherical}} = 0.124C_n^2 k^{7/6} L^{11/6}, \tag{12}$$

$$\sigma_l^2|_{\text{plane}} = 1.23C_n^2 k^{7/6} L^{11/6}, \tag{13}$$

$$\sigma_l^2|_{\text{spherical}} = 0.50C_n^2 k^{7/6} L^{11/6}, \tag{14}$$

where σ_l^2 is the log-irradiance variance, $k = 2\pi/\lambda$ represents the optical wave number, L denotes the distance, and C_n^2 is the altitude-dependent index of refraction structure parameter. The C_n^2 is a main parameter for distinguishing the amount of refractive index fluctuation in the atmospheric turbulence. It is a function of the atmospheric altitude, wavelength, and temperature. There are a number of C_n^2 profile models being proposed in the literature; however, the extensively used one in terms of altitude is the Hufnagle-Valley model given by [3, 5, 35, 39]

$$C_n^2(h) = 0.00594(v_w/27)^2(10^{-5}h)^{10}\exp\left(-h/1000\right) + 2.7\times10^{-16}\exp\left(-h/1500\right) + \hat{A}\exp\left(-h/100\right), \tag{15}$$

where h signifies the altitude in meters (m) and $\hat{A}$ represents the nominal value of $C_n^2(0)$ at the ground level in $m^{-2/3}$. The value of C_n^2 for the FSO links near the ground level is approximately $1.7\times10^{-14}m^{-2/3}$ and $8.4\times10^{-14}m^{-2/3}$ during the daytime and at the night, respectively. Generally, C_n^2 ranges from $10^{-13}m^{-2/3}$ for the strong turbulence to $10^{-17}m^{-2/3}$ for the weak turbulence. Its typical average value is $10^{-15}m^{-2/3}$ [5, 39]. The v_w denotes the root mean square (rms) wind speed (pseudowind) in meters per second (m/s) with the most usual value of 21 m/s, but it can be described by [39]

$$w = \left[\frac{1}{15\times10^3}\int_{5\times10^3}^{20\times10^3} V^2(h)dh\right]^{1/2}, \tag{16}$$

where $V(h)$ is normally defined by the Bufton wind model that can be expressed as [39]

$$V(h) = \omega_s h + V_g + 30\exp\left[-\left(\frac{h-9400}{4800}\right)^2\right] \tag{17}$$

where V_g denotes the ground wind speed and ω_s signifies the *slew rate* that is related to the satellite movement regarding the observer on the ground.

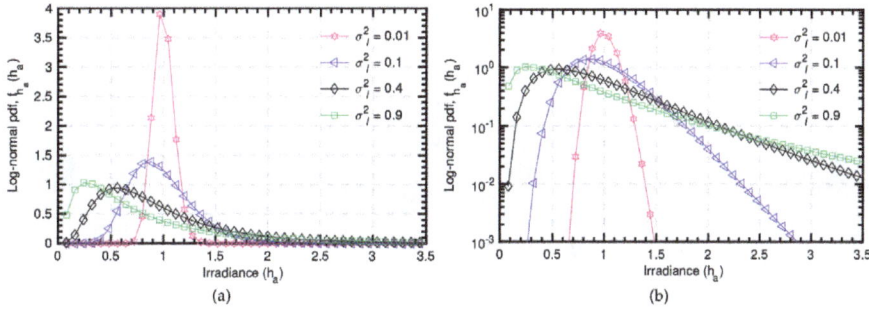

Figure 12. Log-normal pdf for different values of log-irradiance variance using (a) linear scale and (b) logarithmic scale.

We substitute different values of the log-irradiance variance, that is, $\sigma_I^2 \in (0.01, 0.1, 0.4, 0.9)$ in Eq. (11) into Eq. (10) in order to estimate the intensity fluctuation pdf for the LN distribution. The LN pdf plot can be easily achieved by using a linear scale; however, we also present results of a logarithmic scale implementation. This is to ensure uniformity in this chapter because we are going to use the logarithmic scale to present experimental results of irradiance fluctuation in Section 6. Moreover, logarithmic scale has the ability to respond to skewness toward the large values. So, it is more appropriate for illustrating how the measured samples would fit with the distributions at their tails. **Figure 12** shows the LN pdf for different values of log-irradiance variance. In **Figure 12a**, linear scale is employed whereas in **Figure 12b** we use logarithmic scale. It is crystal clear that detailed insight of the tails of the plots is presented in **Figure 12b**. Furthermore, it should be noted that as the value of σ_I^2 increases, the distribution is getting more and more tilted. This signifies the magnitude of the irradiance fluctuation of the system.

3.3.2. Gamma-gamma (ΓΓ) distribution

In most cases, in the strong turbulence regimes in which the LN distribution characterization is not valid, the ΓΓ distribution is normally employed in modeling the scintillation effects. Also, the ΓΓ model can be used in characterizing the fading gains from the weak to the strong turbulence scenarios. The pdf of h_a using the ΓΓ distribution is given by [3, 5, 35, 39]

$$f_{h_a}(h_a) = \frac{2(\alpha\beta)^{(\alpha+\beta)/2}}{\Gamma(\alpha)\Gamma(\beta)} (h_a)^{\frac{(\alpha+\beta)}{2}-1} K_{\alpha-\beta}(2\sqrt{\alpha\beta h_a}), \tag{18}$$

where $\Gamma(\cdot)$ represents the gamma function, $K_\nu(\cdot)$ is the modified Bessel function of the second kind of order ν, α and β are the effective number of large-scale and small-scale eddies of the scattering process, respectively. The parameters α and β are defined, respectively, for the plane wave as [3, 5, 35, 39]

$$\alpha = \left[\exp\left(\frac{0.49\sigma_R^2}{(1 + 1.11\sigma_R^{12/5})^{7/6}} \right) - 1 \right]^{-1}, \tag{19}$$

$$\beta = \left[\exp\left(\frac{0.51\sigma_R^2}{(1 + 0.69\sigma_R^{12/5})^{5/6}} \right) - 1 \right]^{-1}, \tag{20}$$

and for the spherical wave, they can be expressed as [35]

$$\alpha = \left[\exp \left(\frac{0.49\sigma_R^2}{(1 + 0.18d^2 + 0.56\sigma_R^{12/5})^{7/6}} \right) - 1 \right]^{-1}, \tag{21}$$

$$\beta = \left[\exp \left(\frac{0.51\sigma_R^2 (1 + 0.69\sigma_R^{12/5})^{-5/6}}{(1 + 0.9d^2 + 0.62d^2\sigma_R^{12/5})^{5/6}} \right) - 1 \right]^{-1}, \tag{22}$$

where $d \triangleq (kD^2/4L)^{1/2}$, D represents the diameter of the receiver aperture, and σ_R^2 denotes the Rytov variance which is a metric for the strength of the turbulence fluctuations. The σ_R^2 is defined for the plane and the spherical waves, respectively, as [5, 35, 39]

$$\sigma_R^2\big|_{\text{plane}} = 1.23\, C_n^2\, k^{7/6}\, L^{11/6}, \tag{23}$$

$$\sigma_R^2\big|_{\text{spherical}} = 0.492\, C_n^2\, k^{7/6}\, L^{11/6}, \tag{24}$$

Apart from different values of σ_I^2 being employed, that is, $\sigma_I^2 \in (4.0, 1.5, 0.5)$, we substitute different values of the effective number of large- and small-scale eddies of the scattering process, that is, α and $\beta \Rightarrow \alpha \in (4.3, 4.1, 6.0)$ and $\beta \in (1.3, 2.0, 4.4)$ in Eq. (19) and Eq. (20) into Eq. (18) to estimate the $\Gamma\Gamma$ pdf. The results of both linear and logarithmic scales are depicted in **Figure 13**. **Figure 13** shows turbulence regimes that correspond to the weak, moderate, and strong atmospheric scenarios. The results show that an increase in the turbulence from the weak regime to the strong regime leads to corresponding increase in the distribution spreading.

The normalized variance of the irradiance also known as the scintillation index (σ_N^2) can be expressed in terms of σ_x^2 and eddies of the scattering process (α and β), respectively, as [35, 39]

$$\sigma_N^2 \triangleq \frac{\langle h_a^2 \rangle - \langle h_a \rangle^2}{\langle h_a \rangle^2} \tag{25}$$

$$= \frac{\langle h_a^2 \rangle}{\langle h_a \rangle^2} - 1 \tag{26}$$

$$= exp(4\sigma_x^2) - 1 \tag{27}$$

$$= 1/\alpha + 1/\beta + 1/(\alpha\beta). \tag{28}$$

3.3.3. Málaga (M)-distribution

The $\mathcal{M}$-distribution is a generic model that is appropriate for defining the entire turbulent regimes. **Table 4** shows the means of generating the existing atmospheric turbulence models from the $\mathcal{M}$-distribution model [41]. The $\mathcal{M}$-distributed fading model is based on components such as U_L, U_S^C, and U_S^G which represent the line-of-sight (LOS) component, the scattered component by the eddies on the propagation axis that is coupled to the LOS contribution, and the scattered component to the receiver by the off-axis eddies, respectively. The average power of the LOS component as well as that of the total scatter components are $\Omega = E[|U_L|^2]$

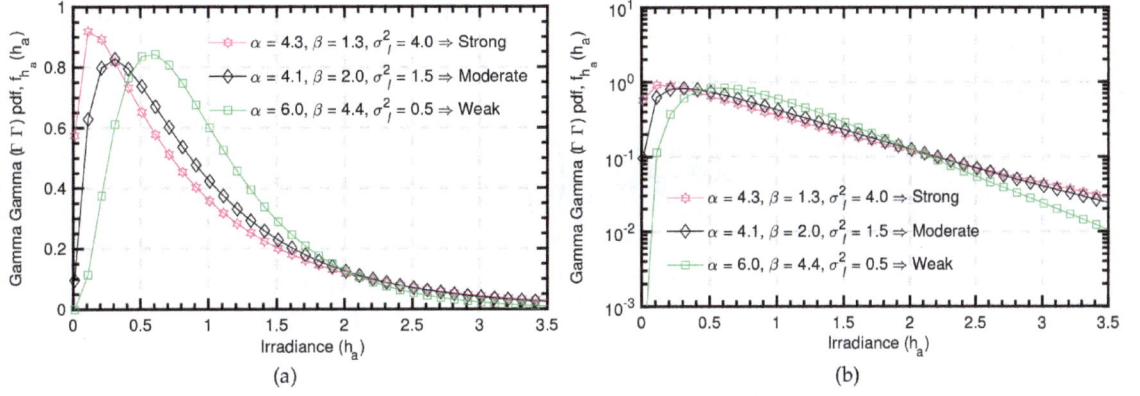

Figure 13. Gamma-gamma pdf for weak-strong turbulence regimes using (a) linear scale and (b) logarithmic scale.

Distribution model	Generation									
	ρ	$\mathrm{Var}[\|U_L\|]$	$\mathrm{Var}[G]$	X	Ω'	$\mathrm{Var}[X]$	γ	Ω	α	β
Rice-Nakagami	0	0								
Gamma	0							0		
Homodyned K (HK)	0		0	γ						
Gamma-gamma	1				1			0		
Shadowed-Rician						0				
Log-normal	0	0					$\rightarrow 0$			
K	0							0		
Exponential	0	0						0	$\rightarrow \infty$	
Gamma-Rician										$\rightarrow \infty$

Adapted from Ref. [41].

Table 4. Approximations required for generating different distribution models from M-distribution model.

and $2b_0 = E[|U_S^C|^2 + |U_S^G|^2]$, respectively. In addition, the average power of the coupled-to-LOS scattering component and that of the scattering component received by off-axis eddies are $E[|U_S^C|^2] = 2\rho b_0$ and $E[|U_S^G|^2] = (1 - \rho)2b_0]$, respectively. The parameter $0 \le \rho \le 1$ represents the amount of scattering power coupled to the LOS component [41, 42]. The pdf of the optical channel gain h_a for the M-distribution can be written as [41, 42]

$$f_{h_a}(h_a) = A \sum_{k=1}^{\beta} a_k h_a^{\frac{\alpha+k}{2}-1} K_{\alpha-k}\left(2\sqrt{\frac{\alpha\beta h_a}{\mu\beta + \Omega'}}\right) \tag{29}$$

where $\mu = E[|U_S^C|]^2 = (1 - \rho)2b_0$, $\Omega' = \Omega + 2\rho b_0 + 2\sqrt{2\rho b_0 \Omega}\cos(\varphi_A - \varphi_B)$, φ_A and φ_B represent the deterministic phases of the LOS and the coupled-to-LOS component, respectively; α denotes a positive parameter that depends on the effective number of large-scale cells of the scattering process, β is a natural number which represents the amount of fading parameter, $K_v(\cdot)$ represents the modified Bessel function of the second kind with order v, A and a_k can be expressed, respectively, as [41, 42]

$$A \triangleq \frac{2\alpha^{\frac{\alpha}{2}}}{\mu^{1+\frac{\alpha}{2}}\Gamma(\alpha)} \left(\frac{\mu\beta}{\mu\beta + \Omega'}\right)^{\beta+\frac{\alpha}{2}} \tag{30}$$

$$a_k \triangleq \binom{\beta \quad -1}{k \quad -1} \frac{(\mu\beta + \Omega')^{1-\frac{k}{2}}}{(k-1)!} \left(\frac{\Omega'}{\mu}\right)^{k-1} \left(\frac{\alpha}{\beta}\right)^{\frac{k}{2}} \tag{31}$$

3.4. Combined attenuation statistics

The pdf of $h = h_\ell h_a h_p$ that constitutes the aforementioned factors of the propagation channel can be defined as [37, 38]

$$f_h(h; w_z) = \int f_{h|h_a}(h|h_a) f_{h_a}(h_a) dh_a \tag{32}$$

where $f_{h|h_a}(h|h_a)$ is the conditional probability given a turbulence state h_a and its distribution can be written as [37, 38]

$$f_{h|h_a}(h|h_a) = \frac{1}{h_a h_\ell} f_{h_p}\left(\frac{h}{h_a h_\ell}\right) = \frac{\gamma^2}{A_0^{\gamma^2} h_a h_\ell} \left(\frac{h}{h_a h_\ell}\right)^{\gamma^2 - 1}, \qquad 0 \leq h \leq A_0 h_a h_\ell. \tag{33}$$

Therefore, $f_h(h; w_z)$ can be written as [37]

$$f_h(h; w_z) = \frac{\gamma^2}{(A_0 h_\ell)^{\gamma^2}} h^{\gamma^2 - 1} \int\limits_{h/A_0 h_\ell}^{\infty} h_a^{-\gamma^2} f_{h_a}(h_a) dh_a. \tag{34}$$

4. Performance analysis

4.1. BER

The channel state distribution $f_h(h; w_z)$ can be calculated by employing an appropriate model for atmospheric turbulence regimes in Eq. (25) as follows:

1. **For a weak turbulence ($\sigma_R^2 < 0.3$)**

    In this regime, $f_{h_a}(h_a)$ follows LN distribution, so, $f_h(h; w_z)$ can be expressed as [37]

$$f_h(h; w_z) = \frac{\gamma^2}{2(A_0 h_\ell)^{\gamma^2}} h^{\gamma^2 - 1} \times \text{erfc}\left(\frac{\ln\left(\frac{h}{A_0 h_\ell}\right) + \mu}{\sqrt{8}\sigma_x}\right) e^{\left(2\sigma_x^2 \gamma^2 (1+\gamma^2)\right)}. \tag{35}$$

2. **For a strong turbulence**

    This regime is characterized by the $\Gamma\Gamma$ distribution, so, $f_h(h; w_z)$ can be written as [37, 38]

$$f_h(h; w_z) = \frac{2\gamma^2 (\alpha\beta)^{(\alpha+\beta)/2}}{(A_0 h_\ell)^{\gamma^2}\Gamma(\alpha)\Gamma(\beta)} h^{\gamma^2-1} \times \int\limits_{h/A_0 h_\ell}^{\infty} h_a^{(\alpha+\beta)/2-1-\gamma^2} K_{\alpha-\beta}(2\sqrt{\alpha\beta\, h_a})dh_a. \tag{36}$$

where $K_\nu(\cdot)$ can be expressed in terms of the Meijer's G-function $G_{p,q}^{m,n}[\cdot]$ as [35]

$$K_\nu(x) = \frac{1}{2}G_{0,2}^{2,0}\left[\frac{x^2}{4}\,\bigg|\,_{(\nu/2),\,-(\nu/2)}^{-}\right]. \tag{37}$$

So, from Eq. (28), $K_{\alpha-\beta}(2\sqrt{\alpha\beta h_a})$ can be expressed as [35]

$$K_{\alpha-\beta}(2\sqrt{\alpha\beta h_a}) = \left(\frac{1}{2}\right)G_{0,2}^{2,0}\left[\alpha\beta h_a\,\bigg|\,_{\frac{\alpha-\beta}{2},\,\frac{\beta-\alpha}{2}}^{-}\right], \tag{38}$$

Therefore, $f_h(h)$ can be written as [38]

$$f_h(h) = \frac{\alpha\beta\gamma^2}{A_0 h_l \Gamma(\alpha)\Gamma(\beta)}\left(\frac{\alpha\beta h}{A_0 h_l}\right)^{(\alpha+\beta/2)-1} \times G_{1,3}^{3,0}\left[\frac{\alpha\beta}{A_0 h_l}h\,\bigg|\,_{-\frac{\alpha+\beta}{2}+\gamma^2,\,\frac{\alpha-\beta}{2},\,\frac{\beta-\alpha}{2}}^{1-\frac{\alpha+\beta}{2}+\gamma^2}\right]. \tag{39}$$

The average BER, $P(e)$, in terms of $P_h(h)$ can be expressed as [38]

$$P(e) = \int_0^\infty P(e|h)f_h(h)dh. \tag{40}$$

Figure 14 shows the average BER in terms of the average SNR for BPSK over different values of turbulence strength. In this analysis, only the atmospheric turbulence effect is assumed

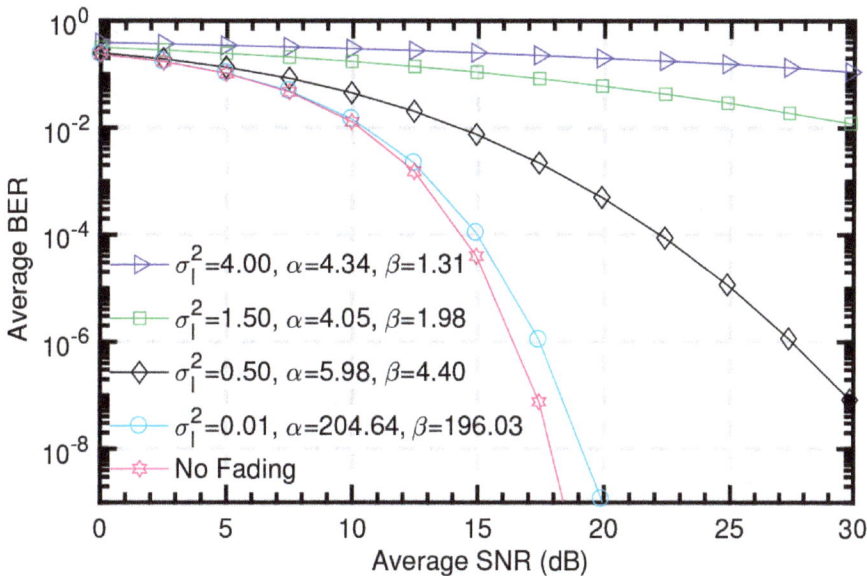

Figure 14. Average BER versus SNR for BPSK under different turbulence conditions.

when Eq. (40) is considered. The atmospheric turbulence parameters $\alpha \in$ (4.34, 4.05, 5.98, 204.64), $\beta \in$ (1.31, 1.98, 4.40, 196.03), and $\sigma_I^2 \in$ (4.00, 1.50, 0.50, 0.01), which represent weak to strong turbulence conditions, are employed in order to estimate the system BER performance. It is observed that the SNR required to achieve a specific BER increases with an increase in the atmospheric turbulence strength. For instance to achieve a BER of 10^{-6} in a channel with $\sigma_I^2 = 0.01$, the required SNR is about 18 dB; however, for fading strength of $\sigma_I^2 = 0.50$, the required SNR increases to 28 dB. Furthermore, at this BER, for $\sigma_I^2 = 0.01$ and $\sigma_I^2 = 0.50$ additional 2 and 12 dB, respectively, are required compared to the ideal channel in which there is no turbulence. This shows that the BER increases as turbulence strength becomes stronger.

4.2. Ergodic channel capacity

The channel capacity is one of the main performance metrics in the design of FSO systems that needs significant attention. The capacity of a multiple-input-multiple-output (MIMO) FSO system with M lasers and N photodetectors in bits/s/Hz can be expressed as [34]

$$C = \log_2 \left[det \left(I_M + \frac{\gamma_{\text{inst}}}{M} R \right) \right],$$ (41)

where $R = \begin{cases} HH^\dagger \text{ if } N < M \\ H^\dagger H \text{ if } N \leq M \end{cases}$, H denotes an $N \times M$ channel state matrix, $(\cdot)^\dagger$ corresponds to the Hermitian transpose, I_M is an $M \times M$ identity matrix, and $\gamma_{\text{inst}} = \eta_e^2 h^2 / N_0$ denotes the instantaneous electrical SNR whose average $\xi_a = \eta_e^2 \mathbb{E} \langle h \rangle^2 / N_0$.

It is worth noting that there are a number of viable approaches for the estimation of ergodic capacity of the MIMO FSO link. One of such is the numerical integration approach.

4.2.1. Numerical integration approach

The ergodic capacity, C_{erg}, of MIMO FSO link can be defined by the expected value of the instantaneous mutual information C, between the transmit and receive apertures. Hence, the C_{erg} of FSO system is a random variable and a function of SNR. The C_{erg} can be expressed as [34, 35, 43]

$$C_{\text{erg}} \triangleq \mathbb{E} \langle C \rangle = \int_0^\infty \log_2 (1 + \gamma_{\text{inst}}) f_{\gamma_{\text{inst}}} (\gamma_{\text{inst}}) d\gamma_{\text{inst}},$$ (42)

where $\mathbb{E}(\cdot)$ denotes the expectation operator and $f_{\gamma_{\text{inst}}} (\gamma_{\text{inst}})$ is the pdf of γ_{inst}.

It should be noted that most of the models for ergodic capacity evaluation are based on numerical integration. However, integration-based approach requires comparatively more computational time. This is even more challenging in the strong turbulence regimes analysis. This consequence can be attributed to the Bessel function that is usually expressed in terms of the Meijer's G-function for easier evaluation [34]. In Refs. [34] and [35], computational-efficient approaches are presented in order to reduce the associated high computational time of the integration-based approach. This enables faster performance evaluation over a wide range of SNR.

4.2.2. Power series approach

Considering the power series representation, the ergodic channel capacity can be expressed as [35]

$$\langle C \rangle = f(\mathcal{X}) = \psi \mathcal{A}(a_0 + a_1 \mathcal{X} + a_2 \mathcal{X}^2 + a_3 \mathcal{X}^3), \tag{43}$$

where $\psi = \pi/4$ and $\mathcal{A} = min\{M, N\}$ represent the minimum number of transmit or receive apertures, $\mathcal{X} = k_1 \gamma$, k_1 is a coefficient whose value is one and the unit is in b/s/Hz, and a_n represents the coefficient of the nth term given by [35]

$$a_0 = 0.13\pi k_1 \mathfrak{R}, \tag{44}$$

$$a_1 = 0.66\mathfrak{R}, \tag{45}$$

$$a_2 = 1.45 \times 10^{-3}\mathfrak{R}, \tag{46}$$

$$a_3 = -1.73 \times 10^{-5}\mathfrak{R}, \tag{47}$$

where $\mathfrak{R} = \pi exp\left(\sigma_R^2 - \sigma_N^2\right)$.

4.2.3. Spatial interpolation lookup approach

The spatial approach is based on B-spline and Barycentric Lagrange interpolation lookup table (B^2LUT). With the B-spline interpolated LUT, the capacity of MIMO FSO system can be expressed as [34]

$$C\left(SNR|_{(M, N, \sigma_N^2)}\right) = \sum_{t=1}^{m_1}\sum_{r=1}^{m_2}\sum_{p=1}^{m_3}\omega(SNR|_{\sigma_N^2}\mathfrak{I}_p, \cdots, \mathfrak{I}_{p+h})$$

$$\times \psi(SNR|_N\mathfrak{D}_r, \cdots, \mathfrak{D}_{r+\ell})\phi(SNR|_M\mathfrak{S}_t, \cdots, \mathfrak{S}_{t+s})\mathfrak{M}_{t, r, p}, \tag{48}$$

where $\mathfrak{I} = (\mathfrak{I}_p, \cdots, \mathfrak{I}_{p+h})$, $\mathfrak{D} = (\mathfrak{D}_r, \cdots, \mathfrak{D}_{r+\ell})$, and $\mathfrak{S} = (\mathfrak{S}_t, \cdots, \mathfrak{S}_{t+s})$ are the knot sequences, $(\mathfrak{M}_{t, r, p} : t = 1, \cdots, m_1; r = 1, \cdots, m_2; p = 1, \cdots, m_3)$ are coefficients array, and $\omega(\cdot), \psi(\cdot), \phi(\cdot)$ are the univariate piecewise B-spline for M, N, and σ_N^2, respectively.

Another main figure of merit for characterizing the communication link performance is the achievable average (ergodic) channel capacity. **Figure 15** shows the average channel capacity of FSO link as a function of average electrical SNR for different values of turbulence strength. The atmospheric turbulence parameters $\sigma_N^2 \in$ (0.0120, 0.1486, 0.2078) result in $C_N^2 \in$ $(6.03 \times 10^{-16}m^{-2/3}, 7.62 \times 10^{-15}m^{-2/3}, 1.09 \times 10^{-14}m^{-2/3})$, which correspond to the weak, moderate, and strong turbulence conditions, respectively. Moreover, the capacity of the nonturbulent channel condition (no fading) is presented for benchmarking. Obviously, the ergodic capacity of FSO link significantly depends on the atmospheric turbulence strength.

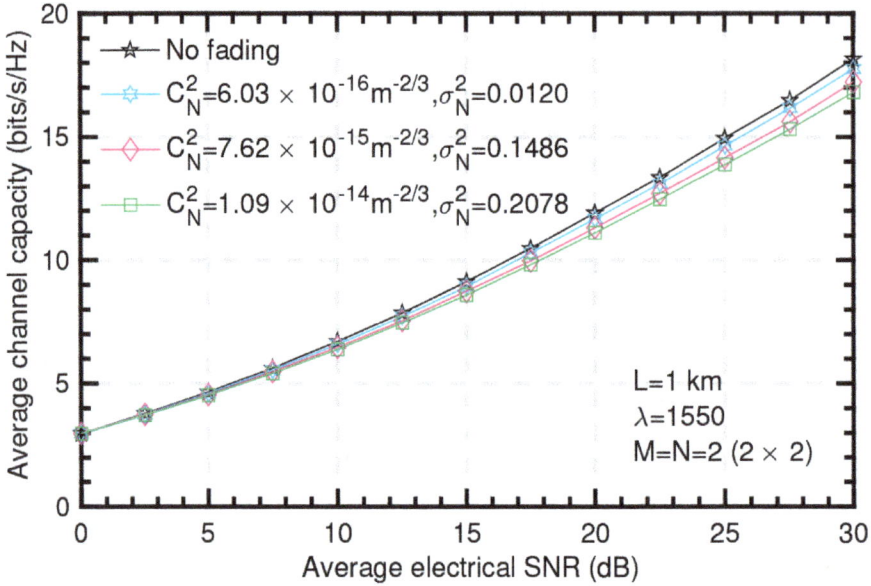

Figure 15. Average channel capacity of FSO link versus average electrical SNR.

Comparatively, the ergodic channel capacity for weak turbulence condition is considerably more than in the cases of moderate and strong turbulence conditions. Moreover, the channel condition with no fading offers the highest capacity. This shows that the atmospheric turbulence-induced fading results in severe impairment on the FSO link performance. This can eventually result in recurrent link failures. Consequently, optical wireless technologies are not as reliable as the conventional optical fiber technologies. In order to address the challenges, innovative technologies can be employed to enhance the system performance.

5. Technologies for performance enhancement

FSO is a promising optical technology that can be employed for different application. However, the tradeoff between the required high data rates and the limitations of atmospheric channel is the major challenge for reliable implementation of FSO technologies in the access networks [35]. Therefore, the problems inhibit the FSO system from being an effective and reliable standalone fronthaul technology. In this section, schemes like hybrid RF/FSO and relay-assisted transmission technologies that can be implemented to enhance the performance of FSO technology in the access networks are presented.

5.1. Hybrid RF/FSO technology

It should be noted that RF wireless technologies that operate above approximately 10-GHz frequencies are adversely affected by rain, whereas fog has insignificant effect on them. On the other hand, FSO systems are highly susceptible to fog, whereas the effect of rain on them is negligible. Therefore, it is of high importance to improve the link reliability to alleviate the adverse effects of the meteorological and weather conditions. An attractive way of addressing the challenge is a simultaneous employment of an RF link and the FSO link for transmission. It is remarkable that fog and rain rarely occur concurrently in nature. Consequently, the two links can function in a complementary way. This concept influences the hybrid RF/FSO scheme. The RF/FSO is a hybrid scheme that combines the benefits of the inherent high transmission capacity of optical technologies and the ease of deployment of wireless links. Moreover, the idea of hybrid RF/FSO system is to concurrently attend to the related drawbacks and take advantages of both technologies. This will help in the reliable transmission of heterogeneous wireless services [8, 24].

In a hybrid RF/FSO technology, there are two parallel links between the transmitter and the receiver. Moreover, subject to deployment scenario and application, both parallel links of the hybrid technology have the capability to transmit data. Nevertheless, based on the weather conditions as well as the EMI levels, either of the links can be used for data transmission [44]. For instance, under adverse atmospheric condition (i.e., fog), the hybrid RF/FSO scheme ensures that the RF link serves as a back-up in case of FSO link outage. However, the resultant data rate of the RF link is less than that of the actual FSO link [8, 24]

5.2. Relay-assisted FSO transmission

A realistic approach for turbulence-induced fading mitigation is spatial diversity technique. In this technique, multiple transmit/receive apertures are employed in order to create and exploit additional degrees of freedom in the spatial domain. The spatial diversity is an appealing technique for fading mitigation because of its typical redundancy. However, the utilization of multiple apertures presents different challenges such as an increase in the system complexity as well as cost. In addition, the distance between the apertures has to be large enough in order to inhibit detrimental effects of spatial correlation. A simplified way of implementing spatial diversity is the dual-hop relaying which has been considerably employed in the RF communication systems. The dual-hop relaying implementation helps substantially in extending the network coverage area as well as improving the quality of the receive signal [45].

Furthermore, the concept of relay-assisted transmission is based on creating a virtual multiple-aperture system in order to realize advantages of MIMO techniques. This is achieved by exploiting both RF and FSO characteristics in order to have an efficient system in a real-life situation. Additionally, a relay-assisted transmission is also known as a mixed RF/FSO dual-hop communication scheme. The dual-hop entails the links from the source to the relay which are RF links and the links between the relay and the destination which are FSO links. In essence, RF transmission is utilized at one hop and FSO transmission is employed at the other. It is remarkable that, in principle, the mixed RF/FSO dual-hop relay scheme is comparatively different from the hybrid RF/FSO technology. In the latter, parallel RF and FSO links are normally used for the same path [45]. Furthermore, in the mixed RF/FSO dual-hop scheme, the main purpose of the FSO link is to enable the RF users to communicate with the backbone network. This helps bridging the connectivity gap between the backbone and the last-mile access networks [42, 46].

The mixed RF/FSO dual-hop model efficiently addresses the last-mile transmission bottleneck of the system. This is achieved by enabling multiplexing of multiple users with RF capabilities as well as their aggregation in a particular high-speed FSO link so as to exploit the inherent optical capacity [42, 45]. Moreover, this implementation stalls any form of interference due to the fact that RF and FSO operate on completely different frequency bands. Therefore, the mixed RF/FSO dual-hop model offers better performance compared to the traditional RF/RF transmission system [45, 46]. In a mixed RF/FSO system, the source comprises multiple RF users, each equipped with an antenna. Furthermore, at the destination, there is an FSO detector that is equipped with an aperture. In addition, the source and the destination are connected by a relay that is usually mounted on a high platform. The relay node performs RF to FSO conversion. Also, the relay has a receive antenna and a transmit aperture that are assigned for the RF signal reception and optical signal transmission, respectively.

6. Channel measurement and characterization

In this section, the performance of an FSO link subjected to a real atmospheric turbulence condition is investigated experimentally. The σ_N^2 is measured from the channel samples obtained so as to determine the degree of atmospheric turbulence and the subsequent effects on the FSO link quality. Furthermore, the C_n^2 can be calculated as explained in Section 3.

6.1. Experimental setup

The experimental setup shown in **Figure 16** is employed in the channel measurement. The setup consists of a point-to-point FSO link that is based on IM/DD technique. The pattern generator uses a pseudorandom binary sequence (PRBS) of length $2^{23} - 1$ bits to generate a 10-Gb/s non-return-to-zero (NRZ) signal. Also, the produced electrical signal is then injected into a JDSU Integrated Laser Mach Zehnder (ILMZ) drives at 1548.51-nm wavelength. A standard single-mode fiber (SSMF) is used to convey the optical output signal launched from the laser to a 3-mm diameter collimator. The optical power at the input of collimator is set to 0 dBm. The collimated laser beam is subsequently transmitted over the FSO channel with a round trip length of 54 m. **Figure 16** inset depicts an outdoor FSO setup employed in the measurement. The overall transmission distance is achieved when the optical signal from collimator 1 passes

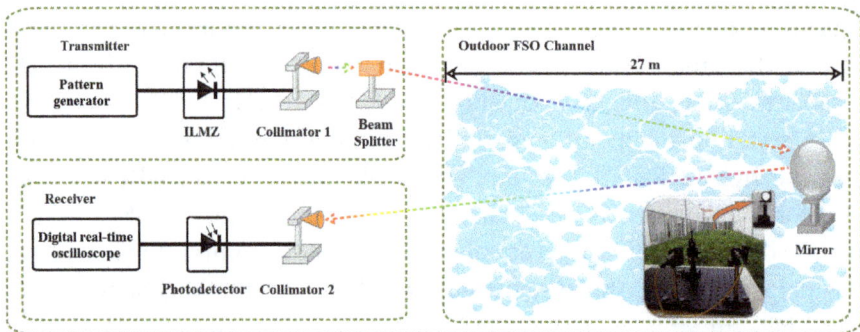

Figure 16. Experimental setup. The inset presents the picture of outdoor FSO setup, ILMZ: Integrated Laser Mach Zehnder.

through the FSO channel to the mirror with a beam diameter of about 2 cm located at the other side of the link that is 27 m long, and then reflected back to the collimator 2 at the receiver with approximately its initial diameter. The collimator is made of a concave mirror, in order to lessen beam scattering and considerably maximize the power transfer. The converged received optical signal at the receiver then focuses on the laser collimator which is coupled to the photodetector by the SSMF. The resulting optical signal is then converted into electrical signal using a 10-Gb/s PIN photodiode. The PIN is followed by a real-time sampling oscilloscope (Tektronix: DPO72004B) with a sample rate of 50 GS/s.

6.2. Experimental results

The results of FSO channel samples collected from November 9 to 20, 2015, for characterization are presented. The data obtained on November 12, 2015, at 01:45 pm and 09:30 pm are analyzed in this work. The recorded weather conditions are as follows:

1. Scenario 1: 01:45 pm

   Temperature, 22°C; wind, 6 mph; humidity, 69%; pressure, 1031 mb; visibility, 10 km; precipitation, 0 mm; and rain rate, 0%.

2. Scenario 2: 09:30 pm

   Temperature, 17°C; wind, 4 mph; humidity, 80%; pressure, 1030 mb; visibility, 9 km; precipitation, 0 mm; and rain rate, 0%.

The FSO channel measurement is realized by injecting an unmodulated optical wave that emanates from the laser into the FSO channel. The FSO channel brings about a path loss with

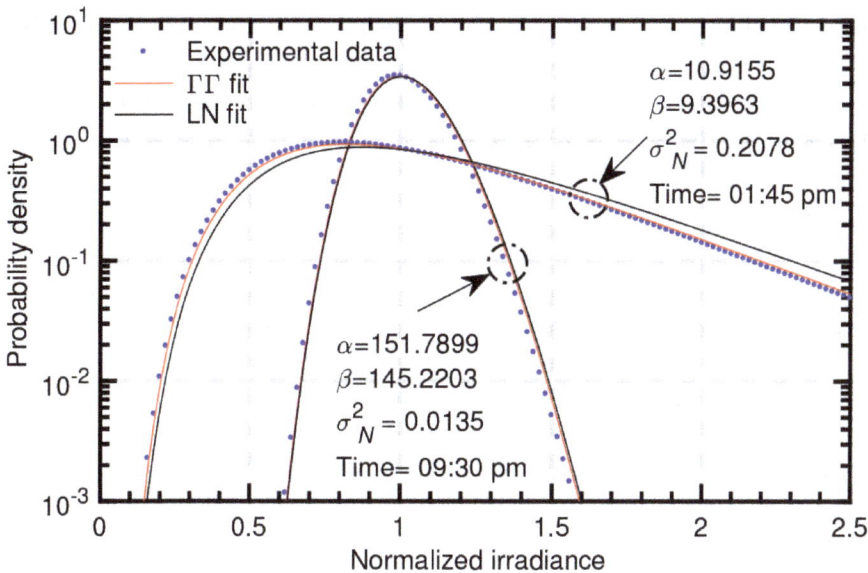

Figure 17. Histogram of normalized irradiance with log-normal and gamma-gamma fits under different scintillation index values.

a mean value of 9 dB, for the setup and the atmospheric conditions. Using statistical means, the resultant signal detected and received by the real-time oscilloscope is analyzed offline using MATLAB®.

The characterization of the refractive-index structure parameter C_n^2 is achieved by fitting the nearest LN and the ΓΓ pdf curves to the pdf of the received data. The fittings are presented in **Figure 17**. The scintillation index σ_N^2 is measured for the two scenarios considered. The values obtained are 0.0135 and 0.2078 for 09:30 pm and 01:45 pm, respectively. For the first scenario with $\sigma_N^2 = 0.0135$, the LN and the ΓΓ fit very well with the measured channel samples σ_N^2. However, when $\sigma_N^2 = 0.2078$, the LN fitting is loose and unable to give an accurate result for the fading model, whereas the ΓΓ fitting still maintains a relatively better result for the fading model. This result shows that the LN model is unsuitable for the strong atmospheric fading characterization.

Furthermore, the estimated values of the refractive-index structure parameters, C_n^2, are $6.7807 \times 10^{-16} m^{-2/3}$ ($\sigma_N^2 = 0.0135$) and $1.0864 \times 10^{-14} m^{-2/3}$ ($\sigma_N^2 = 0.2078$). Therefore, the first ($\sigma_N^2 = 0.2078$) and the second ($\sigma_N^2 = 0.0135$) scenarios correspond to the strong and weak turbulence regimes, respectively.

7. Real-time coherent PON OWC based on dual polarization for the mobile backhaul/fronthaul

In this section, proof-of-concept gigabit-capable long-reach coherent PON and OWC systems with the ability to support different applications over a shared optical fiber infrastructure are experimentally implemented. This is in an effort to demonstrate the FSO application in certain areas in the mobile cellular systems in which physical connections by means of optical fiber cables are impractical or in rural area that lacks fiber infrastructure. Moreover, this is achieved by a reconfigurable real-time DSP reception of a dual-polarization-quadrature phase shift keying (DP-QPSK) signal over the SSMF and FSO systems. It is worth noting that the system is validated by a commercial field-programmable gate array (FPGA) in order to have an open system whose components and protocols conform to standards independent of a particular equipment vendor. In this analysis, we study signal transmission and reception over 100 km of SSMF as well as over a hybrid 100 km of SSMF plus a 54-m outdoor FSO link. We are able to establish the lowest sampling rate that is necessary for digital coherent PON by employing four 1.25 Gsa/s ADCs with an electrical front-end receiver that offers just 1-GHz analog bandwidth. This is realized by implementing a phase and polarization diversity coherent receiver in conjunction with the DP-QPSK modulation formats. This technique is of high importance in order to relax the required electrical digital units at the optical network unit (ONU) toward the RF rates. This scheme also helps in realizing the anticipated data rate for the next-generation coherent optical access networks for the 5G Mobile wireless networks.

7.1. Experimental setup

The experimental setup depicted in **Figure 18a** is used to validate the performance of a PON system with hybrid fiber and FSO link using DP signals. It is worth mentioning that only the

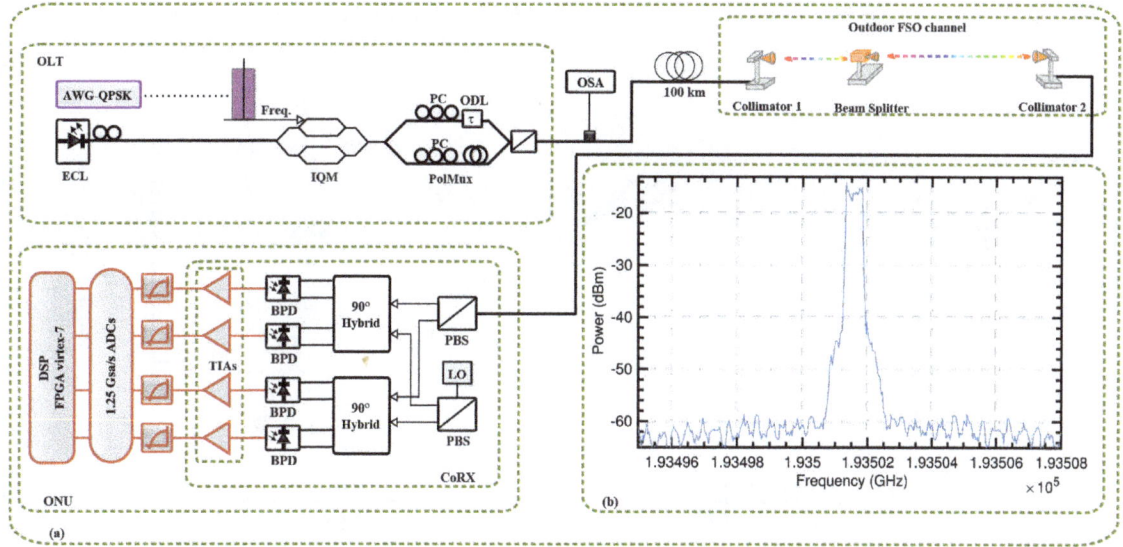

Figure 18. (a) Experimental setup for 20×625 Mbaud DP-QPSK signal; (b) Overall spectrum (PBS: polarization beam splitter; BPD: balanced photodetector; CoRX: coherent receiver).

receiver DSP of the setup is estimated in real time. At the OLT, the light from an external cavity laser (ECL) (<100 kHz linewidth) is injected into an IQ modulator (IQM). The wavelength λ is centered at ~1549 nm. The IQM is driven by a 65-Gsa/s arbitrary waveform generator (AWG) that generates 625 Mbaud signals from a $2^{12} - 1$ pseudo random bit sequence (PRBS). The subsequent signal is then digitally filtered using a raised-cosine (RC) filter with a 0.1-roll-off factor and 32-taps FIR resolution in addition to a simple 3-taps FIR pre-emphasis subsystem. The employed modulation format is the differential DP-QPSK, providing 2.5 Gb/s per end user. The obtained spectrum is shown in **Figure 18b**.

In an effort to emulate the DP system, this signal is divided into two using an optical splitter. Then, we applied a delay of 12 symbols to one of them for an effective decorrelation purposes. Afterward, both polarized signals are multiplexed orthogonally once again using a polarization beam combiner (PBC). This implementation results in dual polarization of the signal. The optical power is managed using variable optical attenuators (VOAs). This subsequent signal is then propagated over 100 km of an SSMF and 54-m FSO. At the receiver side, the signal is coherently detected using a 4×90° optical hybrid by means of a free-running ECL LO with about 100-kHz linewidth tuned to the center channel, λ. The optical signal is converted to the electrical domain using four balanced detectors (BDs) and then amplified by transimpedance amplifiers (TIAs). This results in the in-phase and quadrature components of each polarization. It is worth noting that only an output of the TIA is used. The signal is then filtered using a 1-GHz low-pass filter and sampled by four 8-bit 1.25 Gsa/s ADCs. The digitalized signal is conveyed to a Virtex-7 FPGA, where the entire post-detection 8-bit DSP in real time is implemented. The applied DSP is based on [47]. The bit-error rate (BER) is calculated in real time by bit error counting, averaged between the two polarizations.

In addition, we consider the possibility of an outdoor FSO communication link as part of the system. The employed FSO link setup in this study is similar to that in **Figure 16** that we have

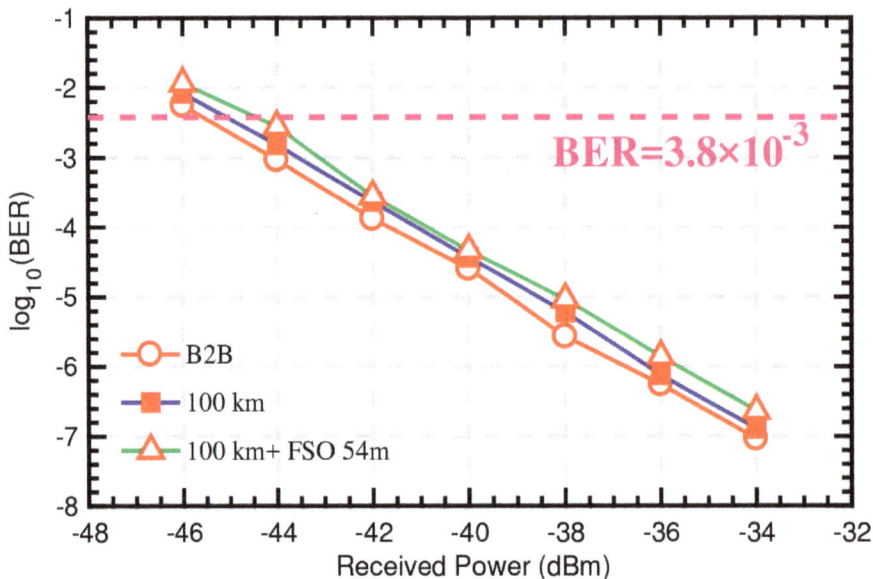

Figure 19. Receiver sensitivity in terms of BER measured for DP-QPSK signals.

discussed in Section 6. The outdoor FSO link experiences a total loss of ~8–9 dBm. At the collimator, the received signal is guided to an integrated phase- and polarization-diverse coherent receiver.

7.2. Experimental results

Figure 19 illustrates receiver sensitivity in terms of BER measured for the DP-QPSK signal. The figure presents results for the back-to-back (B2B) and 100 km of fiber scenarios as well as 100-km plus 54-m FSO. The considered BER limit of 3.8×10^{-3} corresponds to the 7% hard-decision forward error correction (HD-FEC). As shown in **Figure 19**, there is no significant penalty between 100 and 100-km plus FSO in real-time results.

8. Conclusions

In this chapter, we have presented various opportunities of using optical wireless communication technology for addressing the last-mile transmission bottleneck of the fixed/mobile network. We have also discussed various challenges of the OWC and presented different solutions in order to make OWC an efficient technology. In the proof-of-concept experiment, we study the transmission capabilities of a PON based on DP signal in terms of receiver sensitivity. This is implemented with the real-time ONU receiver that is emulated by a commercial FPGA. This helps in facilitating an open system and hence enables interoperability, portability, and open software standards. The transmissions over 100 km of SSMF as well as over a hybrid 100 km of SSMF plus a 54-m outdoor FSO link are successfully realized considering 625 Mbaud DP-QPSK channel.

Acknowledgements

This work is supported by the European Structural Investment Funds (ESIF), through the Operational Competitiveness and Internationalization Programme (COMPETE 2020) under FutPON project [Nr. 003145 (POCI-01 -0247-FEDER-003145)]. Also, it is funded by the Fundação para a Ciência e a Tecnologia under the grants PD/BD/52590/2014 and FRH/BPD/110889/2015.

Author details

Isiaka Alimi*, Ali Shahpari, Artur Sousa, Ricardo Ferreira, Paulo Monteiro and António Teixeira

*Address all correspondence to: iaalimi@ua.pt

Instituto de Telecomunicações, Department of Electronics, Telecommunications and Informatics, Universidade de Aveiro, Aveiro, Portugal

References

[1] Yu C, Yu L, Wu Y, He Y, Lu Q. Uplink scheduling and link adaptation for narrowband internet of things systems. IEEE Access, 2017;**5**:1724–1734

[2] Ejaz W, Anpalagan A, Imran MA, Jo M, Naeem M, Qaisar SB, Wang W. Internet of things (IoT) in 5G wireless communications. IEEE Access. 2016;**4**:10310–10314

[3] Alimi I, Shahpari A, Ribeiro V, Kumar N, Monteiro P, Teixeira A. Optical wireless communication for future broadband access networks. In: 2016 21st European Conference on Networks and Optical Communications (NOC); IEEE; Lisbon, Portugal. Jun 2016. pp. 124–128

[4] Parca G, Tavares A, Shahpari A, Teixeira A, Carrozzo V, Beleffi GT. FSO for broadband multi service delivery in future networks. In: 2013 2nd International Workshop on Optical Wireless Communications (IWOW); IEEE; Newcastle upon Tyne, UK. Oct 2013. pp. 67–70

[5] Ghassemlooy Z, Popoola W, Rajbhandari S. Optical wireless Communications: System and Channel Modelling with MATLAB®. Taylor & Francis; London, New York: 2012

[6] Ghassemlooy Z, Amon S, Uysal M, Xu Z, Cheng J. Emerging optical wireless communications-advances and challenges. IEEE Journal on Selected Areas in Communications. Sep 2015;**33**(9):1738–1749

[7] Alkholidi AG, Altowij KS. Free Space Optical Communications—Theory and Practices, Contemporary Issues in Wireless Communications. InTech; 2014. pp. 159–212

[8] Khalighi MA, Uysal M. Survey on free space optical communication: A communication theory perspective. IEEE Communications Surveys Tutorials. 2014;**16**(4):2231–2258

[9] Bloom S, Korevaar E, Schuster J, Willebrand H. Understanding the performance of free-space optics [Invited]. Journal of Optical Network. Jun 2003;**2**(6):178–200

[10] Kaushal H, Kaddoum G. Optical communication in space: Challenges and mitigation techniques. IEEE Communications Surveys Tutorials. 2017;**19**(1)57–96

[11] Zeng Z, Fu S, Zhang H, Dong Y, Cheng J. A survey of underwater optical wireless communications. IEEE Communications Surveys Tutorials. 2017;**19**(1):204–238

[12] Pompili D, Akyildiz IF. Overview of networking protocols for underwater wireless communications. IEEE Communications Magazine. Jan 2009;**47**(1):97–102

[13] Sevincer A, Bhattarai A, Bilgi M, Yuksel M, Pala N. FIGHTNETs: Smart FIGHTing and mobile optical wireless NETworks—A survey. IEEE Communications Surveys Tutorials. Fourth 2013;**15**(4):1620–1641

[14] Rajagopal S, Roberts RD, Pirn SK. IEEE 802.15.7 visible light communication: Modulation schemes and dimming support. IEEE Communications Magazine. Mar 2012;**50**(3):72–82

[15] Ying K, Yu Z, Baxley RJ, Qian H, Chang GK, Zhou GT. Nonlinear distortion mitigation in visible light communications. IEEE Wireless Communications. Apr 2015;**22**(2):36–45

[16] Yang F, Gao J. Dimming control scheme with high power and spectrum efficiency for visible light communications. IEEE Photonics Journal. Feb 2017;**9**(1):1–12

[17] Pinterest. Color wavelength frequency. Feb 2017. [Online]. Available from: https://www.pinterest.com/pin/408983209884752740/.

[18] Wang M, Wu J, Yu W, Wang H, Fi J, Shi J, Fuo C. Efficient coding modulation and seamless rate adaptation for visible light communications. IEEE Wireless Communications. Apr 2015;**22**(2):86–93

[19] ANDY. Visible Light Communication—VLC & PUREVLC™. February 2017 [Online]. Available from: http://andy96877.blogspot.pt/p/visible-light-communication-vlc-is-data.html

[20] Jan SU, Fee YD, Koo I. Comparative analysis of DIPPM scheme for visible light communications. In: 2015 International Conference on Emerging Technologies (ICET); IEEE; Peshawar, Pakistan. Dec 2015. pp. 1–5

[21] Tsouri GR, Zambito SR, Venkataraman J. On the benefits of creeping wave antennas in reducing interference between neighboring wireless body area networks. IEEE Transactions on Biomedical Circuits and Systems. Feb 2017;**11**(1):153–160

[22] Zhang R, Moungla H, Yu J, Mehaoua A. Medium access for concurrent traffic in wireless body area networks: Protocol design and analysis. IEEE Transactions on Vehicular Technology. Mar 2017;**66**(3):2586–2599

[23] Cahyadi WA, Jeong TI, Kim YH, Chung YH, Adiono T. Patient monitoring using visible light uplink data transmission. In: 2015 International Symposium on Intelligent Signal Processing and Communication Systems (ISPACS); IEEE; Nusa Dua, Indonesia. Nov 2015. pp. 431–434

[24] Uysal M, Capsoni C, Ghassemlooy Z, Boucouvalas A, Udvary E. Optical wireless communications: An emerging technology. Signals and Communication Technology. Springer International Publishing; Switzerland: 2016

[25] Baister G, Gatenby PV. Pointing, acquisition and tracking for optical space communications. Electronics Communication Engineering Journal. Dec 1994;6(6):271–280

[26] Toyoshima M. Trends in satellite communications and the role of optical free-space communications [Invited]. Journal of Optical Networking. Jun 2005;4(6):300–311

[27] D'Amico M, Feva A, Micheli B. Free-space optics communication systems: First results from a pilot field-trial in the surrounding area of Milan, Italy. IEEE Microwave and Wireless Components Letters. Aug 2003;13(8)505–307

[28] Song D-Y, Hurh Y-S, Cho J-W, Lim J-H, Fee D-W, Fee J-S, Chung Y. 4 × 10 Gb/s terrestrial optical free space transmission over 1.2 km using an EDFA preamplifier with 100 GHz channel spacing. Optics Express. Oct 2000;7(8):280–284

[29] Carlson RT, Paciorek S. Environmental Qualification and Field Test Results for the SONAbeam™ 155 and 622. Technical Report, fSONA Communications Corp. 2017; http://www.fsona.com/tech/white_papers/tech_qual-test.pdf

[30] Bandera P. Defining a Common Standard for Evaluating and Comparing Free-Space Optical Products. Technical Report, fSONA Communications Corp. 2017; http://www.fsona.com/tech/white_papers/WHTPAP-Generalized_Link_Margin.pdf

[31] LightPointe. Ultra-Low Latency Point-to-Point Wireless Bridge. White Paper, [Online]. Available from: http://nebula.wsimg.com/793e82b2beac48cb90c347bd86776d12?AccessKeyId=C1431E109BF92B03DF85&disposition=0&alloworigin=1.

[32] Wang Z, Zhong WD, Fu S, Fin C. Performance comparison of different modulation formats over free-space optical (FSO) turbulence links with space diversity reception technique. IEEE Photonics Journal. Dec 2009;1(6):277–285

[33] Navidpour SM, Uysal M, Kavehrad M. BER performance of free-space optical transmission with spatial diversity. IEEE Transactions on Wireless Communications. Aug 2007;6(8):2813–2819

[34] Alimi IA, Abdalla AM, Rodriguez J, Monteiro PP, Teixeira AF. Spatial interpolated lookup tables (FUTs) models for ergodic capacity of MIMO FSO systems. IEEE Photonics Technology Letters. Apr 2017;29(7)583–586

[35] Alimi I, Shahpari A, Ribeiro V, Sousa A, Monteiro P, Teixeira A. Channel characterization and empirical model for ergodic capacity of free-space optical communication link. Optics Communications. 2017;390:123–129

[36] Naboulsi MA, Sizun H, de Fornel F. Fog attenuation prediction for optical and infrared waves. Optical Engineering. 2004;43(2):519–329

[37] Farid AA, Hranilovic S. Outage capacity optimization for free-space optical links with pointing errors. Journal of Lightwave Technology. Jul 2007;25(7):1702–1710

[38] Sandalidis HG, Tsiftsis TA, Karagiannidis GK. Optical wireless communications with heterodyne detection over turbulence channels with pointing errors. Journal of Lightwave Technology. Oct 2009;**27**(20):4440–4445

[39] Andrews FC, Phillips RF. Laser Beam Propagation through Random Media. Press Monographs. SPIE Press; Bellingham, Washington USA. 2005

[40] Kiasaleh K. Performance of APD-based, PPM free-space optical communication systems in atmospheric turbulence. IEEE Transactions on Communications. Sep 2005;**53**(9):1455–1461

[41] Jurado-Navas A, Garrido-Balsells JM, Paris JF, Puerta-Notario A. A Unifying Statistical Model for Atmospheric Optical Scintillation. InTech; Rijeka, Croatia. 2011. pp. 181–206

[42] Yang F, Hasna MO, Gao X. Performance of mixed RF/FSO with variable gain over generalized atmospheric channels. IEEE Journal on Selected Areas in Communication. Sep 2015;**33**(9):1913–1924

[43] Peppas KP, Stassinakis AN, Topalis GK, Nistazakis HE, Tombras GS. Average capacity of optical wireless communication systems over I-K atmospheric turbulence channels. IEEE/OSA Journal of Optical Communications and Networking. Dec 2012;**4**(12):1026–1032

[44] Dahrouj H, Douik A, Rayal F, Al-Naffouri TY, Alouini MS. Cost-effective hybrid RF/FSO backhaul solution for next generation wireless systems. IEEE Wireless Communications. Oct 2015;**22**(5):98–104

[45] Zhang J, Dai F, Zhang Y, Wang Z. Unified performance analysis of mixed radio frequency/free-space optical dual-hop transmission systems. Journal of Lightwave Technology. Jun 2015;**33**(11):2286–2293

[46] Ansari IS, Yilmaz F, Alouini MS. Impact of pointing errors on the performance of mixed RF/FSO dual-hop transmission systems. IEEE Wireless Communications Letters. Jun 2013;**2**(3) 551–354

[47] Ferreira RM, Shahpari A, Reis JD, Teixeira AF. Coherent UDWDM-PON with dual-polarization transceivers in real-time. IEEE Photonics Technology Letters. Jun 2017;**29**(11):909–912

2

Impact of Fiber Duplication on Protection Architectures Feasibility for Passive Optical Networks

Waqas Ahmed Imtiaz, Javed Iqbal, Affaq Qamar,
Haider Ali and Sevia M. Idrus

Additional information is available at the end of the chapter

Abstract

Adaptability of high capacity passive optical network (PON) requires the provision of an efficient fault detection and restoration mechanism throughout the network at an acceptable cost. The readily adapted pre-planned protection strategy relies on component duplication, which significantly increases the cost of deployment for PON. Therefore, it is imperative to determine a suitable component that requires high redundancy and determine the impact of protection for that component on feasibility of PON. Five protection architecture including ITU-T 983.1 Type C, single ring, dual ring, tree- and ring-based architectures with hybrid star-ring topology at the optical distribution network (ODN), are considered to evaluate the impact of fiber duplication in terms of capital expenditure (CAPEX), operation expenditure (OPEX), reliability, and support for maximum number of subscribers. Reliability block diagram (RBD) based analysis shows that desirable 5 nines connection availability is provided by each protection architecture and utilization of ring topology avoids duplication of the fiber but effects the number of subscribers. Furthermore, it is observed that OF duplication at ODN is the main contributor to CAPEX. Collectively hybrid protection architectures provide efficient performance and proves to be a feasible solution for the deployment of survivable PONs at the access domain.

Keywords: passive optical network, protection, network topology, reliability, CAPEX, OPEX

1. Introduction

Exponential growth in Internet traffic has significantly increased the demand for high bandwidth connectivity at both business and residential premises. Internet service providers are deploying passive optical network (PON) at the access domain to provide the required capacity

in terms of reach, bandwidth, and the number of subscribers. In PON, all services are origi-
nated from an optical line terminal (OLT) at the central office (CO). End-face of the OLT is
connected to a 15–20 km feeder fiber (FF) that extends the network toward the subscriber
premises called optical distribution network (ODN). Remote node (RN) receives the FF at
ODN, which houses a 1 : N bidirectional passive optical coupler (POC). N output ports from
the POC are fed into short-branched distribution fibers (DFs) that connect the RN to individual
optical network unit (ONU) transceiver modules [1].

PON has emerged as a promising candidate to resolve the last-mile bottle, owing to its significant advantages like:

- Support for high network capacity in terms of reach, bandwidth, and the number of
  subscribers due to a complete optical fiber (OF) path between OLT and ONU modules.

- Minimum capital expenditure (CAPEX) by sharing FF between OLT and multiple ONUs.

- Reduced operational expenditure (OPEX) through passive components at RN, which
  requires no power, minimum maintenance and planning.

- Smooth service upgradability with existing infrastructures.

- Highly scalable as new subscribers can easily join the network.

- High degree of flexibility, owing to the use of FF between OLT and multiple subscribers.

With the rapid increase in PONs capacity, fault detection and restoration at satisfactory costs have
turned the network reliability to a new challenge for Internet service providers. Each subscriber is
interested in seamless reception of maximum bandwidth at minimum possible cost. However, the
conventional PON architecture has limited protection, which results in significant data loss at the
event of failure in optical components including OF medium. Therefore, it is imperative to devise
an architecture, which is capable of maintaining a seamless flow of upstream and downstream
traffic at required capacity and acceptable costs for a common end subscriber [1, 2].

Two techniques are readily adapted to provide fault detection and restoration in PON, namely
pre-planned and dynamic protection. The latter relies on fault detection and restoration through
diagnosis at the higher levels and dynamically allocates resources at the event of failure. Such
technique requires more time for traffic restoration between OLT and ONU modules, as upper
layer recovery techniques usually utilize routing tables, topology recalculations, and slow convergence time. Yet there is no guarantee for fault restoration at the physical layer [1–3]. Therefore,
for the facilitation of an effective and prompt fault detection and restoration, it is highly desirable
to provide protection measures at the optical layer.

Pre-planned protection utilize an optical-layer approach by providing dedicated backup paths
for components including OF medium. This type of protection is planned at the network design
phase, owing to the fact that topology of PON remains same, and the proposed solution can
address fault restoration at both feeder and ODN. This type of protection provides high reliability at minimum recovery time in the event of failures at both optical components and OF
medium. However, path and resources duplication significantly elevates the CAPEX at the
network deployment phase [4, 5]. Therefore, it is imperative to encompass the following considerations while designing a pre-planned protection architecture.

2. Consideration for PON protection planning

2.1. Network topology

Network topology significantly effects the design, redundancy, and deployment cost for the PON. Two common network topologies are used for the deployment of PON, namely tree and ring [1]. In tree topology, the optical signal sent from OLT is divided into N equal parts and delivered to designated ONU modules through respective DFs. Such deployment can provide the required bandwidth at desired number of subscribers; however, a single cut or failure at the feeder level can cripple the entire network by disconnecting the working OLT from ONU modules. Moreover, failure at the DF can also result in significant data loss and high customer dissatisfaction. Therefore, such topology requires the provision of redundancy at both levels of PON, which is achieved by duplicating the network components.

Ring topology is adapted to minimize the cost incurred by the provision of redundant paths in the conventional PON. It utilizes a single ring-based fiber that connects the OLT directly to all ONU transceiver modules. This significantly reduces the effect of fiber cuts or failures [6, 7]. Ring topology provides the required reliability at acceptable costs; however, use of the POC between OLT and individual ONU module introduces serious power budget issues, which effects network capacity in terms of the number of subscribers [4, 8, 9]. Besides the commonly used ring- and tree-based network architectures, hybrid topologies are readily adapted for the implementation of survivable PON at the access domain. These architectures utilize a combination of tree- and ring-based architectures with subsequent topologies such as tree-ring, tree-star, ring-star, and bus. Hybrid architectures have proved as a promising candidate to provide the required redundancy at desirable network capacity [10–14].

2.2. Resources to be protected

A typical PON primarily comprises two types of resources that require protection for efficient delivery of information between OLT and ONU modules, namely OF medium and optical components. Both significantly effect the flow of upstream and downstream traffic throughout the network. **Figure 1** shows connection availability for PON components based on **Table 1** [15]. It is observed that active and passive devices, such as OLT, ONU, POC, $X : NPOC$, and so on, provide desirable (5 nines) connection availability over the network lifetime, since the rate of failure for these components is significantly low. Furthermore, the mean time to repair (MTTR) for the in-house optical components is minuscule as compared to the on-field components like OF medium that constitutes a major portion of PON architecture and is more prone to failures as shown in **Figure 1**.

Therefore, OF paths require more attention as compared to other components of the networks, in order to ensure seamless transmission of information, minimize the loss of data, service interruption penalty cost, and PON downtime per year [4].

2.3. Number of subscribers

Number of subscribers refer to the amount of users that a PON can accommodate without compromising the reach and provision of nominal bandwidth. It is an important parameter

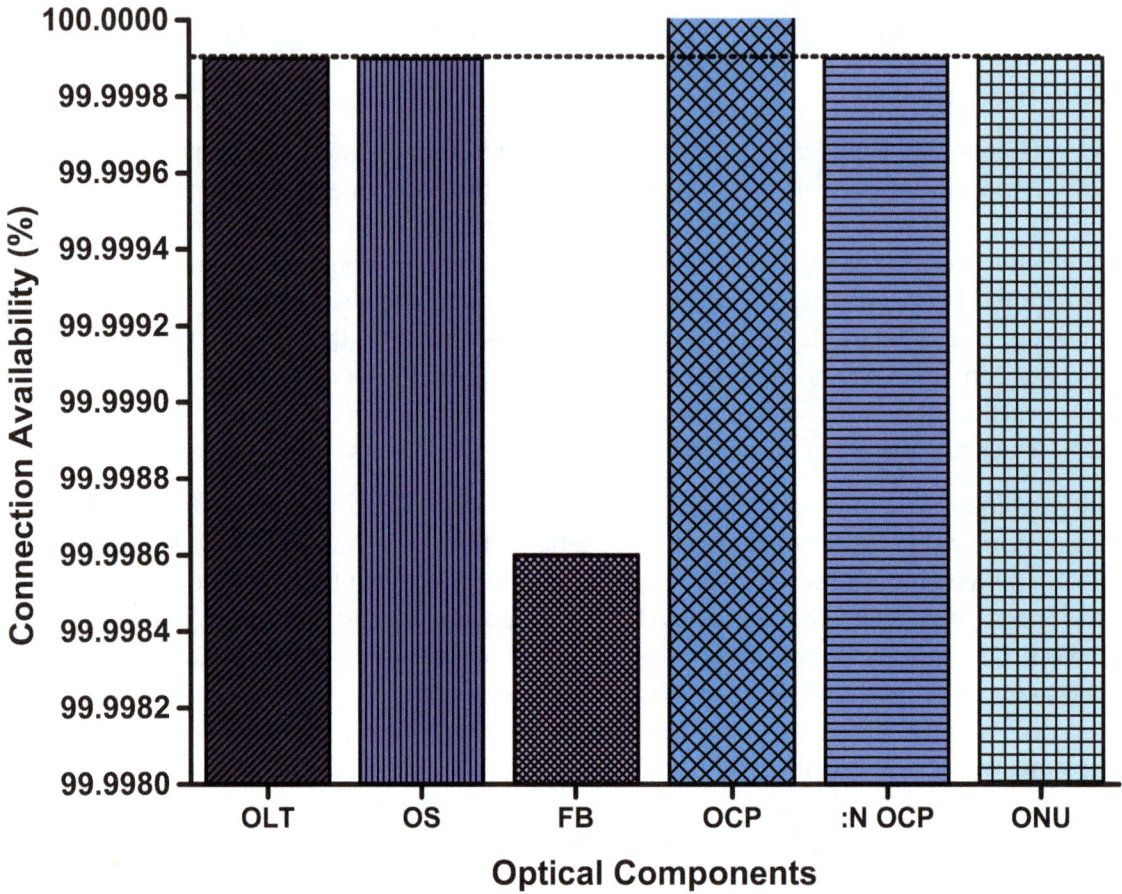

Figure 1. Connection availability of basic optical components in PON.

System components	Unavailability	Cost ($)	Energy consumption
OLT	$5.12e^{-7}$	12100	20 W, 25 W (with EDFA), 2 W (standby)
ONU	$1.54e^{-6}$	350	1 W, 0.25 W (standby)
Optical circulator	$3e^{-7}$	50	
Optical switch	$1.2e^{-6}$	50	
1:2, 2:2 POC	$3e^{-7}$	50	
1:N, 2:N POC	$7.2e^{-7}$	800	
Fiber ($/Km)	$1.3e^{-5}$	160	

Table 1. Components description, connection availability, and cost.

since it is directly associated with the extent and cost of the network. Number of subscribers is primarily effected by the type of topology at both feeder and ODN along with devices at the CO and RN. For example, a typical tree-star topology can accommodate more subscribers as compared to a conventional ring-based architecture due to the use of 1 : N POC. Whereas, the latter utilizes symmetric Y, 1 : 2 or 2 : 2, POC per subscriber, which introduces a power-budget loss of -3 dBm in each symmetric POC, $P_{POC} = 10Log_{10}\left(\frac{1}{2}\right)$ [6, 16]. This significantly effects capacity of the network since nominal received power is required for high bandwidth communication. Therefore, it is imperative to consider these features at the network planning phase, so that the proposed PON can accommodate maximum number of subscribers at desirable capacity and cost.

2.4. Cost and complexity

Deployment/operational costs and feasibility of PONs primarily depend on complexity of the network architecture. For example, some protection mechanics utilize redundant transceivers at both OLT and ONUs, like ITU-T type C and D, in order to avoid $1:1$ or $1 + 1$ switching [17]. Although such techniques provide an abrupt recovery to maintain a smooth flow of information between OLT and ONU modules, they significantly elevate the deployment cost of the network. Since more CAPEX is spent on OLT duplication as compared to the $1:1$ or $1 + 1$ switching, a trade-off must be made between the cost and recovery time at the event of failure. Therefore, it is desirable to minimize the overall system complexity, without compromising the fault detection and restoration time.

3. Protection architectures for PON

Different protection architectures are proposed to facilitate fault detection and restoration in PON. This section discusses five pre-planned protection architectures, which vary in terms of topology, fiber duplication, and devices at both feeder level and ODN.

3.1. ITU-T 983.1 type C architecture

ITU-T 983.1 type C is a pre-planned protection architecture, which provides fault detection and restoration throughout the network with redundant components at both feeder and the distribution levels [17]. The basic type C PON is shown in **Figure 2**, where each component of the network is duplicated to ensure high connection availability and fast restoration time. OLT is placed at the CO and consists of two transceiver modules, where one acts as primary (OLT_p) and another is set as a secondary (OLT_s) module. Under normal mode of operation, OLT_p is responsible for originating and managing services across the network, whereas OLT_s activates in the event of failure at OLT_p module.

Each transceiver module at OLT is connected to a corresponding FF, namely primary (FF_p) and secondary (FF_s). Both fibers extend the network toward the subscribers' premises. Under normal mode of operation, OLT_p is connected to the FF_p. Two FFs are used to provide maximum connection availability and fast restoration time, such that FF_p is immediately replaced with FF_s in the event of failure. The span of each fiber is about 20 km for a standard PON. Both FFs terminate into RN, which serves as chases for two $1:N$ POC modules connected with the corresponding FF at the input port and N DFs at the output port, respectively, where N represents the number of subscribers.

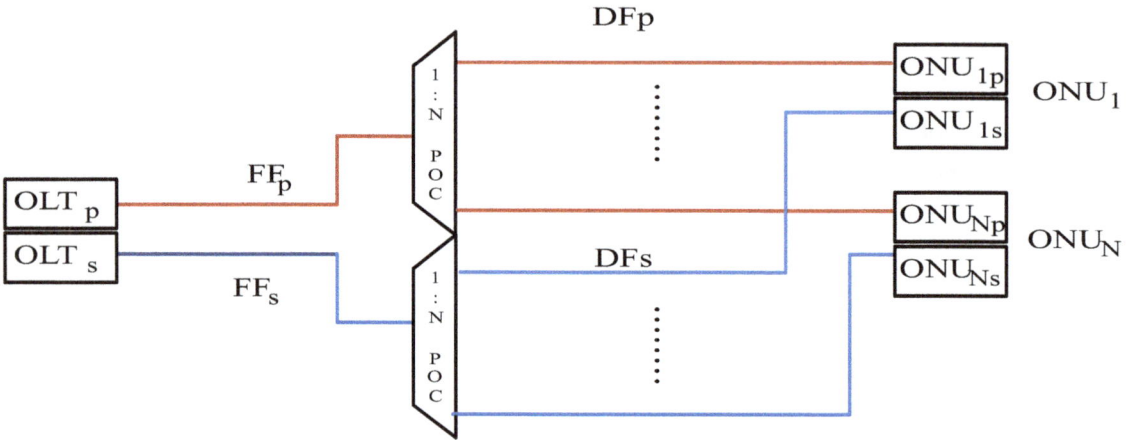

Figure 2. Type C protection architecture.

Consequently, a total of $2N$ DFs are utilized to provide the required protection at the ODN. DFs terminate into ONU with two transceiver modules, (ONU$_p$,ONU$_s$), in order to facilitate abrupt fault detection and restoration. Type C protection architecture duplicates every component including OFs to provide abrupt fault detection and restoration. Furthermore, a two-fiber tree-based topology is laid at the feeder level, whereas a star-based topology is adapted to implement the $2N$ DFs at the ODN.

Number of subscribers in type C protection architecture can be determined through power budget analysis from OLT toward ONU module. Power budgeting in PON ensures an efficient communication between the transmitting and receiving modules. Moreover, it also determines the POC splitting ratio, which translates the number of subscribers in PON, respectively. If P_i represents the power loss across each component i, α is attenuation/km in the OF medium, and R_{sen} represents the receiver sensitivity, then the number of subscribers for type C protection architecture can be determined as:

$$P_T - P_{OLT} - \alpha P_{FF} - P_{POC} - \alpha P_{DF} - P_{ONU} - P_{mis} \geq R_{sen} \tag{1}$$

Equation (1) shows that major power loss occurs across the POC, which is determined by $P_{OPC} = 10\text{Log}_{10}(N)$, where N represents the number of subscribers or splitting ration of $1 : N$ POC. Therefore, to maintain received power $P_{re} \geq R_{sen}$, the value of N must be adjusted to facilitate extended reach, fault detection, and restoration along with high bandwidth connectivity. For example, when the transmitter power is 10 dBm, $\alpha = -0.25 \frac{dB}{km}$, $N = 128$, 25 km fiber, and $P_{mis} = -3$ dBm, the approximate power received at the PIN photo-diode will be $P_{re} = -20$ dBm. Consequently, for $R_{sen} = -25$ dBm, type C architecture can efficiently support 128 subscribers simultaneously accessing the medium.

3.2. Single ring architecture

In order to avoid extensive duplication of the OF medium at both feeder and ODN, ring-based topology is utilized to implement PON with desired connection availability and fault detection/

restoration between OLT and ONU modules [16]. The basic ring-based PON is shown in **Figure 3**, which contains OLT module at the CO. In order to avoid the high deployment cost of OLT module, this architecture employs a single unit, owing to the fact that failure per year of OLT module is minuscule [4, 18]. End-face of the OLT module is connected to a switching arrangement (SA_{co}) that extends the OF medium in both clockwise (CW) and counter clockwise (CCW) directions. SA_{co} serves as chases for a $1:2$ POC_{sa}, and $1:2$ OS_{sa}. POC_{sa} sends and receives the traffic for OLT module. Port 1 splits the optical signal toward the clockwise feeder ring (FR_{cw}), whereas port 2 extends the flow of traffic toward the counter clockwise FR (FR_{ccw}) through a $1:2$ OS_{sa}. This arrangement recovers the flow of information in case of failures at the FR. Under normal working conditions, OS_{sa} is at port a and both upstream and downstream traffic are carried on the FR_{cw}.

FR contains multiple ONU modules $\{ONU_n; n = 1, 2, 3 \ldots N\}$, which are placed directly over the FR through individual RNs (RN_x) housing a $2:2$ $POC_{x,1}$ and a $1:2$ $POC_{x,2}$, respectively. Consequently, the total number of ONU modules is equal to RNs on the FR $X = N$. $POC_{x,1}$ is used to extend the FR to the neighboring ONU modules. Furthermore, it also connects individual ONU_n with FR through $POC_{x,2}$ and controls the flow of traffic in and out of the ONU module.

Network capacity in single ring-based architecture is significantly effected by the utilization of a single POC per subscriber. If N represents the number of RNs, then the total subscribers accessing the medium simultaneously can be determined by:

$$P_T - P_{OLT} - P_{POC_{sa}} - P_{OS_{sa}} - 2 \, XP_{POC_{onu}} - P_{ONU} - P_{mis} \geq R_{sen} \qquad (2)$$

X represents the number of RNs in Eq. (2), which is equal to ONUs N that can access the medium simultaneously. Consequently, for the transmitter power of 10, -3 dBm loss across the POC, $P_{mis} = -3$ dBm, 25 km fiber, and $R_{sen} = -25$ dBm, this architecture can scarcely support 16 subscribers simultaneously accessing the medium [6]. It is observed that fiber duplication is avoided at both levels through a ring-based structure; however, power drop across each POC

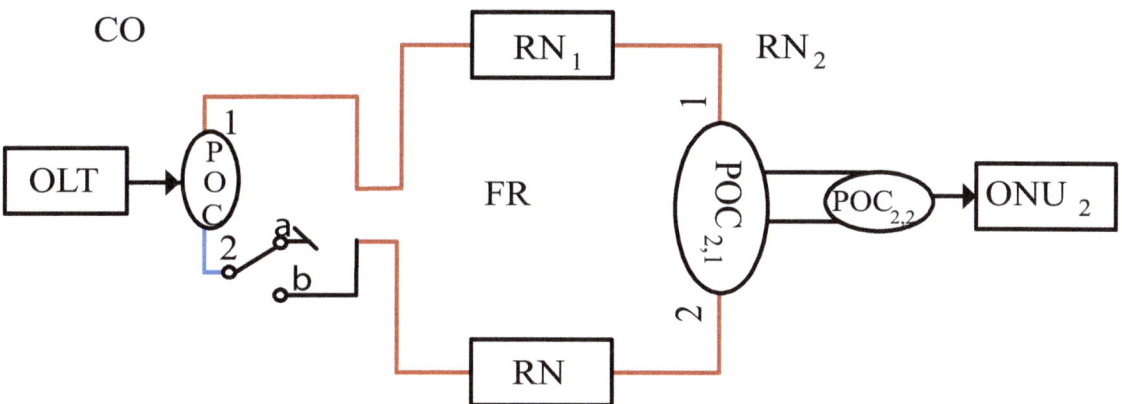

Figure 3. Single ring protection architecture.

introduces serious power budget issues. Consequently, capacity of the PON is effected. It is therefore imperative to utilize a topology that is capable of providing fault detection and restoration without compromising the network capacity and total cost.

3.3. Dual ring protection architecture

Dual ring PON avoids fiber duplication at the feeder level through a single ring-based fiber (FR) as shown in **Figure 4** [9]. Furthermore, several small rings are deployed at ODN to negate the power budget issue. CO contains a single OLT module, which is fed into an optical circulator (OC_{co}) and erbium-doped fiber amplifier (EDFA) module. To maintain passive nature of the PON, EDFA is placed inside the CO as shown in **Figure 4**. Switching arrangement SA_{co} is placed at end-face of the EDFA, which serves as chases for $1:2$ POC_{sa} and $1:2$ OS_{sa}. Port 1 of POC_{sa} splits the down-stream traffic towars FR_{cw}, whereas port 2 extends the flow of traffic towars the FR_{ccw} through a $1:2$ OS_{sa}. In normal mode of operation, all traffic is carried through the FR_{cw}. However, OLT medium access control (MAC) layer flips the switch position when a failure occurs at the FR. This new position of the switch converts the ring-based fiber in two trees-based fibers carrying traffic in clock- and counter clockwise direction between the PoF.

Both FR_{cw} and FR_{ccw} are fed into multiple remote nodes forming several distribution rings (DRs) at the ODN. Each remote node $\{RN_x; x = 1, 2, 3...X\}$ houses two $2:2$ bidirectional POCs, namely ($POC_{x,1}, POC_{x,2}$) as shown in **Figure 4**. The former extends the FR from CW toward CCW direction and the latter carries the traffic from feeder toward the ODN and vice versa. End-face of $POC_{x,2}$ splits the incoming traffic into clock- and counter clockwise DRs (DR_{cw}) and (DR_{ccw}), respectively. Both DR_{cw} and DR_{ccw} are connected to individual ONU

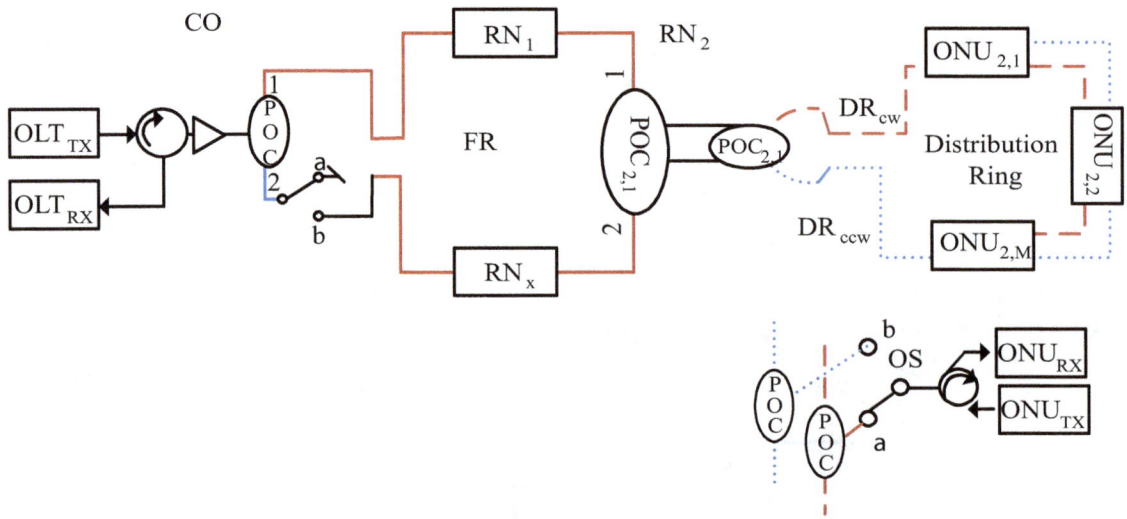

Figure 4. Dual ring protection architecture.

$\{ONU_{x,m}; \; m = 1,2,3...M\}$ module through two 1:2 $POC_{onu_{x,m}}$ and $OS_{onu_{x,m}}$. Under normal working conditions, $OS_{onu_{x,m}}$ is at position a and all traffic is handled through DR_{cw}.

Dual ring architecture utilizes multiple rings at the distribution level to compensate the excessive power drop. Furthermore, EDFA is also employed at the CO to support high capacity transmission. Consequently, the power budget equation from OLT toward ONU module can be written as:

$$P_T - P_{OLT} - P_{OC_{co}} + P_{EDFA} - P_{POC_{sa}} - P_{OS_{sa}} - \alpha P_{FR} - 2XP_{POC_{RN}} - \alpha P_{DR} - MP_{POC_{onu}} - P_{ONU} \geq R_{sen}$$

$$(3)$$

Where X, in Eq. (3), represents the total number of RNs deployed at the FR and M represents the number ONUs per RN. Consequently, the total ONU modules accessing the medium simultaneously become $N = X \times M$. Now, at $P_T = 10$ dBm, 25 km fiber, and $P_{edfa} = 25$ dB, the maximum value of N that can be achieved is 72 with $X = 9$ and $M = 8$ [9]. It is observed that the dual ring architecture supports more ONU modules in comparison with the single ring architecture. Nevertheless, power drops in the ring topology at the distribution level limits the overall capacity of the network.

3.4. Tree-based hybrid protection architecture

This architecture employs a hybrid topology through combination of tree and star-ring architectures at the feeder and ODN, respectively. OLT is placed at the CO, which is fed into an OC_{co}, EDFA, and SA_{co} as shown in **Figure 5** [15]. SA_{co} consists of a $1:2$ OS_{sa}, with port a connected to the FF_p, whereas port b is fed into FF_s. In normal working conditions, OS_{sa} is at port a and all upstream and down-stream traffic is sent and received through the FF_p. However, in case of failures or cuts at the FF_p, OLT MAC layer flips the switch position and resumes the flow of traffic through the FF_s.

Since a tree-based topology is adapted at the feeder level, the required protection is provided by duplicating the long-span fiber. End-face of both FF_p and FF_s terminates into RN that serves as chases for a $2:N$ POC. Output of the $2:N$ POC is connected to a series of dedicated DFs connecting N ONU transceiver modules as shown in **Figure 5**. In order to avoid extensive duplication of the DFs for protection, a ring-based topology is adapted by connecting intermediate ONU modules, respectively. Consequently, each ONU_n module, where $\{n = 1,2,3...,N\}$, consists of $1:2$ POC_n and $1:2$ OS_n. Port 1 of the POC is connected to respective DF, whereas port 2 is fed into a $1:2$ OS_n. Port 3 extends the DF toward a DR that provides the necessary protection in case of failure at the DF.

OS_n is used to connect OC_n and ONU_n with transmission media at the distribution level. Port "a" of OS_n connects ONU_n with the DF, whereas port "b" is used to connect ONU_n with the redundant DR in case of failure. Consequently, the required protection is achieved by avoiding the extensive duplication of the DFs. Under normal mode of operation, OS_n is at port "a" and all services are delivered through the DF, whereas in case of failure the affected ONU_n MAC

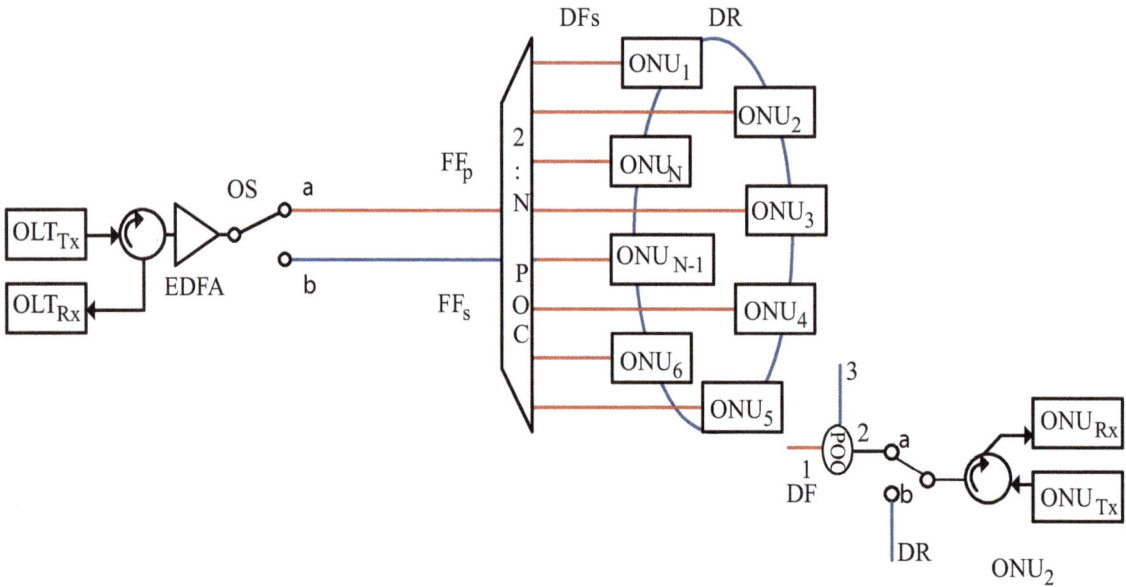

Figure 5. Tree-based hybrid protection architecture.

layer flips the switch position and transfers the flow of traffic to the DR between the adjacent ONU module.

The downstream power budget for tree-based hybrid protection architecture can be written as:

$$P_T - P_{OLT} - P_{OC_{co}} + P_{EDFA} - P_{OS_{co}} - \alpha P_{FF} - 10\text{Log}_{10}N - \alpha P_{DF}$$
$$- P_{POC_{onu}} - P_{OS_{onu}} - P_{OC_{onu}} - P_{ONU} \geq R_{sen} \tag{4}$$

It is observed from Eq. (4) that for $P_T = 10$ dBm, 25 km fiber, $P_{edfa} = 25$ dB, and $R_{sen} = -25$ dBm, the tree-based hybrid protection architecture can efficiently support 128 subscribers simultaneously accessing the medium [15].

3.5. Ring-based hybrid protection architecture

This architecture is formed by the combination of a single ring topology at the feeder level and a star-ring architecture at the ODN. A single OLT is placed at the CO, which is connected to an OC_{co}, EDFA, and SA_{co} as shown in **Figure 6** [18]. This architecture utilizes a single ring-based fiber to provide the required high bandwidth connectivity and reliability at the feeder level. Port 1 and 2 of the POC_{sa} extend traffic toward the FR_{cw} and FR_{ccw}, respectively. A 1:2 OS_{sa} is connected with port 2 of the POC_{sa} to provide the required fault recovery in the event of failure. Port a of the OS_{sa} is connected to ground, whereas port b is fed into the FR_{ccw}. Under normal mode of operation, OS_{sa} is at port a and all traffic is sent and received through the FR_{cw}.

FR is implemented at each RN by combining FR_{cw} and FR_{ccw} through a special arrangement. This significantly reduces the power budget penalty and cost of the overall architecture. RN

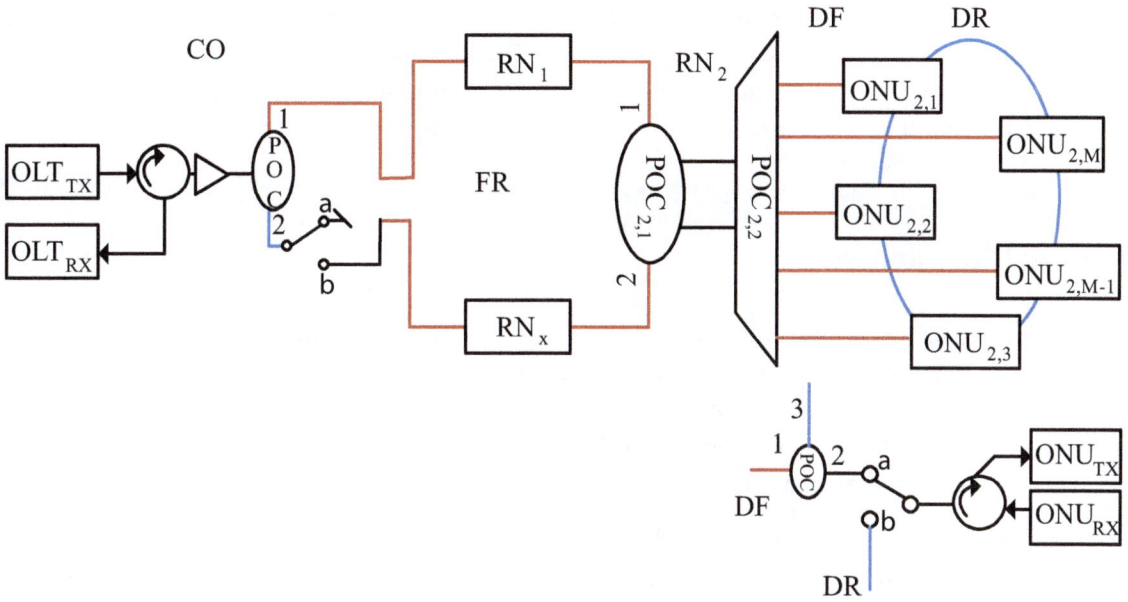

Figure 6. Ring-based hybrid protection architecture.

in the ring-based hybrid network $\{RN_x;\ x = 1, 2, 3...X\}$ consists of two POCs, namely $(POC_{x,1}, POC_{x,2})$. Where $POC_{x,1}$ and $POC_{x,2}$ are bidirectional couplers with $2:2$ and $2:M$ ports respectively. FR is formed by connecting both CW and CCW paths through port 1 and 2 of the $POC_{x,1}$ as shown in **Figure 6**. $POC_{x,1}$ further extends the FR toward the ODN through a $2:M$ $POC_{x,2}$, where M is the number of ONUs connected to each $POC_{x,2}$ through dedicated DFs. If X represents the total number of RNs deployed over the FR and M is the number ONUs per RN, then the proposed system can support a total of $N = X \times M$ ONUs.

ONU module, $\{ONU_{x,m};\ m = 1, 2, 3...M\}$, starts with a 1:2 $POC_{x,m}$ with three ports as shown in **Figure 6**. Port 1 is used to connect each ONU to its dedicated DF, whereas port 2 is fed into 1:2 $OS_{x,m}$. Port 3 is used to form a backup DR, which ensures immediate survivability of traffic between ONUs and OLT. Port a of the $OS_{x,m}$ connects $ONU_{x,m}$ with corresponding DF, under normal working conditions, and port "b" is fed into the DR.

The number of subscribers in ring-based hybrid protection architecture can be determined as:

$$P_T - P_{OLT} - P_{OC_{co}} + P_{EDFA} - P_{POC_{sa}} - P_{OS_{sa}} - \alpha P_{FR} - XP_{POC_{x,1}} - 10\mathrm{Log}_{10}(M) - \alpha P_{DF}$$
$$- P_{POC_{onu}} - P_{OS_{onu}} - P_{OC_{onu}} - P_{ONU} \tag{5}$$

If N represents the total number of ONUs, then the value of X can be written as:

$$X \le \frac{P_T - R_{sen} - 16.25 - 10\mathrm{Log}_{10}M + P_{EDFA}}{3} \tag{6}$$

Analysis of Eq. (6) shows that capacity of the network increases when more ONUs are placed per RN. Furthermore, it is observed that for $P_T = 10$ dBm, $P_{edfa} = 25$ dB, and $R_{sen} = -25$ dBm, the ring-based hybrid protection architecture can efficiently support 128 subscribers simultaneously accessing the medium [18].

4. Reliability analysis

Network reliability analysis is an important tool that is used to determine the overall connection availability for a given protection architecture. Reliability block diagrams (RBDs) are commonly used to determine the overall connection availability of PON, owing to its significant advantages like, accuracy, simplicity, visual impact, and flexibility. This section analyzes network reliability in terms of connection availability of the selected protection architectures through RBDs.

RBD represents each component including the OF medium as a functional block connected in series or parallel combination with adjacent blocks. Series and parallel connectivity represent the unprotected and protected components in protection architecture, respectively. The characteristic parameter of each RBD block is the asymptotic unavailability (U_i) of the components that represent their probability of failures. Consequently, if I represents the total number of components in PON, then overall connection availability A is given by:

$$A = 1 - \sum_{i=1}^{I} U_i \tag{7}$$

where U_i is determined by

$$U_i = 1 - \frac{\text{MTBF}}{\text{MTBF} + \text{MTTR}} \tag{8}$$

MTBF in Eq. (8) is the mean time between failures, which is used to represent the number of failures per million hours for an optical component [19]. MTTR represents the mean time to repair, which is the time required for reparation or replacement of faulty hardware modules [5, 20, 21]. In order to determine the overall connection availability for each architecture, RBDs with series and parallel combination of functional blocks are extracted as shown in **Figure 7**.

If $\sum_{i=1}^{I} U_i$ is the summation of i components unavailability in a protection architecture, then connection availability for each survivable PON can be written as:

$$A_{[C]} = 1 - \left[\left(U_{OLT} \times U_{OLT_{pt}} \right) + \left(U_{FF} \times U_{FF_{pt}} \right) + \left(U_{1:N\,POC} \times U_{1:N\,POC_{pt}} \right) + \left(U_{DF} \times U_{DF_{pt}} \right) \right.$$
$$\left. + \left(U_{ONU} \times U_{ONU_{pt}} \right) \right]$$

$$\tag{9}$$

$$A_{[SR]} = 1 - \left[U_{OLT} + U_{POC} + U_{OS} + \left(U_{FR} \times U_{FR_{pt}} \right) + U_{POC} + U_{POC} + U_{ONU} \right] \tag{10}$$

$$A_{[DR]} = 1 - \left[U_{OLT} + U_{POC} + U_{OS} + \left(U_{FR} \times U_{FR_{pt}} \right) + U_{POC} + U_{POC} + \left(U_{DR} \times U_{DR_{pt}} \right) \right.$$

$$\left. + \left(U_{POC} \times U_{POC_{pt}} \right) + U_{OS} + U_{ONU} \right] \tag{11}$$

$$A_{[HT]} = 1 - \left[U_{OLT} + U_{OS} + \left(U_{FF} \times U_{FF_{pt}} \right) + U_{2:N\,POC} + \left(U_{DF} \times U_{DR_{pt}} \right) + U_{POC} + U_{OS} + U_{ONU} \right] \tag{12}$$

$$A_{[HR]} = 1 - \left[U_{OLT} + U_{POC} + U_{OS} + \left(U_{FR} \times U_{FR_{pt}} \right) + U_{POC} + U_{2:N\,POC} \right.$$

$$\left. + \left(U_{DF} \times U_{DR_{pt}} \right) + U_{POC} + U_{OS} + U_{ONU} \right] \tag{13}$$

Figure 8 shows the overall connection availability of the selected protection architectures based on **Table 1**. It is observed that maximum 5 nines connection availability is provided by all architectures, owing to the efficient utilization of redundant components and OF medium throughout the network. Maximum connection availability is provided by type C protection architecture, which duplicates entire PON including OLT and ONU modules. Ring-based architecture at both feeder and ODN, which avoids the duplication of light-wave path, is also observed to provide the required availability. A single failure at the ring-based fiber converts

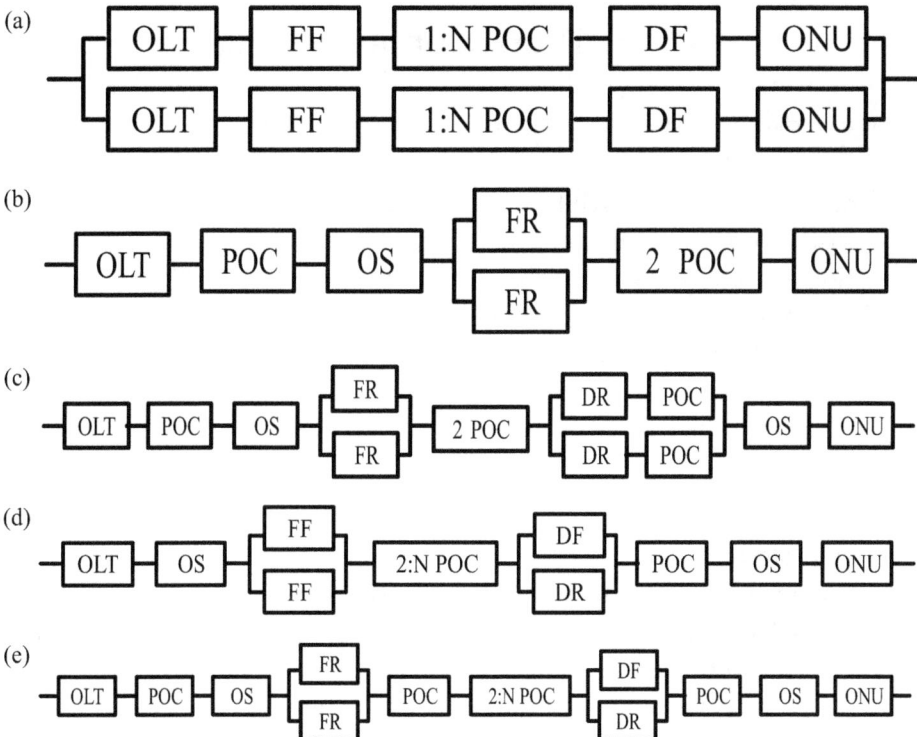

Figure 7. Reliability block diagrams for (a) type C, (b) single ring (SR), (c) dual ring (DR), (d) tree-based hybrid (HT), and (e) ring-based hybrid (HR) protection architectures.

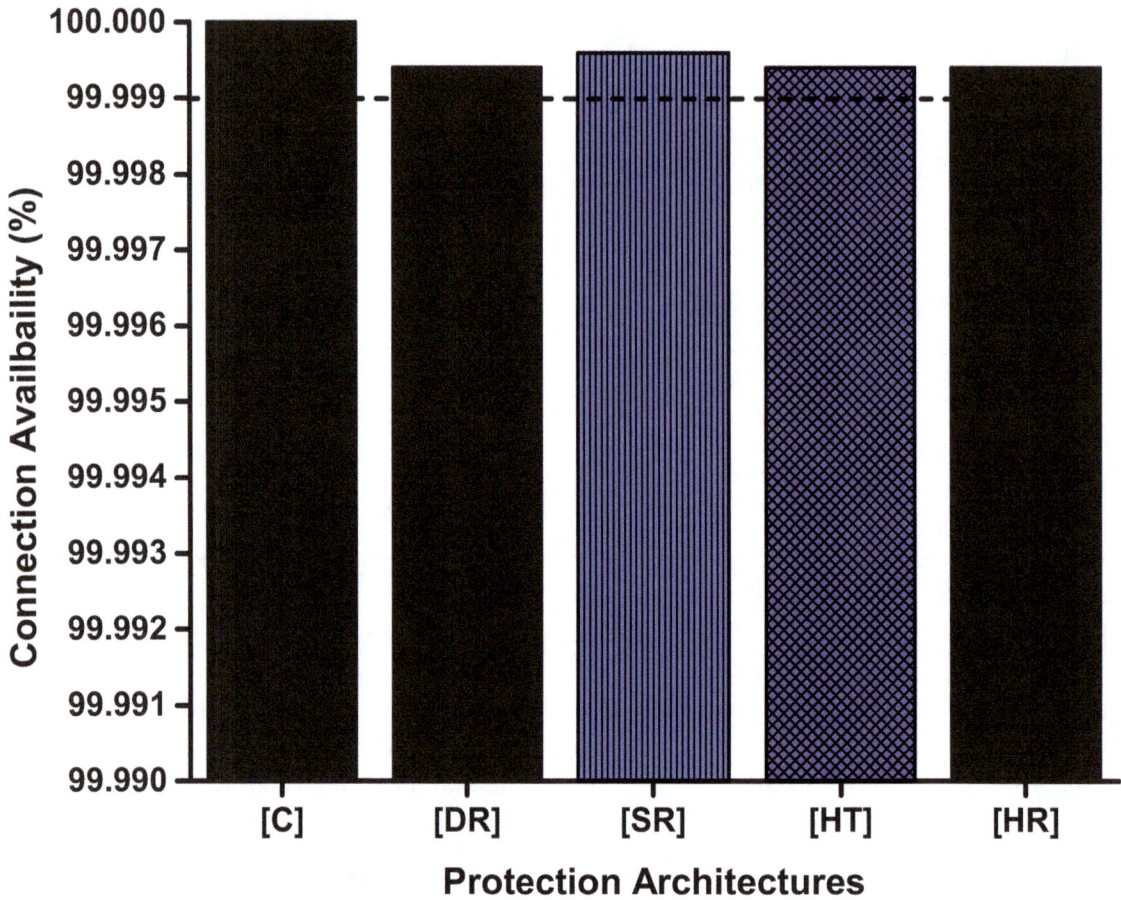

Figure 8. Overall connection availability of selected protection architectures.

the network into two tree-based architectures with the flow of traffic in both clockwise and counter clockwise directions. Thus, redundancy is achieved without duplicating the entire fiber. Furthermore, it is observed that desirable connection availability can be maintained without duplicating the transceiver modules, which further helps in reducing the overall cost of deployment for such architectures.

5. Cost analysis

Cost figures for protection architectures are an important parameter, showing the economic benefits for a common end user at the access domain. This section determines the overall expenditure for the selected architecture while using component costs in **Table 1**. Cost figures for PON can be categorized as CAPEX and OPEX. CAPEX includes the investment utilized for the deployment of PON at access domain and is calculated by computing the cost expenditure on network devices along with the OF infrastructure. OPEX includes the cost incurred on the network operations from the time of deployment till replacement by a new technology [4, 15, 22, 23].

OPEX primarily includes the cost required for repairing the faulty, failed equipment (including OF medium), service interruption penalty cost that is commonly applicable for business subscribers, and energy consumed by active components at transceiver modules of OLT and ONU. Reparation cost is determined by multiplying the total downtime/year with resources required to remove the fault, which includes the number of technicians along with their wages and miscellaneous charges. Service interruption penalties include the expenditure that is spent on the fine defined in service level agreement (SLA) between network operators and subscribers. Power consumption by each component is determined by multiplying the unit price of electricity with the sum of energy consumption of all active components over the network life span [15, 23].

Analysis is performed for residential customers only, and life span of each network is taken as 20 years. Furthermore, following specifications are considered for fair analysis of the selected protection architectures

- OLT can support 16, 32, 64, and 128 subscribers based on the type of adapted topology.

- Length of FF = 20 km, DF = 5 km, and DR = 1 km between adjacent nodes in selected architectures.

- EDFA cost is considered in dual ring, tree- and ring-based hybrid architectures.

- Digging cost for OF medium is ignored due to high variation.

- No service interruption penalty cost is considered.

- Repairing cost is 1000 $/h.

- Per hour cost of electricity is taken to be 0.25 $/kWh.

If N is the total number of ONUs in each PON and X represents the number of RNs in ring-based topologies at the feeder level, then the CAPEX equations (based on RBDs) for selected protection architectures can be written as:

$$C_C = (OLT + OLT) + (20 \times FF \times FF) + (2 \times 1 : N\ POC) + (5 \times N \times DF \times DF)$$
$$+ (N \times ONU \times ONU) \tag{14}$$

$$C_{SR} = OLT + POC + OS + (25 \times FR) + (2X \times POC) + (N \times ONU) \tag{15}$$

$$C_{DR} = OLT + POC + OS + EDFA + (20 \times FR) + (2X \times POC) + (N \times DR_{cw})$$
$$+ (N \times DR_{ccw}) + (2N \times POC) + (N \times OS) + (N \times ONU) \tag{16}$$

$$C_{HT} = OLT + OS + EDFA + (20 \times FF \times FF) + (2 : N\ POC) + (5 \times N \times DF)$$
$$+ (N \times DR) + (N \times POC) + (N \times OS) + (N \times ONU) \tag{17}$$

$$C_{HR} = OLT + POC + OS + EDFA + (20 \times FR) + (X \times POC) + (X \times 2 : N\ POC)$$
$$+ (5 \times N \times DF) + (N \times DR) + (N \times POC) + (N \times OS) + (N \times ONU) \tag{18}$$

Figure 9 shows the CAPEX for each protection architecture at different number of subscribers by referring to **Table 1**. It is observed that ring-based architecture requires minimum CAPEX when compared to conventional tree and hybrid techniques. However, due to the power budget and capacity limitations of single and dual ring-based architectures, their analysis is performed till 16 and 64 number of subscribers, respectively. Overall analysis shows that the deployment cost for each architecture decreases as the number of subscriber increases. It is evident from the fact that total cost incurred is distributed among the number of subscribers, which results in reduced CAPEX per user as the network capacity increases. Furthermore, it is shown that type C architecture requires the highest cost, for all subscribers, in comparison with other schemes due to the extensive duplication of optical components and OF medium throughout the network.

Figure 9 shows that hybrid protection architectures provide nominal performance, for all subscribers, in terms of the deployment cost as they avoid extensive duplication of the OF medium. **Figure 10** further elaborates the CAPEX required for the deployment of light wave path in type C, tree- and ring-based hybrid protection architectures for 128 subscribers in the

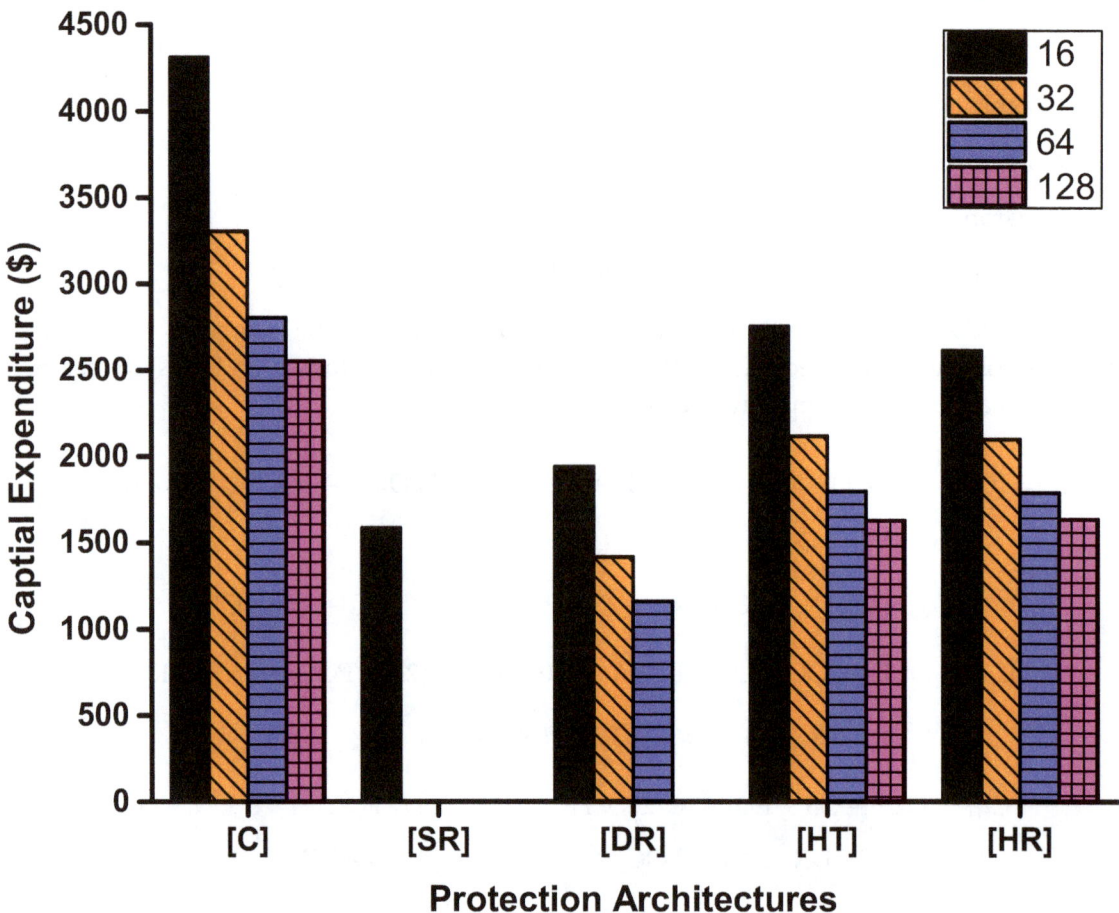

Figure 9. CAPEX at different number of subscribers.

network. It is observed that the main contributor for CAPEX, besides OLT module, is the DF as each subscriber requires a dedicated path for high speed connectivity. Furthermore, it is shown in **Figure 10** that type C requires the highest deployment cost for DFs due to the duplication of each fiber between RN and corresponding ONU module. On the other hand, the deployment of star-ring topology at ODN significantly reduces the CAPEX, on redundancy, for hybrid protection architectures.

It is also observed that the CAPEX in hybrid protection architecture varies with the type of topology at the feeder level. Since ring-based architecture requires multiple RNs, the cost of deployment increases with the number of subscribers due to increase in RNs. Consequently, tree-based hybrid architecture requires less cost when compared to ring-based scheme for 64 and 128 subscribers. However, the difference is minuscule.

Figure 11 shows the total expenditure for protection architectures with high network capacity in terms of the number of subscribers. It is evident from the fact that each architecture provides the desired connection availability, which significantly reduces their downtime/year. Hence, minuscule amount is spent on reparation cost over the network life span. Type C protection architecture requires highest OPEX due to the duplication of transceiver modules at both OLT

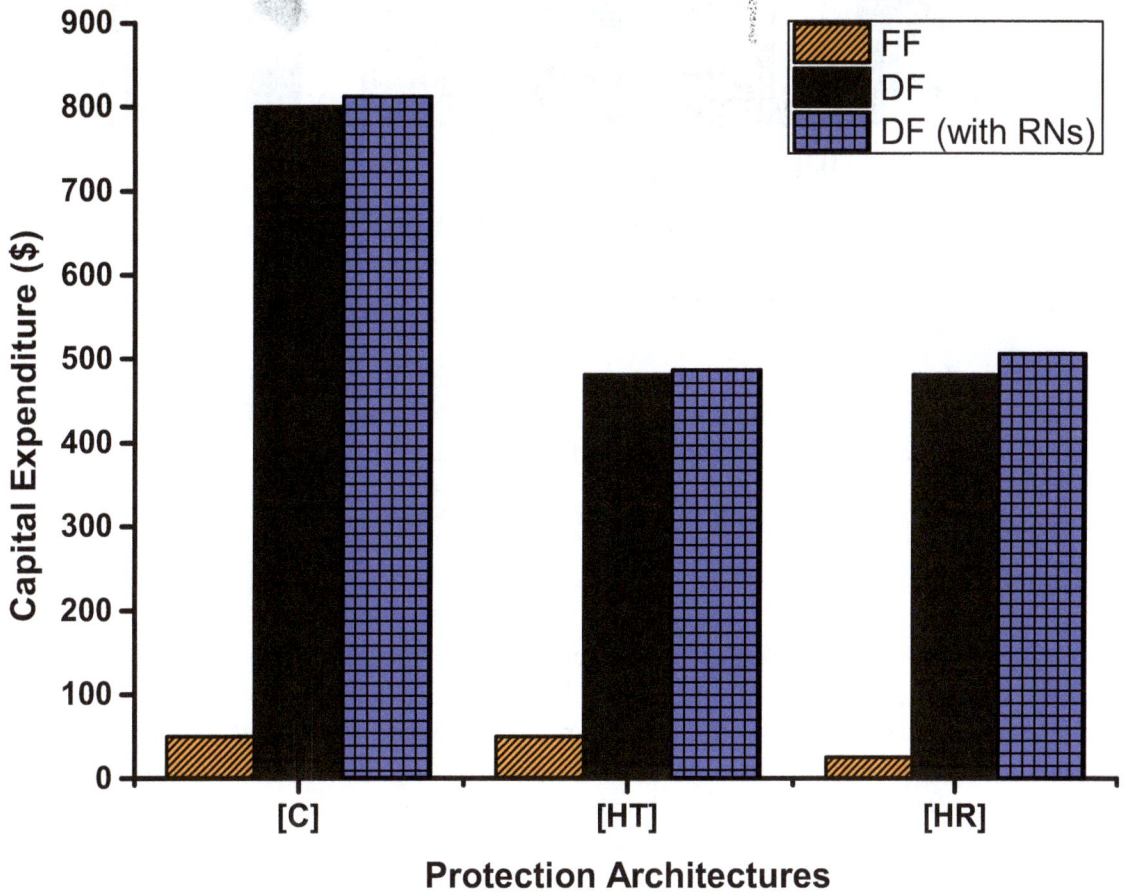

Figure 10. CAPEX for OF infrastructure at 128 subscribers.

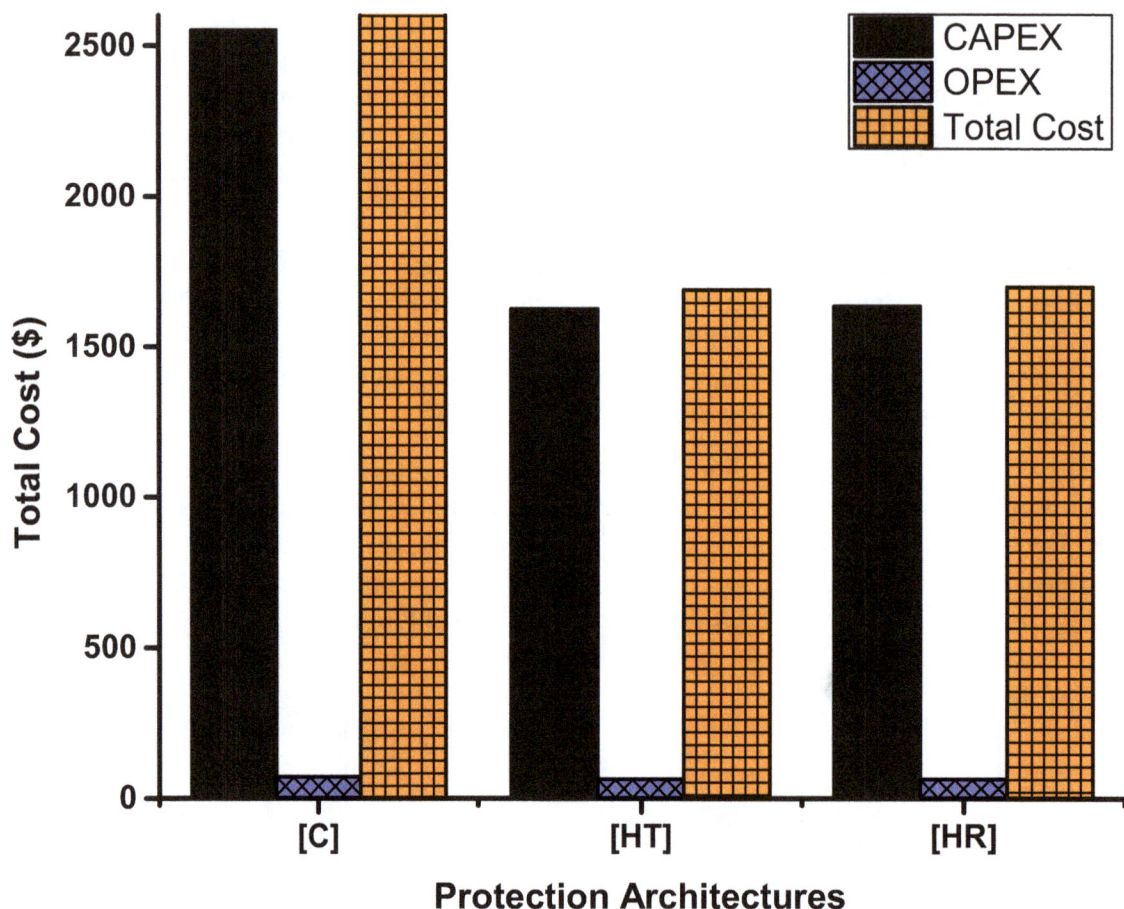

Figure 11. Total cost including CAPEX and OPEX at 128 subscribers.

and ONUs, which consumes more energy as compared to hybrid architectures with single OLT and ONU modules. Consequently, hybrid architecture requires minimum cost, collectively, as compared to the conventional protection scheme.

6. Conclusion

This chapter analyzes the effect of fiber duplication, for protection, on feasibility of PON. Increase in the capacity and adaptability of PON demands for an efficient fault detection and restoration mechanism, which is able to provide the required connection availability at minimum cost. Among all components in the PON, OF medium requires significant protection due to its high rate of failure. Furthermore, it is an on-field component and constitutes a major portion of PON; therefore, it is imperative to provide pre-planned redundant paths in order to ensure swift recovery of failures at the OF medium throughout the network. Five protection architectures are considered, which mainly emphasize on the type of topology laid through the OF medium at both feeder level and ODN. ITU-T 983.1 type C and single ring-based architectures are considered to encompass the conventional tree-star and ring topology, respectively. For further analysis, dual ring architecture and tree- and ring-based hybrid protection architectures for PON are analyzed to study the effect of hybrid topologies on PON feasibility in terms of the number of subscribers, reliability, and cost. It is observed that ring-based architectures provide desirable connection availability without duplicating the OF medium; however, the use of multiple POCs to extend the FR significantly elevates the power budget issues. Consequently, the number of subscribers that can access the medium simultaneously is reduced. Hybrid topologies formed by the variation of tree, ring, and star topology provide nominal performance, in terms of all parameters, as compared to the survivable PONs based on conventional topologies. Furthermore, it is shown that duplication of the OF medium at the ODN network is more critical in comparison with the feeder level duplication. Nevertheless, both tree- and ring-based hybrid pre-planned protection architectures provide desirable network capacity and connection availability at minimum cost.

Author details

Waqas Ahmed Imtiaz[1]*, Javed Iqbal[2], Affaq Qamar[1], Haider Ali[1] and Sevia M. Idrus[3]

*Address all correspondence to: w.imtiaz.1985@ieee.org

1 Abasyn University, Peshawar, Pakistan

2 Sarhad University of Science and Information Technology, Peshawar, Pakistan

3 Faculty of Electrical Engineering, Universiti Teknologi Malaysia, Malaysia

References

[1] Cedric F Lam. *Passive optical networks: principles and practice*. Academic Press, 2007. ISBN 978-0123738530.

[2] Mozhgan Mahloo, Jiajia Chen, Lena Wosinska, Abhishek Dixit, Bart Lannoo, Didier Colle, and Carmen Mas Machuca. Toward reliable hybrid wdm/tdm passive optical networks. *IEEE Communications Magazine*, 52(2):273–280, 2014. doi: 10.1109/MCOM.2014.6736740.

[3] Hussein T Mouftah and Pin-Han Ho. *Optical networks: architecture and survivability*. Springer Science & Business Media, 2012. ISBN 978-1-4615-1169-4.

[4] Lena Wosinska, CHEN Jiajia, and Claus Pop Larsen. Fiber access networks: reliability analysis and swedish broadband market. *IEICE Transactions on Communications*, 92 (10):3006–3014, 2009. doi: 10.1587/transcom.E92.B.3006.

[5] Mozhgan Mahloo. *Reliability versus cost in next generation optical access networks*. Doctoral Dissertation, School of Information and Communication Technology, KTH Royal Institute of Technology, 2013, ISSN 1653-6347 ; 1303.

[6] Pavel Lafata and Jiri Vodrazka. Application of fiber ring for protection of passive optical infrastructure. *Radioengineering*, 22(1):357–362, 2013, ISSN: 1805-9600.

[7] Chien-Hung Yeh and Sien Chi. Self-healing ring-based time-sharing passive optical networks. *IEEE Photonics Technology Letters*, 19(13–16):1139–1141, 2007. doi: 10.1109/LPT.2007.900155.

[8] Byung Tak Lee, Mun Seob Lee, and Ho Young Song. Simple ring-type passive optical network with two-fiber protection scheme and performance analysis. *Optical Engineering*, 46(6), 2007. doi: 10.1117/1.2746929.

[9] Waqas Ahmed Imtiaz, Yousaf Khan, and Khalid Mahmood. Design and analysis of self-healing dual-ring spectral amplitude coding optical code division multiple access system. *Arabian Journal for Science and Engineering*, 41(9):3441–3449, 2016. doi: 10.1007/s13369-015-1988-z.

[10] Haibin Chen, Chaoqin Gan, Maojun Yin, and Cuiping Ni. A single-star multi-ring structure of self-healing wavelength division multiplexing optical access network. *Fiber and Integrated Optics*, 33(1–2):4–16, 2014. doi: 10.1080/01468030.2013.879435.

[11] Yang Qiu and Chun-Kit Chan. A novel survivable architecture for hybrid wdm/tdm passive optical networks. *Optics Communications*, 312:52–56, 2014. doi: 10.1016/j.optcom.2013.09.005.

[12] Yeh C.H., Chow C., and Liu Y.L. Self-protected ring-star-architecture tdm passive optical network with triple-play management. *Optics Communications*, 284:3248–3250, 2011. doi: 10.1016/j.optcom.2011.03.032.

[13] Rastislav Róka. Optimization of traffic protection schemes for utilization in hybrid passive optical networks. *International Journal of Application or Innovation in Engineering and Management*, 5(9):107–116, 2016. ISSN 2319-4847.

[14] Elaine Wong, Carmen Mas Machuca, and Lena Wosinska. Survivable hybrid passive optical converged network architectures based on reflective monitoring. *Journal of Lightwave Technology*, 34(18):4317–4328, 2016. doi: 10.1109/JLT.2016.2593481.

[15] Waqas Ahmed Imtiaz, Yousaf Khan, and Waqas Shah. Cost versus reliability analysis of tree-based hybrid protection architecture for optical code division multiple access system. *Journal of Engineering and Applied Sciences*, 35(1):47–53, 2016.

[16] Chen-Hung Yeh and Sien Chi. Self-protection against fiber fault for ring-based power-splitting passive optical networks. *Optical Engineering*, 47(2): 020501–020501, 2008. doi: 10.1117/1.2841702.

[17] Rec ITU-T. G. 983.1. study group 15: broadband optical access systems based on passive optical networks (pon). *ITU-T, Rec. G*, 983, 1998.

[18] Waqas A Imtiaz, Pir Mehar, Muhammad Waqas, and Yousaf Khan. Self-healing hybrid protection architecture for passive optical networks. *International Journal of Advanced*

Computer Science and Applications (IJACSA), 6(8):144–148, 2015. doi: 10.14569/IJACSA. 2015.060819.

[19] Susan Stanley. Mtbf, mttr, mttf & fit explanation of terms. *IMC Network*, 1–6, 2011.

[20] Patrick O'Connor and Andre Kleyner. *Practical reliability engineering*. John Wiley & Sons, 2012. ISBN 978-0-470-97982-2.

[21] David J Smith. *Reliability, maintainability and risk 8e: Practical methods for engineers including reliability centred maintenance and safety-related systems*. Elsevier, 2011. ISBN 978-00809690 22.

[22] Álvaro Fernández and Norvald Stol. Capex and opex simulation study of cost-efficient protection mechanisms in passive optical networks. *Optical Switching and Networking*, 17:14–24, 2015. doi: j.osn.2015.01.001.

[23] Jiajia Chen and Lena Wosinska Wosinska. Cost vs. reliability performance study of fiber access network architectures. *IEEE Communications Magazine*, 48:56–65, 2010. doi: 10.1109 /MCOM.2010.5402664.

3

Dielectric Resonator Nantennas for Optical Communication

Waleed Tariq Sethi, Hamsakutty Vettikalladi,

Habib Fathallah and Mohamed Himdi

Additional information is available at the end of the chapter

Abstract

Dielectric resonator antennas (DRA) are ceramic based materials that are nonmetallic in nature. They offer high permitivity values (ε_r: 10-100). DRAs' have made their mark in various applications specially in the microwave and millimeter wave (MMW) spectrum, and are making encouraging progress in the THz band, because of their low conduction losses and higher radiation efficiencies compared to their metallic counterparts. With the advancements in nano fabrication, metallic antennas designed in the THz band have taken an interest. These antennas are termed as optical antennas or nantennas. Optical antennas work by receiving the incident electromagnetic wave or light and focusing it on a certain point or hot spot. Since most of the antennas are metallic based with Noble metals as radiators, the conducting losses are huge. One solution that we offer in this work is to integrate the nantennas with DRs. Two different DR based designs, one triangular and other hexagonal, are presented. Both the antennas operate in the optical C-band window (1550 nm). We design, perform numerical analysis, simulate, and optimize the proposed DR nantennas. We also consider array synthesis of the proposed nantennas in evaluating how much directive the nantennas are for use in nano network applications.

Keywords: dielectric resonator (DR), optical nanoantenna, hexagonal DR nantenna, triangular nantenna, optical communication C-window

1. Introduction

Over the last few years, with the introduction of various portable and wireless handheld devices, a surge in mobile and internet data traffic has been observed. This drastic increase is also effected by the way our society creates, shares and consumes information on a regular basis. These abundant data and information sharing also demand for increase in delivery time. High data rate communication with compact size of device is the new norm in technology. Researchers and industry are working at a fast pace to fulfil their consumers demands. It is estimated that wireless data rates are getting doubled every year and are quickly approaching the provided capacity of the wired communication systems [1–5]. Following this trend, systems working at higher data rates are needed. Although millimetre waves and 60 GHz radio [6–9] currently provide a solution and are being implemented in 5G applications, still researchers have to think ahead of time. The electromagnetic spectrum has a lot of bands to offer in terms of wide bandwidth. One section of the spectrum that has not been explored completely in the Terahertz (THz) band can offer communication in the Terabit-per-second (Tbps) domain. These (Tbps) links can be realised over the next 10 years. Terahertz band communication [10–14] is intended as a key wireless technology to satisfy this growing demand of bandwidth-hungry devices with requirements of higher data rates. Since THz band is in the early exploring phase, a lot of revisions and new standards have to be designed for the systems operating at these higher bands of the electromagnetic spectrum (ultraviolet band-infrared band). Special considerations have to be made in the design of transmission and receiving portion of the THz system [15].

Antenna is an important component of any wireless system. For the antenna to be designed at this higher end of the spectrum (THz band), special tools are needed from its realisation to characterisation. The antennas designed at this spectrum are termed as optical nanoantennas [16]. Optical nanoantennas work on the operating principle that the electromagnetic (EM) wave or the light wave received can be controlled or placed into localised energy spots pertaining to the design of the nanoantenna. This property of the optical nanoantennas have gained immense interest from the research point of view as it can be applied to various fields of applications, such as spectroscopy, sensing, photodetection, metasurfaces, medicine, photovoltaics and energy-harvesting applications[17–20]. With the advancements in nanotechnology fabrication, the nanoantennas designed at THz band can be realised. Nanotechnology is a fast-growing research area that marked use of machines that can fabricate nanocomponents. It is considered as an enabling technology for a set of applications in biomedical, environmental and military fields as shown in **Figure 1**. Being inherently simple and performing primitive operations only, nanomachines in isolation are not expected to manage advanced tasks. To enable more complex applications such as intra-body drug delivery or cooperative environment sensing, the exchange of information and commands between networking entities and/or external controller is required. The need for coordination and information sharing naturally leads towards the concept of nanonetworks. One promising way to enable networking capabilities is to use wireless communications between nanomachines [21] made possible with optical nanoantennas fabricated on these machines.

Optical nanoantennas present some similarities with their radio frequency (RF) counterparts, yet there still exists some major difference. The main challenge arises from the fabrication tolerances at nanoscales and from the drastic deviation of metals from perfect conductors to lossy plasmonic materials at optical frequencies [22]. This is usually described as the well-known dispersive plasmonic effects and results in a significant decrease of radiation efficiency caused by conduction losses. These losses can be theoretically explained via different models at

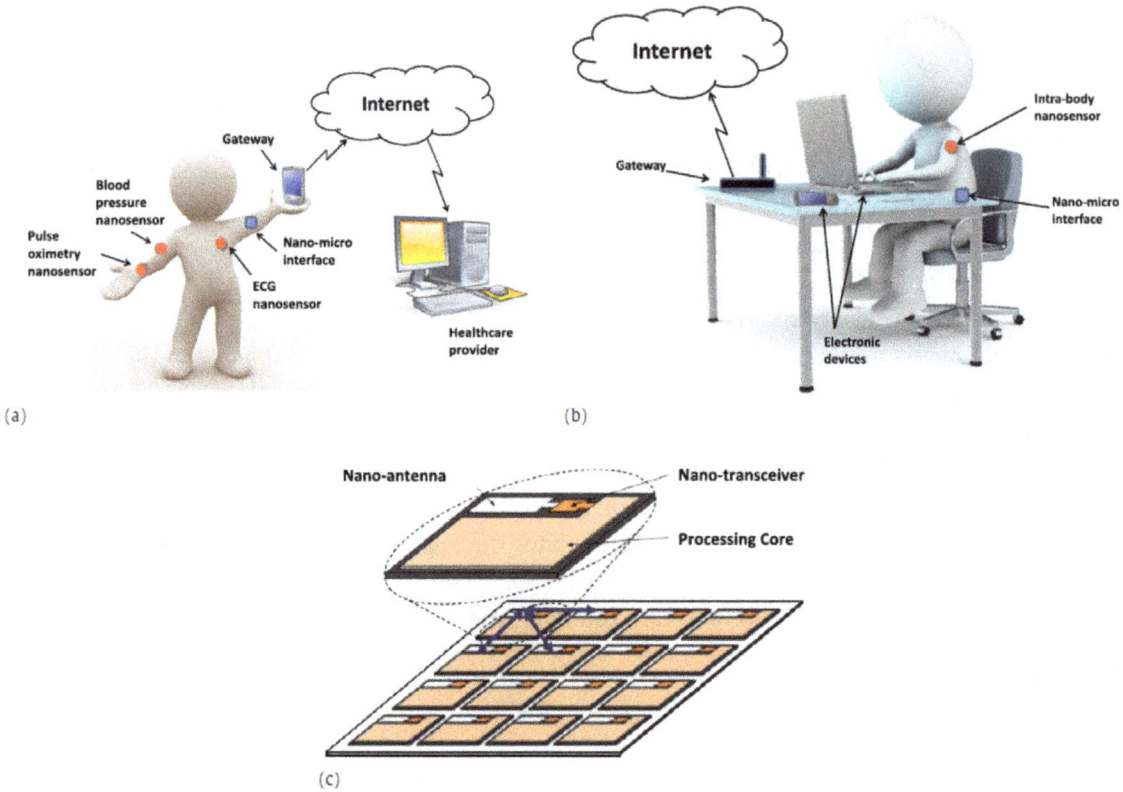

Figure 1. Applications for terahertz band (nanoscale antennas) [15].

specific frequencies with mostly known models such as Drude and Lorrentz model [23, 24]. One solution that is proposed in this chapter to cope with these metallic losses at high frequencies and to fully utilise the properties of metallic nanostructures is to design resonators or absorbers made with dielectric-based materials such as ceramics.

The idea of using antennas based on dielectric resonators (DR) was first proposed by Long et al. in 1983 [25], and since then the research into this novel idea is steadily increasing among the antenna researchers. The operation of DRAs as radiators exploit the 'radiation losses' of dielectric resonators made of moderate to high relative permittivity ($5 < \varepsilon_r < 100$) when excited in their lower order resonant modes in an open environment. Over the years, various aspects of DR antennas have been published and designs patented [26, 27]. Most notable publications are seen at the antennas operating as filters and circuits integrators at the microwave regime due to their compact size and high resistibility to losses. One striking feature of DR resonators is that they are immune to ohmic losses, which drastically appear at higher frequencies (bands above millimetre waves till far infrared), making them suitable candidates for optical nanoantennas. DRs are very efficient radiators compared to metallic antennas even at higher frequencies [28]. In this context, the work presented in this chapter focuses on integrating the resilient DR with metallic nanoelements for operating at THz band. The losses incurred due to appearing of plasmonics inside the metal elements are subsided with the integration of various DR-shaped resonating elements. The results presented in the later sections validate the selection of DR as a suitable candidate and replacement to traditional metallic resonators working at higher THz band.

2. State-of-the-art antennas

Optical nanoantennas are an attractive area for research in the field of optics and nanophotonics. With the advent of nanofabrication machines, the antennas designed at the lower RF and microwave millimetre wave (MMW) domains can now be scaled up to the THz domain. Thus, some of the properties and analysis obtained for antenna designs at the lower band can be applied to higher bands. To study the effects and analyse the current distributions on the optical nanoantennas, a new branch of physics emerged known as nanophotonics. Nanophotonics studies the transmission and reception of optical signals by submicron and even nanometre-sized objects. For nanooptics, it is important to efficiently detect and direct the transmitting signals for optical information between nanoelements. The sources and detectors of radiation in nanooptics are nanoelements themselves, their clusters and even individual molecules (atoms, ions). Nano objects functioning as antennas must exhibit high radiation efficiency and directivity [29, 30]. Most of the nanoantennas existing in the literature are based on plasmonic metallic structures (**Figure 2**). In Ref. [31], dipole nanoantennas exhibit the electric field localisation at certain spots, whereas the bowtie nanoantennas presented in Ref. [32] present broadband characteristics; and in Ref. [33], Yagi-Uda displays high directive nature which can be useful for nanonetwork communication among nanoscaled devices. Similarly in Refs. [34, 35], plasmonic nanoantennas provide enhanced and controllable light-matter interactions and strong coupling between far-field radiation and localised sources at the nanoscale. In Refs. [36, 37], magneto-plasmonic response of the nanoantennas is observed when ferromagnetic metals are driven not only by light but also by external magnetic fields. The authors observed that the magneto-plasmonic nanoantennas enhance the magneto-optical effects, which introduces additional degrees of freedom in the control of light at the nanoscale. However, regardless of various advantages of plasmonic nanoantennas associated with their small size and strong localisation of the electric field, such nanoantennas have large dissipative losses resulting in low radiation efficiency.

To alleviate this conduction loss phenomenon, we propose a combination of DR and plasmonic metallic-based optical nanoantennas. Literature review shows some of the existing designs that perform well when working with DRs. Since DR offers high dielectric constant and refractive index values of the materials, the losses are minimised and subsided when integrated or placed with metallic resonators. In rest of the sections, we detail the proposed DR-based nanoantenna designs with simulated results. The enhancement in the directivity of the nanoantenna is also discussed by implementing an array structure of 1×2 ETDRNA elements. Also, the tunability of the nanoantenna array is discussed. Finally, the chapter ends with the conclusion section discussing the presented results based on selected DR design geometries.

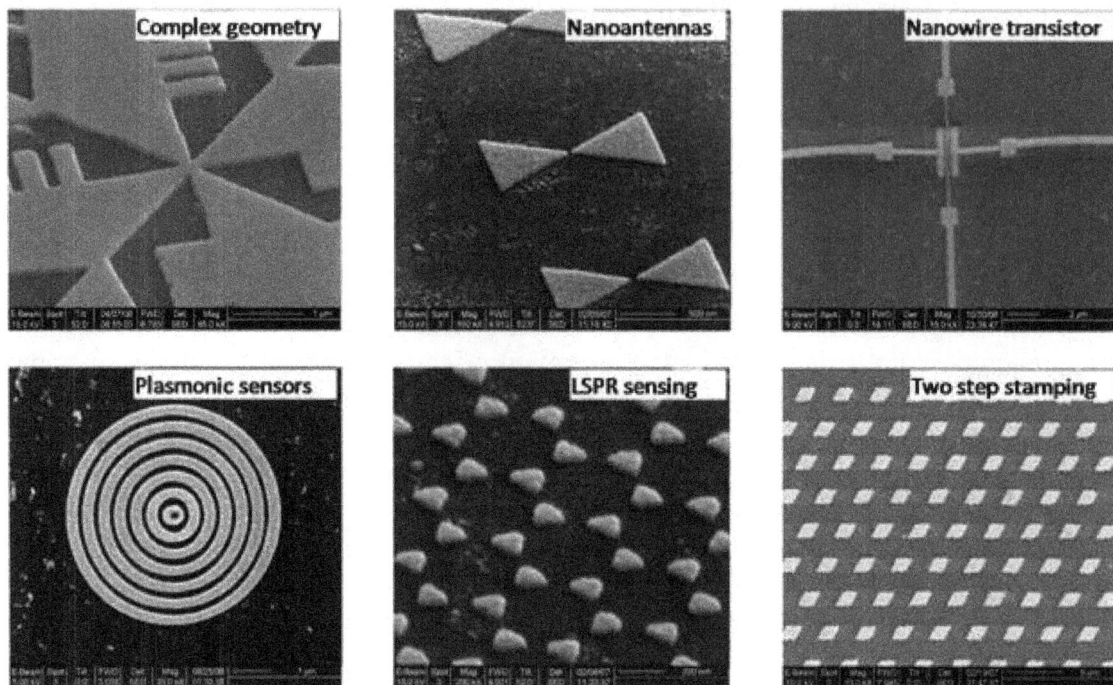

Figure 2. Different types of plasmonic nanoantennas: (a) complex connection, (b) bowtie, (c) dipole nanowires, (d) spiral sensor, (e) bowtie sensors and (f) rectangular patches [38].

Although we know that realization of any device is the ultimate proof of its operation, but at these very high THz frequencies it is difficult to have them fabricated as per limitations in fabrication resources. Secondly, our proposed designs are very small in dimensions (as will be discussed in later sections) to be realized currently with the exiting nano fabrication tools. Therefore we present the simulated designs using two different DR based resonators that we think will be viable candidates for designers and researchers who are in need of antenna designs having minimum losses and better radiation characteristics in the optical communication C-band at 1550 nm.

3. Equilateral triangular dielectric resonator nantenna

In this section, we present the simulated design and analysis of an equilateral triangular dielectric resonator nantenna (ETDRNA). The proposed nantenna is composed of a multilayer 'Ag-SiO$_2$-Ag' structure with noble metal silver (Ag) working as a feed transmission line. The dielectric triangular is made of silicon (Si) material and is excited via coupling mechanism from the feed line. The antenna yields a wide impedance bandwidth of 2.58% (192.3–197.3 THz) with a high directive radiation pattern of 8.6 dBi at 193.5 THz (1550 nm) with an end-fire radiation pattern.

3.1. Antenna geometry

Figure 3 shows the configuration (cross-sectional and front view) of the proposed equilateral triangular dielectric resonator nantenna (ETDRNA). The nanoantenna is designed to operate as a receiving antenna that can capture energy from the free space. The operating band of interest lies in the standard optical communication band at a wavelength of 1550 nm, which corresponds to central frequency of 193.5 THz. From the geometric configuration presented in **Figure 3**, the design follows a basic multilayer substrate approach. A silicon substrate (Si) having an oxide layer (O_2) is sandwiched between two conducting materials layers. The SiO_2 substrate has properties of thickness of $h_1 = 0.150$ μm, $\varepsilon_r = 2.09$ and loss tangent $tan\ \delta = 0$ [39]. The partial conducting material below the substrate acts as a ground plane. Its dimensions and

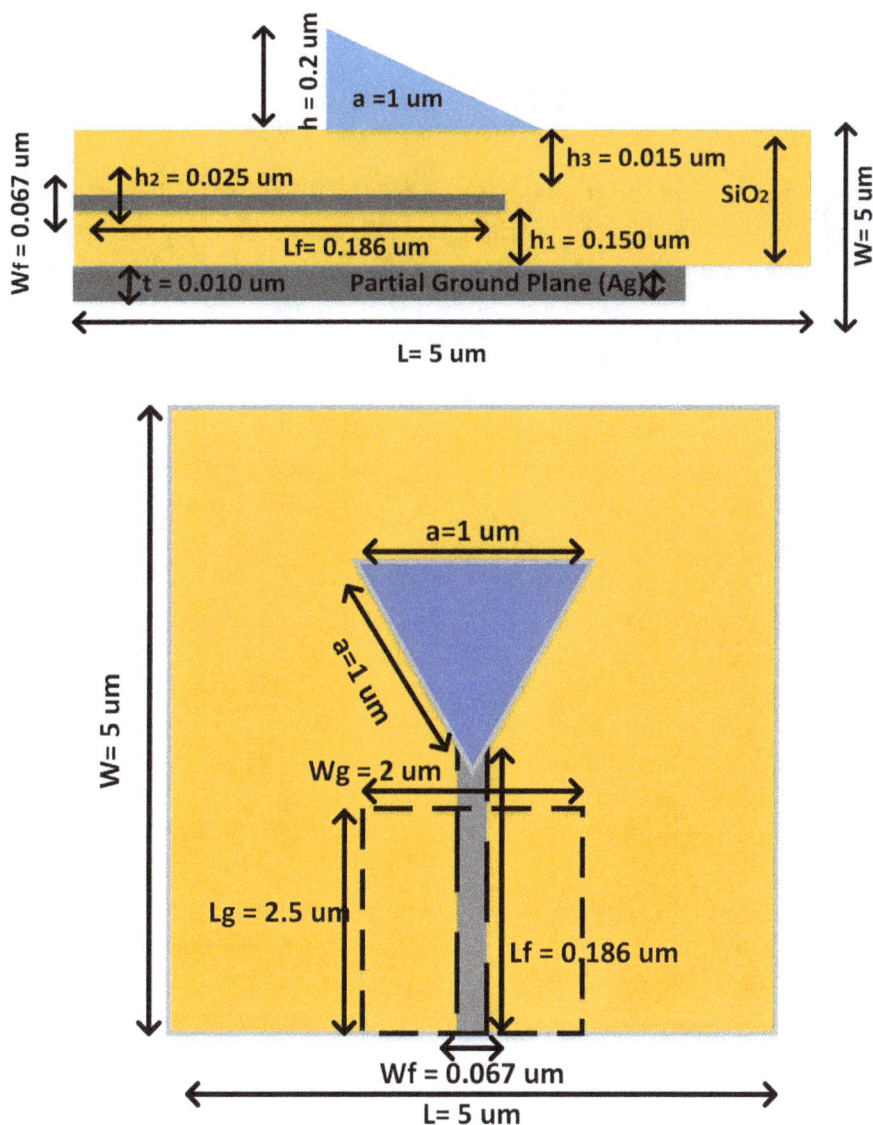

Figure 3. Antenna geometry of proposed nantenna design based on equilateral triangular DR (ETDRNA): (a) cross-sectional view; (b) front view.

thickness are $W_g \times L_g$ having a thickness of $t = 0.010$ μm. The nanoantenna is fed via a feed line placed on the top side of the substrate. It has geometric dimensions and thickness of $W_f \times L_f$; h_2 = 0.025 μm. The ground and the nanostrip are made up of noble metal silver (Ag). The dimensions of the substrate are taken as $W \times L = 5 \times 5$ μm^2. The resonator, made from (Si), placed on top of the feed line is made from dielectric material. It has the shape of an equilateral triangle with properties as $\varepsilon_r = 11.9$ and estimated loss tangent $\tan \delta = 0.003$ at 100 THz [40]. The nanoantenna is excited and matched considering 50 Ω impedance source. In order to control the matching at the central frequency of 193.5 THz and to achieve a wide bandwidth with acceptable radiation patterns, the same (SiO$_2$) substrate material with thickness $h_3 = 0.015$ μm has been introduced between the equilateral triangle and the nanostrip. The dimensions of the equilateral triangular dielectric are calculated from Eq. (1) [41, 42].

$$f_{mnl} = \frac{c}{2\sqrt{\varepsilon_r}} \left[\sqrt{\left(\frac{4}{3a}\right)^2 + \left(\frac{p}{h}\right)^2} \right]^{1/2} \tag{1}$$

where 'a' is the side length of the equilateral triangular DRA, ε_r is the dielectric constant of the DRA, 'h' is two times the height of the triangular DRA to account for the image effect of the ground plane and $p = 1$ for the fundamental mode [42]. For a low-profile triangular DRA, we have $a \gg h$, and therefore Eq. (2) demonstrates that the frequency is predominantly determined by the height of the DRA:

$$fr = \frac{c}{4h\sqrt{\varepsilon_r}} \tag{2}$$

where h and ε_r are the height and dielectric constant of triangular DRA.

Metals working in the optical regime are faced with another loss. This loss appears as negative permittivity, therefore complex permittivity ε_{Ag} of the metals, in our case silver (Ag), is calculated from Eq. (3) explained by the Drude model which is based upon kinetic theory of electron gas in solids [39]:

$$\varepsilon_{Ag} = \varepsilon_o \left\{ \varepsilon_\propto - \frac{f_{p^2}}{[f(f + i\gamma)]} \right\} = -129.17 + j3.28. \tag{3}$$

where $\varepsilon_o = 8.85 \times 10^{-12}$ [F/m], $\varepsilon_\propto = 5$, plasmonic frequency $f_p = 1.41e^{16}$ rad/s, f = central frequency and collision frequency $\gamma = 2.98e^{13}$. The plasmonic frequency, which appears after the photon and free electron gas collision, defines the collective motion of the electrons and can be expressed as follows:

$$f_p = \sqrt{ne^2/\varepsilon_0 m} \tag{4}$$

where n is electron concentration, e is the free electron charge (1.6×10^{-19} C), ε_0 is the free space (vacuum) permittivity (8.854×10^{-12} F/m) and m is the electron effective mass. From

Eq. (4), the behaviour of arriving EM wave to the metal can be deduced. For $f < f_p$ corresponds to and exponential decay field, EM wave will be reflected back and will not propagate through the metal. On the other hand, if $f > f_p$, the EM wave will behave as a travelling wave and will pass through the metals. Similarly, the collision or damping factor describes the losses within the metal and can be expressed as:

$$\gamma = \frac{e}{\mu m} \tag{5}$$

where μ is mobility of free carriers. The proposed model has taken into account the conductive and dielectric losses and has been simulated in commercially available EM simulator CST MWS 2014 based on FIT numerical technique using optical template.

3.2. Design simulation and optimisation

To get an understanding on the working principle of the proposed ETDRNA, various parameters of the nanoantenna were extensively optimised. In order to study the effects of the antenna performance in terms of bandwidth and directivity, the following parameters were observed and analysed.

a.   **Nanostrip feed:** The nanostrip feed line placed on top of the SiO$_2$ substrate was optimised in terms of its length and width. The traditional empirical formulas [43] were used as a starting point for the nanostrip design. The nanostrip acts like a coupling resonator that excites the triangular dielectric place on an upper SiO$_2$ substrate with height h_3. Traditionally at RF frequencies, the length of the transmission lines are characterised to the wavelengths (λ) of incoming and outgoing radiations. However, working at the optical frequencies, the incident waves reflect less and penetrate more in to the substrate passing through the metal atoms. This phenomenon is known as plasmonic affect, and it gives rise to free plasmonic gaseous atoms. To deal with this new phenomenon at optical frequencies, we use shorter effective wavelength (λ_{eff}) compared to traditional wavelengths (λ), which depends on material characteristics given by Eq. (6) for length of a transmission line [44]:

$$\frac{m\lambda_{eff}}{2} = L(\lambda_o) \tag{6}$$

where Eq. (6) shows the relationship between the free space wavelength (λ_o) and the effective wavelength (λ_{eff}) and the order of resonance (m). Here, effective wavelength is given by:

$$\lambda_{eff} = \frac{\lambda_o}{n_{eff}} \tag{7}$$

Typical values of n_{eff} has been measured to be in the range of 1.5–3 [45]. In our case for the silver nanostrip feed line, we use $n_{eff} = 2.8$ which resulted the minimum resonating length of to be 0.27 µm. The length L_f of the nanostrip was optimised from 0.1 to 0.27 µm with the best optimised value producing required resonance at 193.5 THz was at $L_f = 0.186$ µm as shown in **Figure 4a**. The effect of the width 'W_f' of the nanostrip was also examined by extensive parametric studies. Initial values were taken from the empirical formulas [43] and optimisation was done from 0.02 to 0.28 μm. **Figure 4b** shows the best optimised value achieved at resonance of −22 dB with W_f = 0.067 μm.

b. **Partial ground plane:** The ground plane plays an important role in controlling the bandwidth and radiation characteristics of any designed antenna. At the nanoscale geometry, we simulated and observed its parameters effect on our nanoantennas resonance behaviour. We started off initially with a finite ground plane that achieved a good radiation pattern with an acceptable bandwidth. The ground plane was then optimised and a partial section of it was used with optimised dimensions $L_g \times W_g$ = 0.5 μm × 2 μm. **Figure 5a** and **b** shows the effects of varying the ground plane in terms of its length and width. The optimised results produce a wide impedance bandwidth of 2.5% (192.3–197.3 THz) at a centre frequency of 193.5 THz. This makes our proposed nanoantenna covers all the standard optical transmission widow (C-band), with a directivity of 8.6 dB.

c. **Height of triangular DR:** Since the height of the triangular DR predominately determines the resonance frequency as according to Eq. (2), the height h of the DR was optimised from 0.1 to 0.5 μm. **Figure 6** shows the best optimised value of h = 0.3 μm having a resonance at −23 dB.

d. **Rotation of triangular DR:** In order to study the effects of bandwidth, frequency shift and directivity of the nanoantenna design, the proposed silicon-based triangular DR was rotated on its axis. The rotation was from 0° to 360° with an angular spacing of 40°. **Figure 7** shows the angular rotation of the triangular DR. The tip of the triangle was initially aligned at 0° shown in green colour. The DR was then rotated along the counterclockwise direction with varying angles. It was observed that with the rotation of the DR, the bandwidth remained the same at 2.5% but the resonant frequency shifted to other bands (200–205 THz) in the frequency range from (180–220 THz) as shown in **Figure 8a**. Since the triangle is an equilateral one, the angular rotation produces the same shifts at

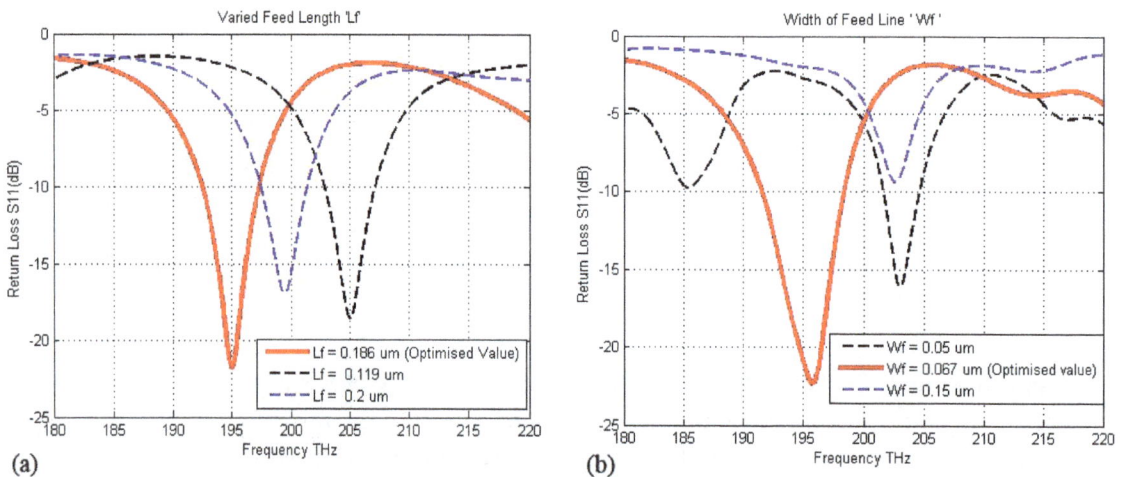

Figure 4. Optimized parameters: (a) length of feed line; (b) width of feed line.

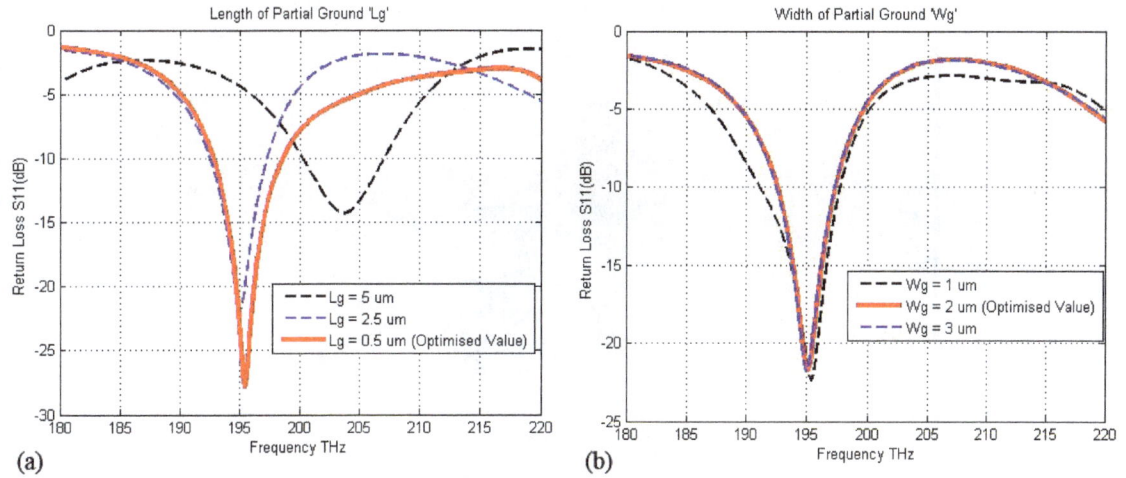

Figure 5. Optimized parameters: (a) length of ground plane; (b) width of ground plane.

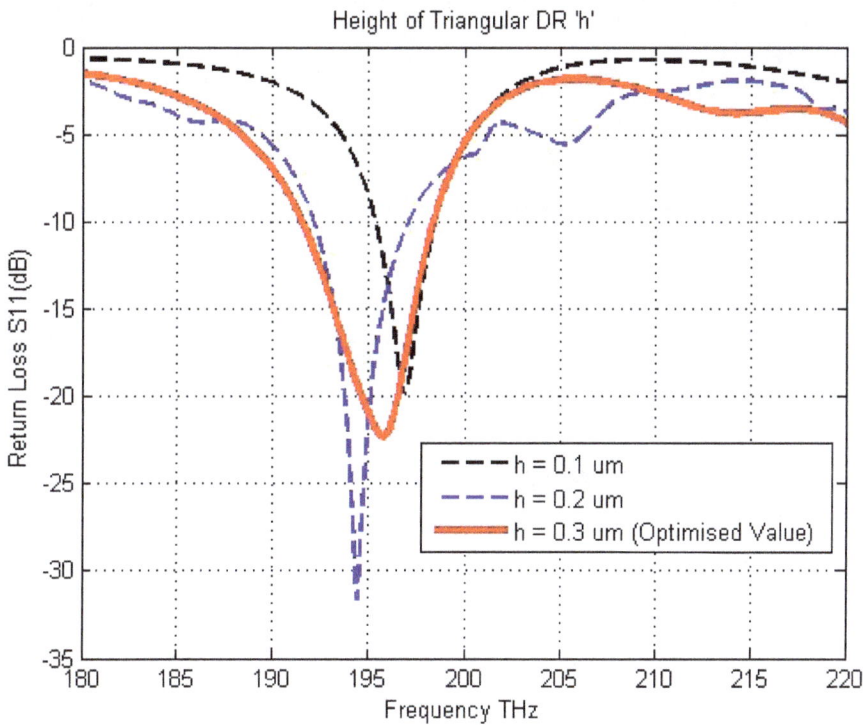

Figure 6. Optimized parameters: height of triangular DR.

other angles, that is, the shift will be the same at $(0 = 120 = 240 = 360°)$ as shown in **Figure 8b**. The directivity was also affected with the rotation of the triangle as shown in **Figure 8b**. It is clear that the effect of the rotation of the triangular DR lowers the directivity to nearly 3 dBi.

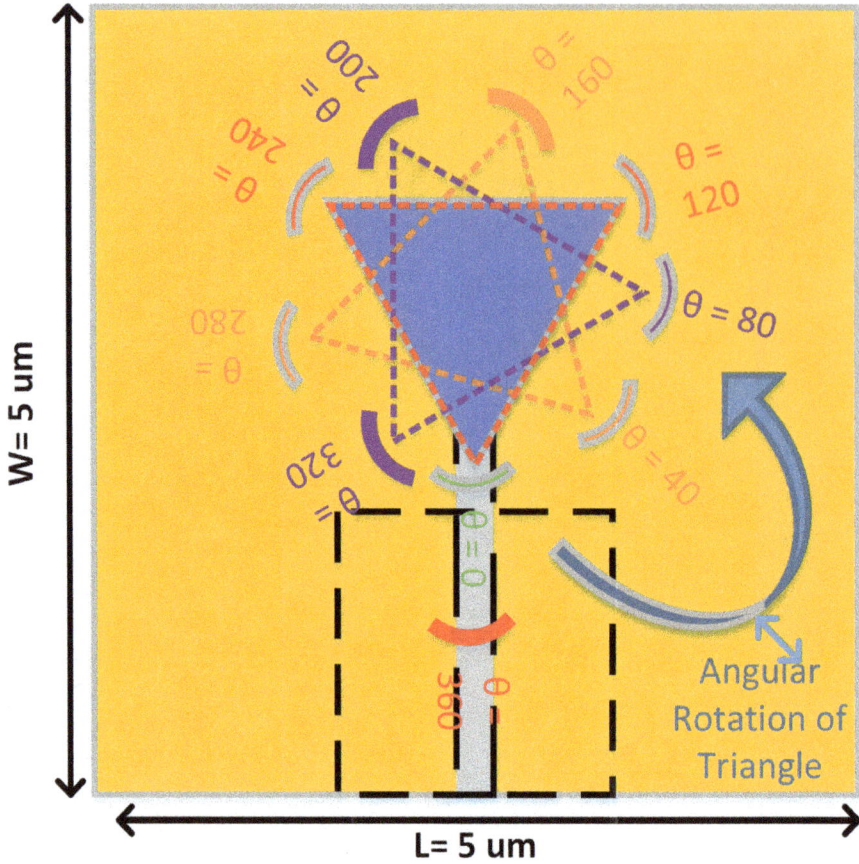

Figure 7. Angular rotation of triangular DR.

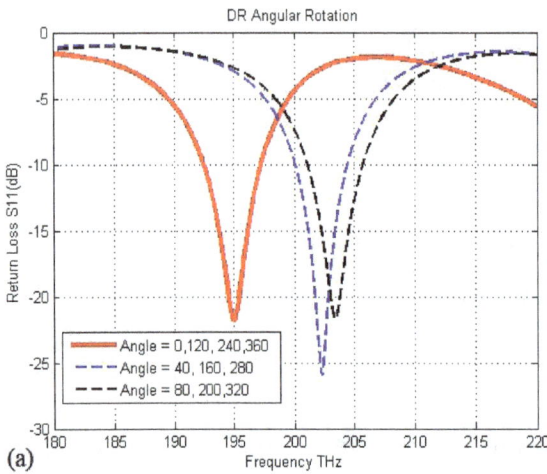

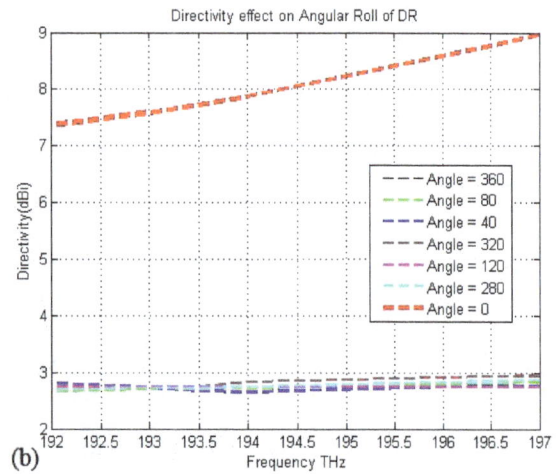

Figure 8. Optimized parameters: (a) effect of angular rotation of resonant frequency; (b) effect of angular rotation on directivity.

Parameters	Value (μm)
Feed length L_f	0.186
Feed width W_f	0.067
Ground length L_g	2.5
Ground width W_g	2
Height of triangular DR h	0.2
Area of triangular side a	1
Rotation angle θ	0^0

Table 1. Optimised parameters of ETDRN.

After extensive optimisation of stated parameters above and analysing the results achieved from these optimisations, the best geometric parameters that achieve an impedance bandwidth of 2.5% (192.3–197.3 THz) and a directivity of 8.6 dB are listed in **Table 1**. It is also observed that the simple ETDRNA structure can act as a tunable resonator when rotated around its axis resulting in usage of applications that work in the wavelengths in the range of (1463–1500 nm). The proposed design, if facility exists, can be fabricated via the known techniques involved in nanofabrication technology, that is, e-beam lithography, photolithography and chemical vapour deposition. In our case, the fabrication will follow a bottom-up approach where the SiO_2 substrate will have silver deposited on its surface.

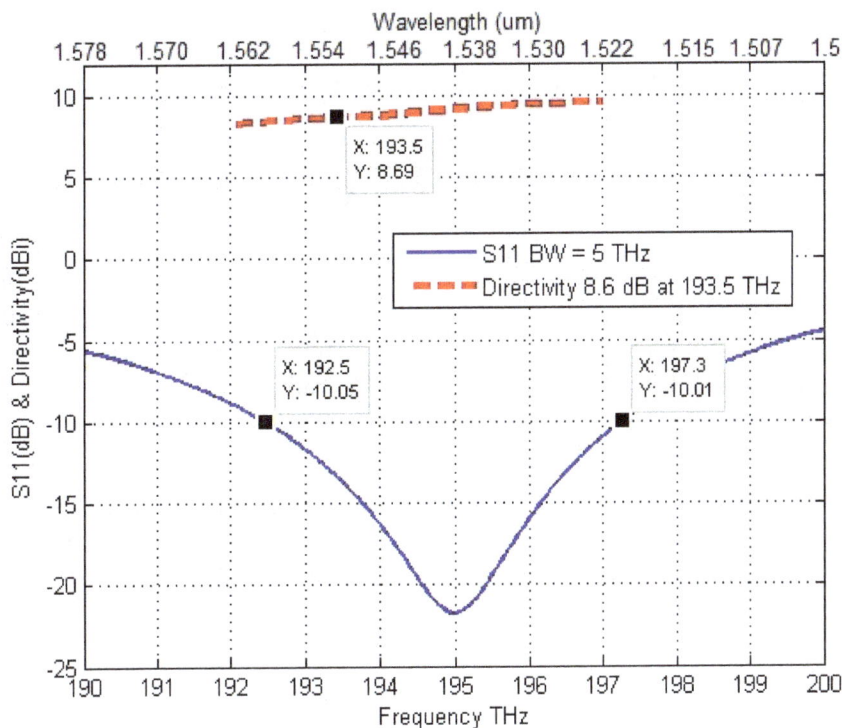

Figure 9. Simulated return loss and directivity of ETDRNA.

3.3. Results and discussion

In this section, we present the simulation results. **Figure 9** shows the return loss (S_{11}) and directivity of the proposed ETDRNA. The three-dimensional (3D) radiation patterns of the nanoantenna at 192, 193.5 and 197 THz are shown in **Figure 10(a–c)**. The maximum dip of −22 dB is achieved from the resonance of the nanoantenna at the central frequency of 1936.5 THz. The nanoantenna covers some part of the S-band while most part is covered for the C-band optical communication window. 3D radiation patterns provide the proof of the ETDRNA radiating in end-fire pattern. At present, the nanofabrication technology is limited and the proposed design is a theoretical one, yet we believe that our contribution in the fast-growing field of nantennas, with the proposed ETDRNA design, will prove itself to be a promising candidate for next-generation energy harvesting and green sustainable solution applications based on nanotechnology designs.

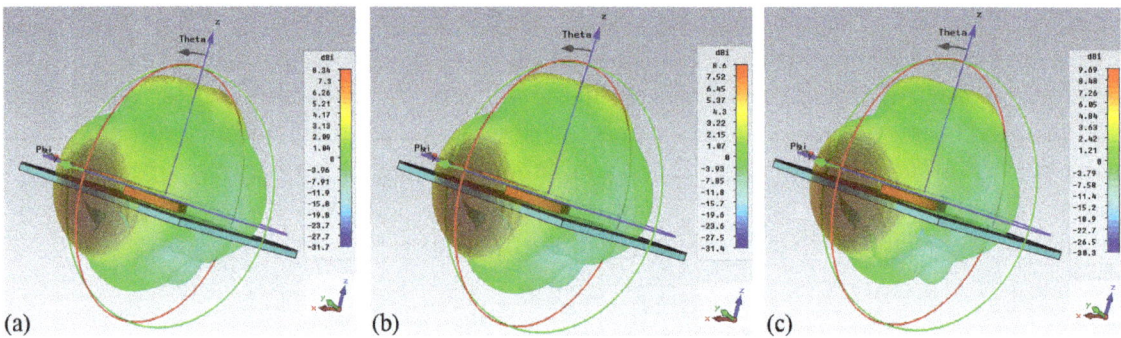

Figure 10. 3D end-fire radiation pattern at: (a) 192 THz, (b) 193.5 THz, (c) 197 THz.

4. Hexagonal dielectric resonator nantenna

In this section, we present another nantenna design based on dielectric resonator material. The shape of this DR is in hexagonal form with the material chosen as silicon (Si). The proposed design works as a loading element. The structure is again in multilayer form having (SiO_2) sandwiched between two silver (Ag) sheets. The radiating element is an equal-sided hexagonal-shaped (Si) dielectric loaded material. The whole nantenna structure is excited via a nanostrip transmission line made from a noble silver metal (Ag) whose conductive properties are calculated using the Drude model. The antenna achieves an impedance bandwidth of 3.7% (190.9–198.1 THz) with a directivity of 8.6 dBi at the frequency of interest. The obtained results make the proposed nantenna a possible solution for future nanophotonics and nanoscale communication devices.

4.1. Antenna geometry

In this section, we present another dielectric resonator (DR) design that takes the shape of a hexagon. The proposed nanoantenna works utilised the loading properties of ceramic dielectric silicon and is termed as hexagonal dielectric loaded nantenna (HDLN). It is also designed to operate at the central frequency of 193.5 THz which corresponds to a wavelength of 1550 nm. The cross-sectional and front view of the proposed HDLN is in **Figure 11(a)** and **(b)**. The design is based on a multilayer structure with (SiO$_2$) substrate sandwiched between two noble metals each made from silver (Ag). The properties of substrate are: thickness of h_1 = 0.150 μm, ε_r = 2.1 and loss tangent $tan\ \delta$ = 0.003 at f = 100 THz [39]. The ground layer, made from silver, below the substrate has partial form with properties as thickness of t = 0.010 μm and dimensions $L_g \times W_g$ = 1.95 × 2 μm. The feeding line is a nano silver strip on top of the substrate with parameters: thickness h_2 = 0.025 μm, W_f = 0.067 μm and L_f = 0.186 μm. The substrate dimensions are taken as $W_s \times L_s$ = 5 × 5 μm^2. The hexagonal dielectric is made of (Si), with ε_r = 11.9 and estimated loss tangent, $tan\ \delta$ = 0.0025. To achieve a further increase in the bandwidth with minimum resonance losses, a small substrate with thickness h_3 = 0.015 μm made from (SiO$_2$) has been introduced between the hexagon and the nanostrip. The dimensions of hexagonal dielectric are calculated from Eq. (8) [43] by inscribing the hexagon inside a circle and equating the areas of both designs, thus giving an optimised equal side lengths of hexagon as s = 1 μm and thickness ($\lambda_g/4 < h < \lambda_g/2$) h = 0.377μm;

$$\pi a_e^2 = \frac{3\sqrt{3}}{2} s^2 \tag{8}$$

where a_e = area of the circle and s = side of the hexagon. Since at optical frequencies, metals appear with a negative permittivity; therefore, complex permittivity 'ε_{Ag}' of silver (Ag) calculated from Eq. (9) was explained by the Drude model [39]:

$$\varepsilon_{Ag} = \varepsilon_o \left\{ \varepsilon_\alpha - \frac{f_{p^2}}{[f(f + i\gamma)]} \right\} = -128 + j3.28 \tag{9}$$

where ε_o = 8.85 × 10^{-12} [F/m], ε_α = 5, plasmonic frequency f_p = 2175 THz, f = central frequency and collision frequency γ = 4.35 THz. **Figure 11(b)** illustrates the antenna operating in the transmitting (Tx) mode by means of propagation vector orientation (k). The magnetic and electric field distributions of the hexagonal dielectric and nanostrip waveguide, along with the wave propagation in the y-axis, are also shown. Optical nantennas can be excited with a few known techniques ; (1) coupling of light using the so called nanotapers [46–47] since nano antennas cannot handle much power because of their small footprints, this makes them ideal candidates for being excited by micro lasers such as micro disks and photonic crystal lasers and (2) by reducing the reflection induced power loss by using slot dielectric waveguides [48].

4.2. Design simulation and optimisation

In this section, we make use of the optimisation techniques available to us from the simulator. We investigate the performance of each parameter involved in the design geometry of the proposed nanoantenna as shown in **Figure 11**. In order to study the impact on the antenna performance in terms of bandwidth, the following parameters have been studied and analysed.

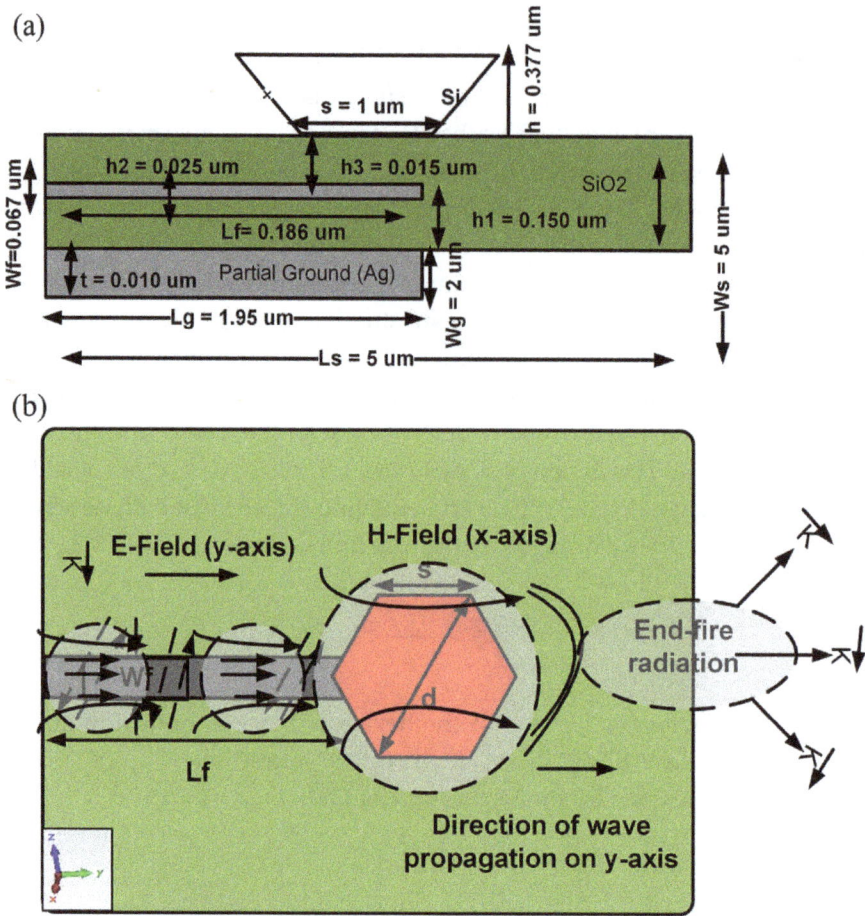

Figure 11. Geometry of hexagonal dielectric loaded nantenna (HDLN): (a) cross-sectional view; (b) front view with field vectors.

a. **Nanostrip feed:** Properties of conducting materials change when working at optical frequencies. The silver nanostrip line used to feed the nanoantenna was analysed in terms of Drude model. The nanostrip acts like a coupling resonator that excites the hexagonal dielectric, placed on an upper SiO_2 substrate with height h_3. Traditionally at RF frequencies, the length of the transmission lines is characterised to the wavelengths (λ) of incoming and outgoing radiations. However, working at the optical frequencies most of the incident light is transparent through the metals. This gives rise to plasmonic-free electrons, thus the feed line is analysed considering shorter effective wavelength (λ_{eff}), which depends on the material properties [44], refractive indexes and Eqs. (6) and (7). In our simulations, the optimised dimensions of the feed line produced values of length L_f to be 0.27 μm. The length L_f of the nanostrip stub was optimised from 0.1 to 0.27 μm with the best optimised value producing required resonance at 193.5 THz was at $L_f = 0.186$ μm as shown in **Figure 12(a)**. Similarly, the width 'W_f' of the nanostrip was also examined and optimisation was done from 0.02 to 0.28 μm. **Figure 12(b)** shows the best optimised value achieved at resonance of −22 dB with $W_f = 0.067$ μm.

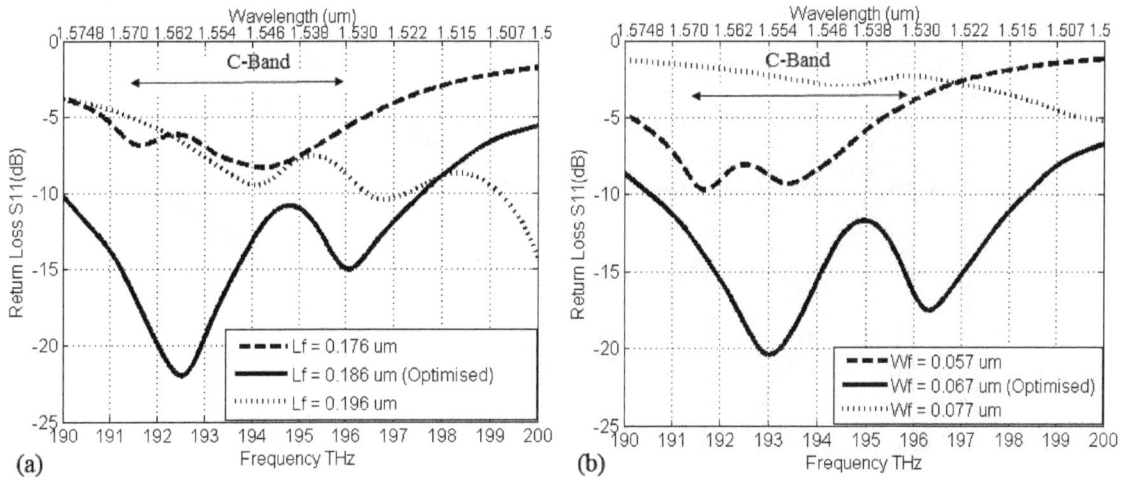

Figure 12. Optimized parameters: (a) length of feed line; (b) width of feed line.

b. **Partial ground plane:** The effect of the ground plane was studied and its optimisation gave dimewnsions of $L_g \times W_g = 1.95$ μm × 2 μm. **Figure 13(a)** and **(b)** shows the effects of varying the partial ground plane in terms of its length and width. The optimised results produce a wide impedance bandwidth of 3.7% (190.9–198.1 THz) at a centre frequency of 193.5 THz, covering all of the standard optical transmission widow (C-band).

c. **Height of hexagonal DR:** The wide impedance bandwidth achieved is also affected by the height of the hexagonal DR. The height h of the DR was optimised within the range ($\lambda_g/4 < h < \lambda_g/2$) [3]. **Figure 14** shows the best optimised value of $h = 0.37$ μm having a resonance at -23 dB.

4.3. Results and discussion

To compare the properties and results of our proposed HDLN nanoantenna, we first simulated a hexagonal dielectric resonator antenna at the lower frequency band [41]. Observations were

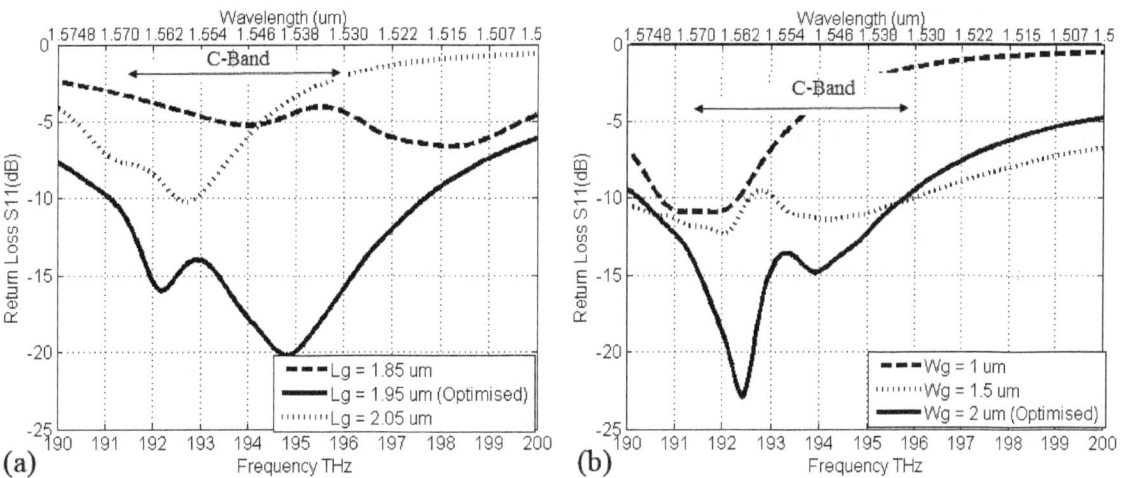

Figure 13. Optimized parameters: (a) length of ground plane; (b) width of ground plane.

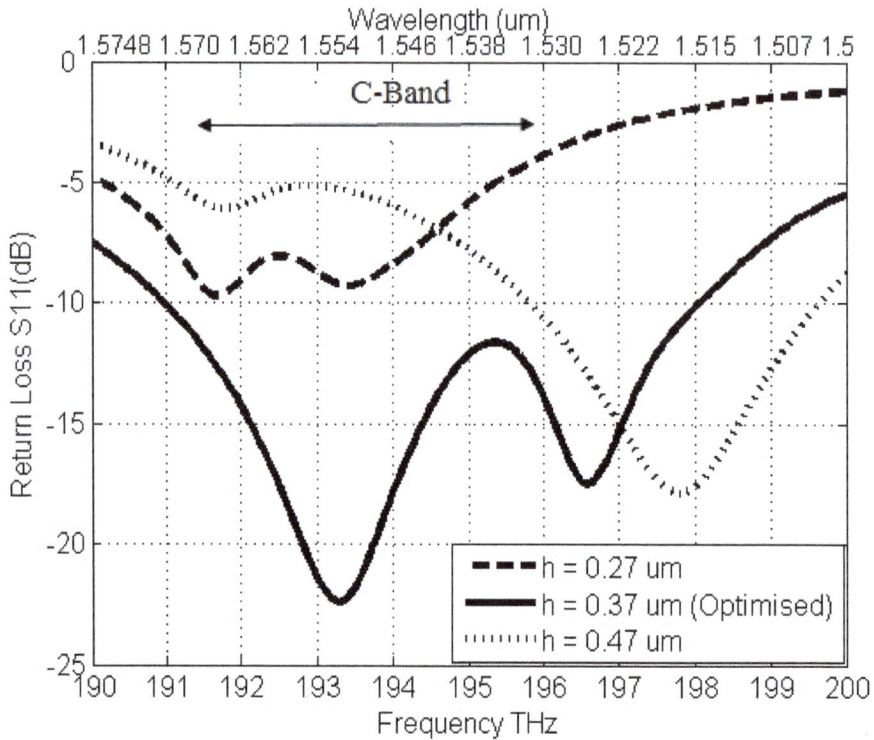

Figure 14. Optimised parameters: height of hexagonal DR.

made in terms of plane-wave propagation in the transmission lines to the radiating structures of the two antennas with results shown in **Figure 15(a)** and **(b)**, respectively. From **Figure 15 (a)**, it can be observed that the E-field propagation or the power propagation in the transmission line is following the fringing effects in order to radiate the hexagonal structure operating in the microwave domain. Whereas the proposed nantenna structure depicted in **Figure 15(b)** shows the E-field propagation in the nanotransmission line follows a travelling wave effect. It is also observed that the hexagonal DR elements for both the cases exhibit different properties. At the microwave domain, the hexagonal DR as shown in **Figure 15(a)** works as a resonator,

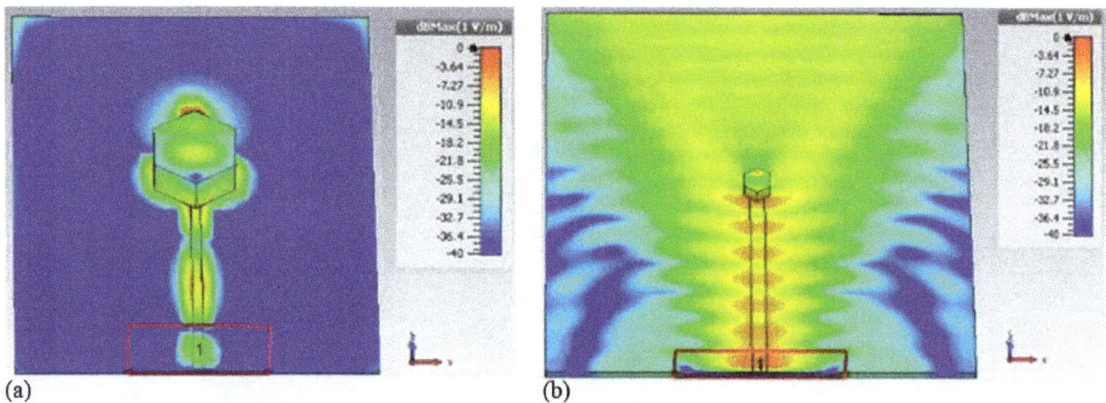

Figure 15. E-field propagation: (a) fringing effects at lower frequency; (b) travelling wave effect at THz spectrum.

whereas the DR at the nanoscale structure shown in **Figure 15(b)** exhibits loading properties which benefit the nantenna to operate as a lens and thus achieve more directivity.

The return loss (S_{11}) and directivity of the proposed HDLN with respective wavelength and directivity axis is shown in **Figure 16**. After extensive optimisation, the nantenna achieves an impedance bandwidth of 3.7% (190.9–198.1 THz) with a directivity of 8.6 dBi, making it useful for nanoscale fabrication due to its robustness against fabrication tolerances.

Typically, the modes of hexagonal DR [49] are derived from the cylindrical dielectric resonator, which has three distinct types: TE (TE to z), TM (TM to z) and hybrid modes. The TE and TM modes are asymmetrical and have no azimuthal variation. On the other hand, fields produced by hybrid modes are azimuthally dependent. Hybrid mode is further divided into two sub-groups of HE and EH. The modes generated by the hexagonal dielectric nantenna are represented in terms of magnitude of electric field distribution on its surface as shown in **Figure 17**, at the centre frequency of 193.5 THz. The mode analysis was done via EM simulator CST MWS.

From the infinite modes available [50], in our simulation as shown in **Figure 17**, we observed the nano hexagonal dielectric antenna producing $HE_{20\delta}$ mode between the achieved wide impedance band. The subscript in the modes represents the variation of fields along azimuthal, radial and z-direction of the cylindrical axis. It is observed from the figure that the magnitude of electric field variation is produced on the azimuthal direction with no variation in the radial direction, thus giving a mode excited at $HE_{20\delta}$. Also, the intensity is highest at the azimuthal

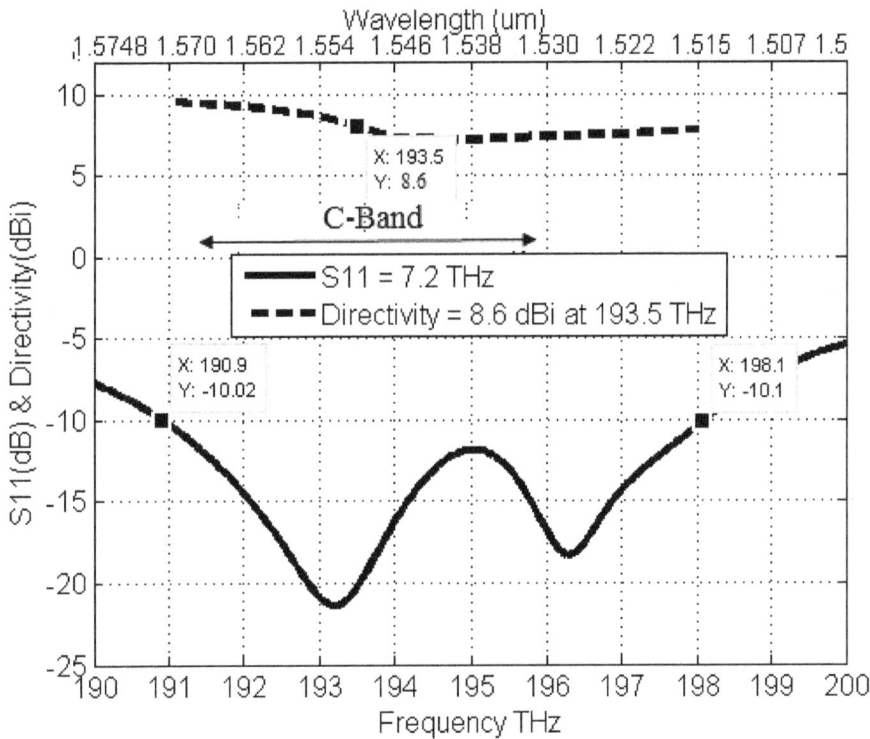

Figure 16. Simulated return loss and directivity of proposed HDLN.

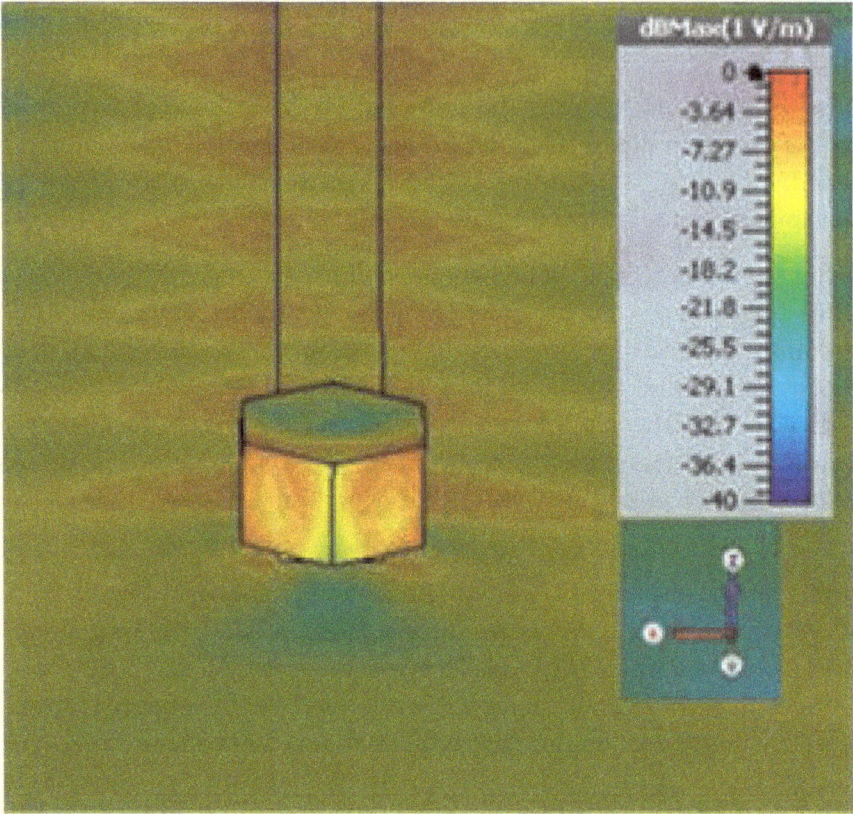

Figure 17. E-field distribution as shown via the magnitude at 193.5 THz with $HE_{20\delta}$ mode.

plane resulting in a radiation pattern towards the end-fire direction. The 3D radiation patterns of the nanoantenna at 191, 193.5 and 198 THz are shown in **Figure 18(a–c)**. The directivity of the antenna at the centre frequency is 8.6 dBi.

Keeping in mind the state-of-the-art nantenna designs and limited availability of nanofabrication equipment and facilities worldwide, we believe our proposed theoretical HDLN design will prove itself to be a promising communication device for applications based on nanotechnology.

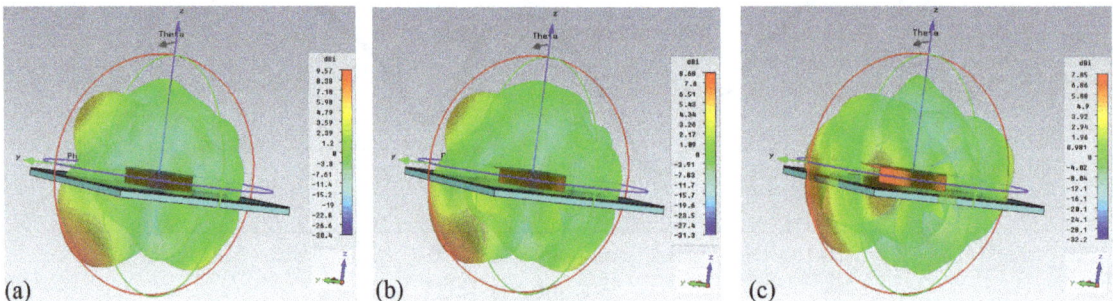

Figure 18. 3D radiation pattern: (a) 191 THz, (b) 193.5 THz and (c) 198 THz.

5. Array synthesis

In this section, we present the array synthesis done on one of the two proposed nantenna designs based on DR element. The ETDRNA was opted for array optimisation. The ETDRNA was fed via a 1×2 corporate feed network working at optical C-band (1.55 μm). Numerical results prove that the proposed nantenna exhibits a directivity of 9.57 dB with an impedance bandwidth of 2.58% (189–194 THz) covering the standard optical C-band transmission window. Furthermore, by selecting the appropriate orientation of the triangular dielectric resonators, the proposed nantenna structure can be tuned to operate at the higher or lower optical bands offering a threshold value of directivity and bandwidth Δf. By tuning the nantenna, they achieve an increase in bandwidth of 4.96% (185.1–194.7 THz) and directivity also improves to 9.7 dB. The wideband and directive properties make the proposed nantenna attractive for a wide range of applications including broadband nanophotonics, optical sensing, optical imaging and energy-harvesting applications.

5.1. Antenna array geometry

An array of 1×2 configuration for the ETDRNA is presented. The front and side view is shown in **Figure 19**. The geometry of the proposed nanoantenna utilised the same parameters as the proposed ETRDN in Section 3. The gap between the elements to counter mutual

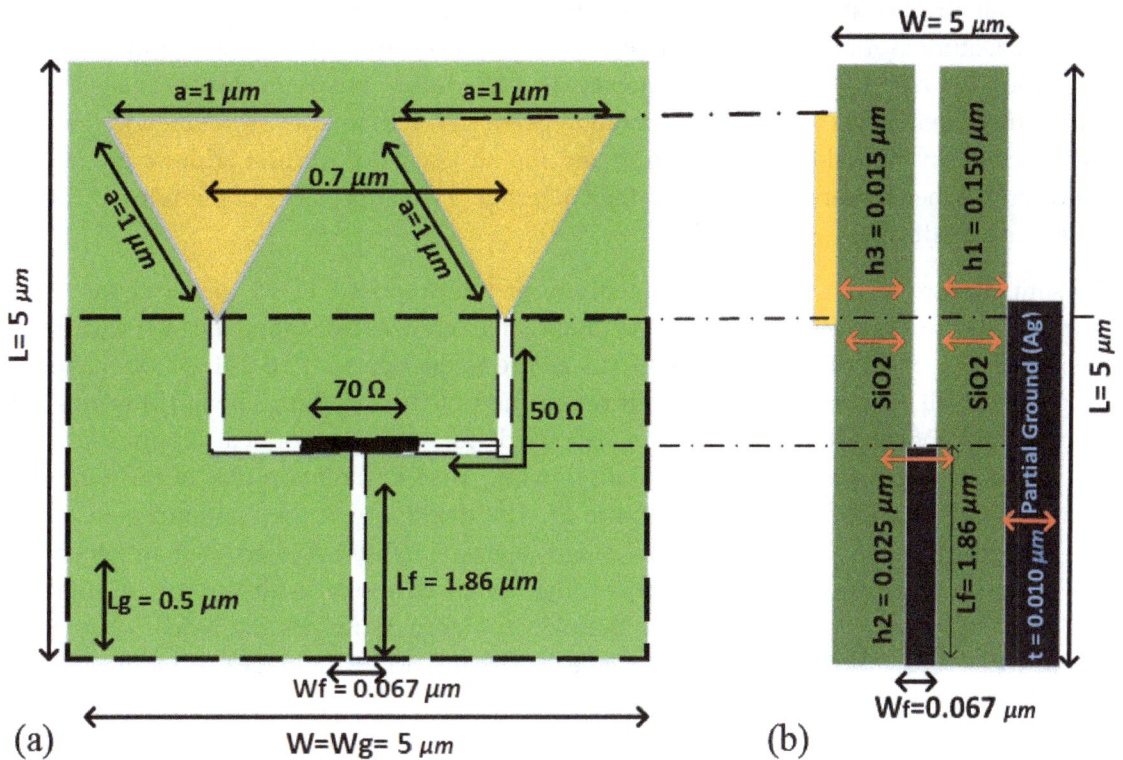

Figure 19. Geometry of proposed 1×2 array: (a) front view; (b) side view.

coupling is around ($\lambda_{eff}/2$) at the central frequency of 193.5 THz. The dimensions of the single element equilateral triangle dielectric based on frequency dependence can be calculated from Eqs. (1) and (2) previously presented in Section 3.

In order to increase the directivity of the proposed nantenna, arrays with a corporate feed network has been utilised for best power transfer from the source to the radiating ETDRNA. **Figure 19(b)** shows the corporate feed network along with appropriate feed line (50 and 70 Ω) markings. The optimised width of the 70 Ω feed line is 0.045 µm, whereas the length is 0.76 µm. For the 50 Ω feed line, the width and length are W_f and L_f, respectively. The shaded region in black shows the partial ground and feed lines to be on the backside of the substrate. The optimum distance between the triangular plasmonic resonators achieve minimal mutual coupling at 0.7 µm centre to centre. The properties of the noble metals are explained as per Drude Model used throughout the sections for the proposed nantenna designs.

5.2. Results and discussion

In this section, we investigate the proposed nanoantenna arrays results in terms of two important features: (1) directivity improvement and (2) tunability. Directivity is very important because it measures the power density the antenna radiates in the direction of its strongest emission versus the power density radiated by an ideal isotropic radiator (which emits uniformly in all directions) radiating the same total power, and on the other hand tunability is an attractive feature as it makes antenna operational in various frequency bands simultaneously as well as resonating at its centre frequency. In fact, during our performance analysis, we learned the importance of the angle of rotation between the triangular structure and the feed line direction. In the following paragraph and the next section, we first address the performance study when this angle is zero and then investigate the importance of this when different angles are introduced.

The simulated return loss (S_{11}) and directivity of the proposed 1 × 2 nanoantenna array are presented in **Figure 20**. The proposed nantenna achieves an impedance bandwidth of 2.58% with $S_{11} < -10$ dB (189–194 THz) at a centre frequency of 193.5 THz (1550 nm) with a dip at −15 dB. The antenna covers all the portion of the standard C-band transmission window in optical domain and can be used for relevant optical applications in nanonetworks and high-speed optical data transfer. The 3D radiation plot of the nantenna resonating at 193.5 THz is shown in **Figure 21**. The directivity of the nantenna is 9.57 dBi, 0.97 dB improvement compared to the single element ETDRNA discussed in Section 3. CST MWS studio has been used to acquire the optimised results with verification done by another EM simulator HFSS. Examining the plot in **Figure 21** reveals the E-field component having main lobe direction at 45°, side lobe levels at −4.1 dB and beam width of 20.1°. Similarly, the H-field component in **Figure 21** has main lobe direction of 180°, side lobe levels of −1.5 dB and beam width of 127.2°. Although the nantenna achieved a directive nature, the losses associated with high side lobe levels are to be expected due to substrate selection and working at higher THz frequencies.

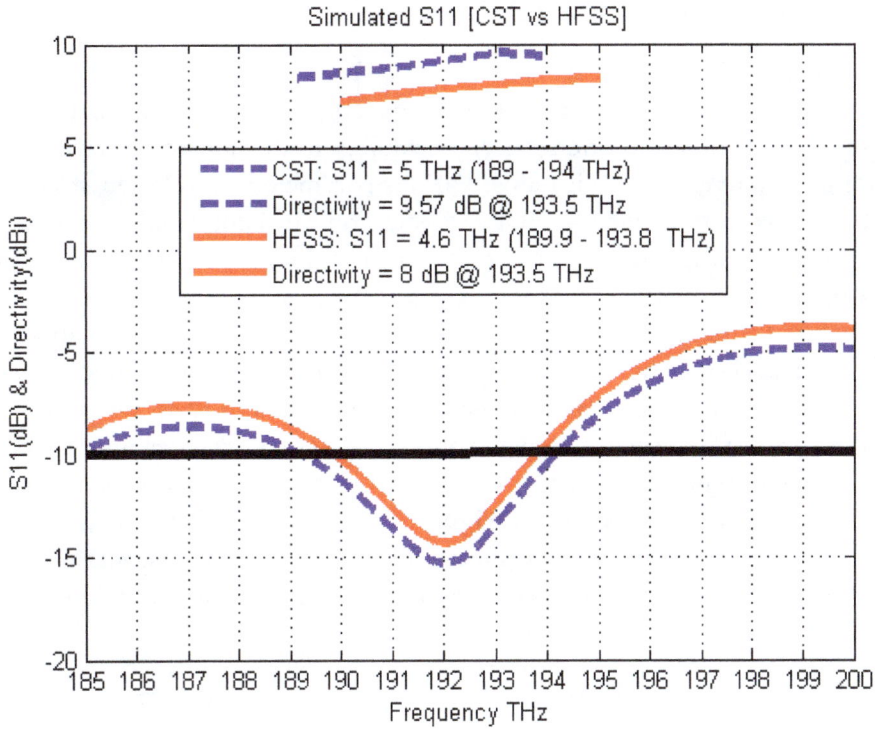

Figure 20. Return loss and directivity of 1×2 ETDRNA array.

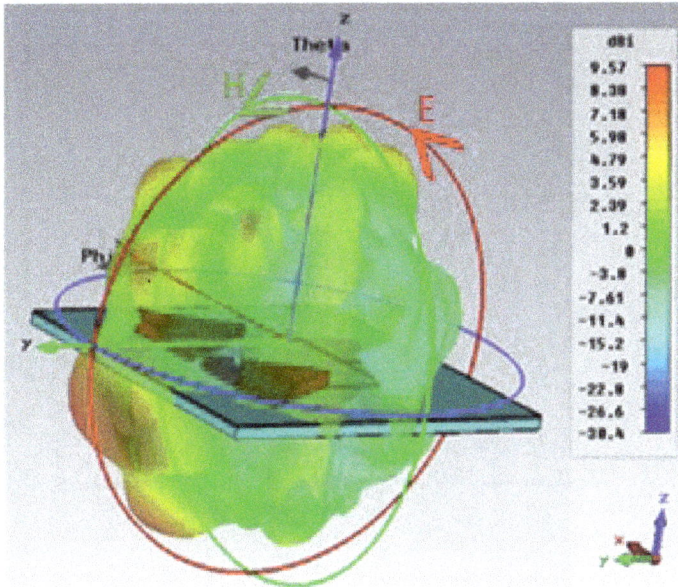

Figure 21. 3D radiation pattern at 193.5 THz.

5.3. Tunability of proposed 1 × 2 ETDRNA array

Achieving an antenna design for a specific frequency band (wide, notch or filter) with tunable (or changeable) centre frequency is an important challenge task. Our proposed nantenna array structure achieves this task and offers a large flexibility to the antenna in terms of operating frequency for a wide range of applications. Various parameters and design strategies such as loading slots, lumped components, switches, diodes and capacitors are introduced in order to achieve this feature. Although we have utilised a corporate feed network and the antenna geometry is based on nanoscale dimensions, which means more parametric study, but our simple and featured equilateral shape of the DR elements enable us to make natural yet simple tunability. **Figure 22** shows the rotation angles of the two equilateral triangular DR elements (T1 and T2). Since it is an equilateral triangle, the rotation angles were swept from ($-180°$ to $180°$) with a step size of $10°$. The rotation angle is defined as counterclockwise movement from the bottom tip of the triangle. We apply three scenarios of rotations: (1) rotate T1 while T2 is fixed (**Figure 22a**), (2) fix T1 and rotate T2 (**Figure 22b**) and (3) simultaneously rotate T1 and T2 in the same counterclockwise direction (**Figure 22c**). In all the three scenarios, as shown in **Figure 22**, the simulation of rotation was done for the whole $360°$ but only five rotating angles

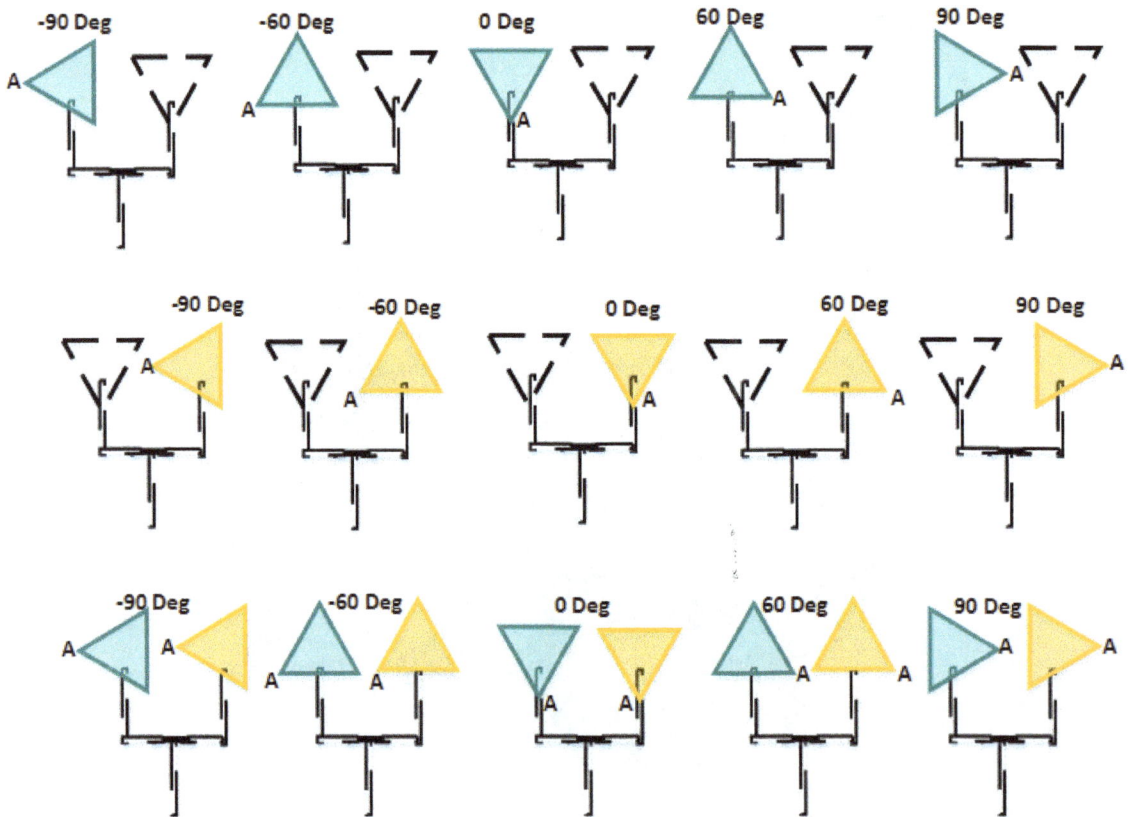

Figure 22. Rotation scenario of triangular DRs for spectral shifts: (a) rotate triangle T1 (left); (b) rotate triangle T2 (right) and (c) rotate both T1 and T2 simultaneously $\pm 90°$.

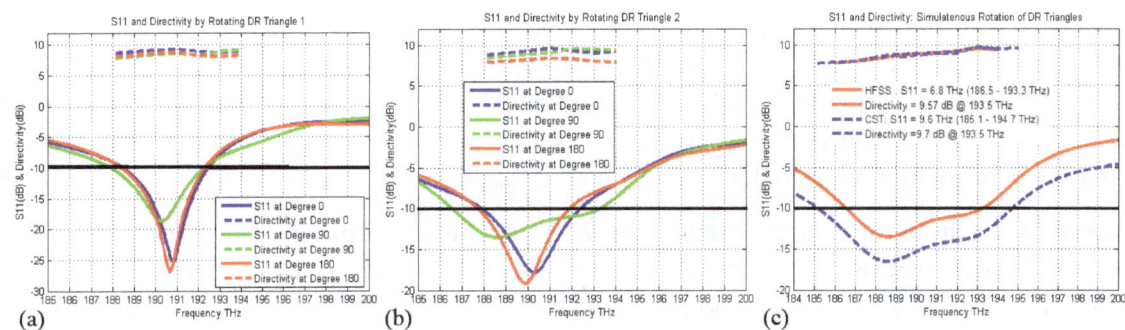

Figure 23. Return loss and directivity as per rotation: (a) T1 rotated; (b) T2 rotated and (c) both T1 and T2 simultaneously rotated.

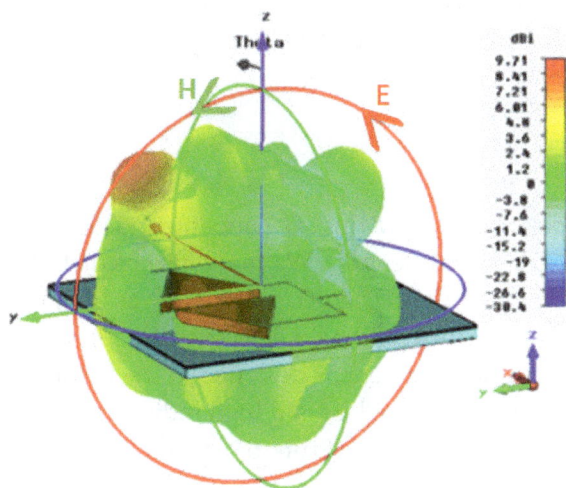

Figure 24. 3D radiation pattern of directivity at 193.5 THz after appropriate selection of DR triangular angles ($-90°$ and $90°$).

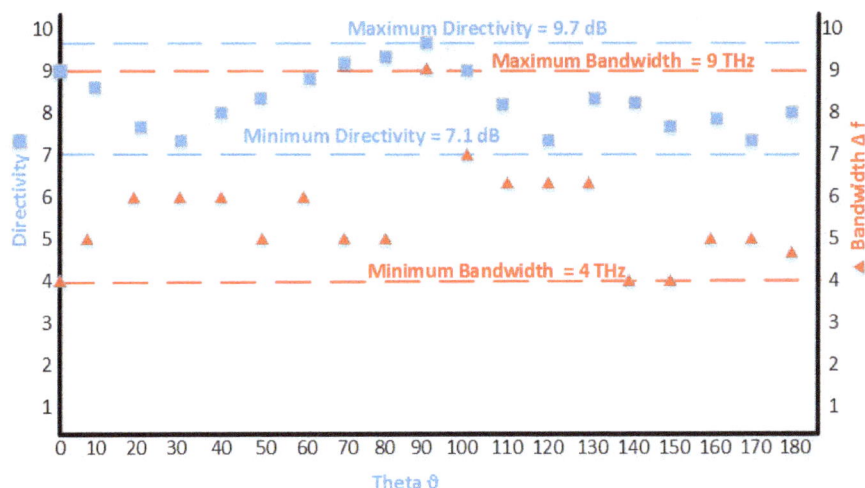

Figure 25. Effect of tunability on directivity and fractional bandwidth.

have been shown, that is, 0°, +60°, +90°, −60°, −90°, for purpose of simplicity. For our case of centre frequency at 193.5 THz, the best resonance with improved bandwidth of 4.96% (185.1–194.7 THz) and a directivity (9.7 dBi) is achieved when we opted to select the angles of T1 and T2 to be −90° and 90°, respectively. **Figure 23** displays the results in terms of s-parameters and directivity for rotation of triangles T1 and T2.

Figure 24, details of the E-field (yz-axis), shows main lobe at a direction of 45°, side lobe levels of −4.5 dB and beam width of 19°, whereas the H-field (xz-axis) has a main lobe direction of 115°, side lobe levels of −2.6 dB and beam width of 99.8° with a lot of losses in terms of high side lobe levels at a higher frequency. **Figure 25** shows the tunability effects of simulataneous rotation of triangular DRs in terms of the minimum and maximum values achieved for directivity and also the bandwidth difference Δf at −10 dB resonance in the frequency band (185–195 THz). The minimum directivity achieved is 7.15 dB and the maximum achieved is 9.71 dB when the triangles are tuned. Similary, the difference of bandwidth Δf ranges from a minimum to maximum of 4 to 9 THz, respectively.

6. Conclusion

In this chapter, we proposed two new optical nanoantenna designs working at centre frequency of 193.5 THz (1550 nm wavelength). We made use of dielectric resonators, which are ceramic based having very high dielectric permittivity, to assist the metallic designs mostly utilised in making plasmonic nanoantennas. Dielectric resonator (DR)-based antennas have made their mark and importance of antennas designed at the RF and MMW spectrum. Due to their unique feature of minimum surface loss and high radiation efficiency, they are considered nominal candidate to be used at higher spectrum bands, that is, THz.

Keeping with the offered characteristics of DR, we suggested two nanoantenna designs with triangular- and hexagonal-shaped DR materials. Both the designs are based on multilayer technology where the ground and transmission lines are made from noble metal silver. The properties of silver at optical domain are discussed and defined on the basis of Drude Model. The first design ETDRNA achieved an impedance bandwidth of 2.58% (192.3–197.3 THz), whereas the second nanoantenna design with hexagonal-shaped DR, HDLN, offered an impedance bandwidth of 3.7% (190.9–198.1 THz). Both the nanoantennas achieved high directivity of 8.6 dBi with end-fire radiation pattern. Array synthesis was also performed in order to compare and observe how much improvement is possible. The 1×2 ETDRNA achieved an improvement of 0.97 dB compared to the original directivity. Also the array antenna offers tunability in simple manner, compared to other methods discussed, making the proposed nanoantenna adaptable to many other optical frequency bands of interest for various optical communications.

Acknowledgements

This research work was supported by King Saud University, Deanship of Scientific Research and College of Engineering Center.

Author details

Waleed Tariq Sethi[1,2]*, Hamsakutty Vettikalladi[3], Habib Fathallah[2] and Mohamed Himdi[1]

*Address all correspondence to: wsethi@ksu.edu.sa

1 Institute of Electronics and Telecommunications of Rennes University (IETR), University of Rennes, France

2 KACST Technology Innovation Center in Radio Frequency and Photonics for the e-Society (RFTONICS), King Saud University, Riyadh, Saudi Arabia

3 Electrical Engineering Department, King Saud University, Riyadh, Saudi Arabia

References

[1] Cherry S. Edholm's law of bandwidth. IEEE Spectrum. 2004;**41**(7):58–60

[2] Li QC, Niu H, Papathanassiou AT, Wu G. 5G network capacity: Key elements and technologies. IEEE Vehicular Technology Magazine. 2014;**9**(1):71–78

[3] Stephan J, Brau M, Corre Y, Lostanlen Y. On the Effect of Realistic Traffic Demand Rise on LTE-A HetNet Performance. 2014 IEEE 80th Vehicular Technology Conference (VTC2014-Fall). Vancouver, BC; 2014. pp. 1–5

[4] Gregori M, Gómez-Vilardebó J, Matamoros J, Gündüz D. Wireless content caching for small cell and D2D networks. IEEE Journal on Selected Areas in Communications. 2016;**34**(5):1222–1234

[5] Mustafa IB, Uddin M, Nadeem T. Understanding The Intermittent Traffic Pattern Of HTTP Video Streaming Over Wireless Networks. 2016 14th International Symposium on Modeling and Optimization in Mobile, Ad Hoc, and Wireless Networks (WiOpt) (2016): n. pag. Web. 6 March 2017

[6] TU/e to develop 5G technology with European grant [Internet]. Available from: https://www.tue.nl/en/university/news-and-press/news/13-05-2016tueto-develop-5g-technology-with-european-grant/

[7] Valdes-Garcia A, Yong SK, Xia P. 60Ghz Technology For Gbps Wlan And Wpan. 1st ed. Hoboken, NJ: Wiley; 2013

[8] Chester E. 4.6Gbps Wi-Fi: How 60Ghz Wireless Works—And Should You Use It? Ars Technica. N.p., 2017. Web. 6 March 2017

[9] Researchers Demonstrate World's First 5G, 100 To 200 Meter Communication Link Up To 2 Gbps". Phys.org. N.p., 2017. Web. 6 March 2017

[10] Kürner T, Priebe S. Towards THz communications-status in research, standardization and regulation. Journal of Infrared, Millimeter, and Terahertz Waves. 2014;**35**(1):53–62

[11] Nagatsuma T, Ducournau G, Renaud CC. Advances in terahertz communications accelerated by photonics. Nature Photonics. 2016;**10**(6):371–379

[12] Petrov V, Pyattaev A, Moltchanov D, Koucheryavy Y. Terahertz band communications: Applications, research challenges, and standardization activities. Lisbon: 2016 8th International Congress on Ultra Modern Telecommunications and Control Systems and Workshops (ICUMT); 2016. pp. 183–190

[13] Lukasz L, Brzozowski M, Kraemer R. Data Link Layer Considerations for Future 100 Gbps Terahertz Band Transceivers. Wireless Communications and Mobile Computing. Vol. 2017. Article ID 3560521, 11 pages, 2017

[14] Yilmaz T, Akan OB. On the use of low terahertz band for 5G indoor mobile networks. Computers and Electrical Engineering. 2015;**48**:164–173

[15] Akyildiz IF, Jornet JM, Han C. Terahertz band: Next frontier for wireless communications. Physics Communications. 2014;**12**:16–32. DOI: 10.1016/j.phycom.2014.01.006 ()

[16] Obayya S, Areed NFF, Hameed MFO, Hussein M. Optical nano-antennas for energy harvesting. In: Innovative Materials and Systems for Energy Harvesting Applications. Hershey, PA, USA: IGI Global; 2015. p. 26

[17] Yang Y, Li Q, Qiu M. Broadband nanophotonic wireless links and networks using on-chip integrated plasmonic antennas. Scientific Reports. 2016;**6**:19490. DOI: 10.1038/srep19490

[18] Yousefi L, Foster AC. Waveguide-fed optical hybrid plasmonic patch nano-antenna. Optics Express. 2012;**20**(16):18326–18335

[19] Hong SG, et al. Bioinspired optical antennas: Gold plant viruses. Light: Science & Applications. 2015;**4**:267–272

[20] Horiuchi N. Optical antennas: Reconfigurable resonance. Nature Photonics. 2016;**10**(5): 2680–2685

[21] Jornet JM, Akyildiz AF. Channel modeling and capacity analysis for electromagnetic wireless nanonetworks in the terahertz band. IEEE Transactions on Wireless Communications. 2011;**10**(10):3211–3221

[22] Novotny L, van Hulst N. Antennas for light. Nature Photonics. 2011;**5**:83–90

[23] Palik ED. Handbook of Optical Constants of Solids. New York: Academic Press; 1998

[24] Johnson PB, Christy RW. Optical constants of the noble metals. Physical review B, APS Physics. 1972;**6**(12):5056–5070

[25] Petosa A, Ittipiboon A. Dielectric resonator antennas: A historical review and the current state of the art. IEEE Antennas and Propagation Magazine. 2010;**52**:91–116

[26] Luk KM, Leung KW, editors. Dielectric Resonator Antennas. Baldock, England: Research Studies Press; 2003

[27] Petosa A. Dielectric Resonator Antenna Handbook. Norwood, MA: Artech House; 2007

[28] Keyrouz and Caratelli D. Dielectric resonator antennas: Basic concepts, design guidelines, and recent developments at millimeter-wave frequencies. International Journal of Antennas and Propagation. 2016;**10**:0–20

[29] Belov PA, et al. Superdirective all-dielectric nanoantennas: Theory and experiment. IOP Conference Series: Materials Science and Engineering. 2014;**67**:012008–012013

[30] Devi I, et al. Modeling and design of all-dielectric cylindrical nanoantennas. Journal of Nanophotonics. 2016;**10**(4):046011–046011

[31] Andryieuski A, Malureanu R, Biagi G, Holmgaard T, Lavrinenko A. Compact dipole nanoantenna coupler to plasmonic slot waveguide. Optics Letters. 2012;**37**:1124–1126

[32] Suh JY, Huntington MD, Kim CH, Zhou W, Wasielewski WR, Odom TW. Extraordinary nonlinear absorption in 3d bowtie nanoantennas. Nano Letters. 2012;**12**:269–274

[33] Maksymov IS, Miroshnichenko AE, Kivshar YS. Actively tunable bistable optical yagi-uda nanoantenna. Optics Express. 2012;**20**:8929–8938

[34] Patel SK, Argyropoulos C. Plasmonic nanoantennas: Enhancing light-matter interactions at the nanoscale. EPJ Applied Metamaterials. 2015;**2**:03140–03174

[35] Giannini V, et al. Plasmonic nanoantennas: Fundamentals and their use in controlling the radiative properties of nanoemitters. Chemical Reviews. 2011;**111**(6): 3888–3912

[36] Wu CH, et al. Multimodal magneto-plasmonic nanoclusters for biomedical applications. Advanced Functional Materials. 2014;**24**(43):6862–6871

[37] Maksymov IS. Magneto-plasmonic nanoantennas: Basics and applications. Reviews in Physics. 2016;**1**:36–51

[38] Professor Nicholas Fang's Research Group @ MIT. Web.mit.edu. N.p., 2017. Web. 6 March 2017

[39] Johnson PB, Christy RW. Optical constants of the noble metals. Physical Review B. 1972;**6**(12):4370–4379

[40] Sinha R, Karabiyik M, Al-Amin C, Vabbina PK, Guney DO, Pala N. Tunable room temperature THz sources based on nonlinear mixing in a hybrid optical and THz micro-ring resonator. Scientific Reports. 2015;**5**:9422

[41] Lo HY, Leung KW, Luk KM, Yung EKN. Lowprofile equilateral-triangular dielectric resonator antenna of very high permittivity. Electronics Letters. 1999;**35**(25):2164–2166

[42] Kishk AA. A triangular dielectric resonator antenna excited by a coaxial probe. Microwave and Optical Technology Letters. 2001;**30**(5):340–341

[43] Balanis C. Antenna Theory: Analysis and Design. New York, NY,USA: John Wiley & Sons; 2005

[44] Olmon RL, Raschke MB. Antenna-load interactions at optical frequencies: Impedance matching to quantum systems. Nanotechnology. 2012;**23**(44):444001

[45] Neubrech F, et al. Resonances of individual lithographic gold nanowires in the infrared. Applied Physics Letters. 2008;**93**(16)163105

[46] Hattori HT, Li Z, Liu D, Rukhlenko ID, Premaratne M. Coupling of light from microdisk lasers into plasmonic nanoantennas. Optics Express. 2009;**17**(23):20878–20884

[47] Li Z, Hattori HT, Fu L, Tan HH, Jagadish C. Merging photonic wire lasers and nanoantennas. Journal of Lightwave Technology. 2011;**29**(18):2690–2697

[48] Hattori HT, Li Z, Liu D. Driving plasmonic nanoantennas with triangular lasers and slot waveguides. Applied Optics. 2011;**50**(16):2391–2400

[49] Hamsakutty V, Mathew KT. Hexagonal dielectric resonator antenna: A novel DR antenna for wireless communication [Ph.D. thesis]. Department of Electronics, Cochin University of Science and Technology, Dyuhti, India, 2007.

[50] Mongia RK, Bhartia P. Dielectric resonator antennas: A review and general design relations for resonant frequency and bandwidth. International Journal of Microwave and Millimeter-Wave Computer-Aided Engineering. 1994;**4**(3):230–247

4

Next-Generation Transport Networks Leveraging Universal Traffic Switching and Flexible Optical Transponders

Bodhisattwa Gangopadhyay, João Pedro and
Stefan Spälter

Additional information is available at the end of the chapter

Abstract

Recent developments in communication technology contributed to the growth of network traffic exponentially. Cost per bit has to necessarily suffer an inverse trend, posing several challenges to network operators. Optical transport networks are no exception to this. On one hand, they have to keep up with the expectations of data speed, volume, and growth at the agreed quality-of-service (QoS), while on the other hand, a steep downward trend of the cost per bit is a matter of concern. Thus, the proper selection of network architecture, technology, resiliency schemes, and traffic handling contributes to the total cost of ownership (TCO). In this context, this chapter looks into the network architectures, including the optical transport network (OTN) switch (both traditional and universal), resiliency schemes (protection and restoration), flexible-rate line interfaces, and an overall strategy of handover in between metro and core networks. A design framework is also described and used to support the case studies reported in this chapter.

Keywords: optical transport network, flexible line interfaces, protection and restoration, network planning

1. Introduction

The exponential growth of consumer demands and machine-to-machine network traffic coupled with the downward trend in revenue per bit transported is challenging network operators to adopt a strategy which tackles a twofold problem. The dual nature of the problem, on one hand lies in selecting a network architecture/technology which can efficiently transport traffic originating from multiple sources, be it time division multiplexing (TDM) or Packet, and on the other hand to make use of an increasingly flexible and heterogeneous optical layer, where the characteristics of the optical light paths to be set up (e.g., modulation format, spectral width) are customized to the specific path properties.

Today, typical transmission networks are a layered combination of dense wavelength division multiplexing (DWDM) equipment (the lowest layer above the optical fiber layer), a subwavelength aggregation and grooming layer and an internet protocol (IP) layer. These layers form server/client relationships and are independent of each other. From a technological point of view, the functions of each layer are very different. Higher layer equipment is typically more expensive per bit transported because it needs to do more processing, so the use of the layers must be carefully balanced to deliver cost-optimized networks. The introduction of coherent 100G (100 Gigabits per second) optical transport was a key catalyst which offered massive performance gains over incumbent technologies to exploit more capacity from a single fiber. Telco operators achieved considerable gains as the cost per bit started going down with the introduction of 100G. But this was not the end, as they eyed newer improved scalable architectures to strengthen network operations and total cost of ownership (TCO). These next-generation network architectures aimed at efficiently grooming and aggregating sublambda data streams resulting in cost-optimized well-packed 100G wavelengths which would allow telco operators to survive within challenging capital expenditure (CAPEX) and operational expenditure (OPEX) cost targets in the near future. However, due to their relatively compact and short reach topologies, an abundance of optical fiber and the requirement to interconnect network elements at 10G rates or less, metro networks still mostly relied on a direct-detect 10G optimized optical transport infrastructure, though this situation is bound to change dramatically [1].

Importantly, the abundant deployment of 100G coherent systems in core networks has been attained at the expense of relatively costly line interfaces, performing electrical-optical (EO) and optical-electrical (OE) conversions. In the meanwhile, optical transport network (OTN) emerged as a key building block to complement the capacity gains unleashed by 100G and coherent optics. Efficiency, predictability, and reliability of the transport world and the agility, programmability of the packet world were blended into the ITU G.709 standard (OTN protocol) [2] and thus became an automatic choice. The demanding service level agreements (SLAs) of private E-LANs, E-Lines, and other packet traffic along with wavelength services from the near future could be met by the many features that OTN was offering. Moreover, traffic processing on IP packet level (layer 3) is much more expensive (per GByte) than switching the same amount of traffic in optical channel data unit (ODU) containers by OTN switches. Therefore, OTN provides a more cost-effective platform for subwavelength services to be multiplexed not only at their source node, but also at selected intermediate nodes and, as a result, reduce the amount of expensive (WDM) line interfaces used without having to resort to expensive router equipment to perform this task [3].

Nevertheless, a continuous steep inverse trend between the data volume and cost per bit being carried and the explosive growth of the traffic between data centers has supported the development of higher-order modulation formats, namely, 8-quadrature amplitude modulation

(8-QAM) and 16-QAM, which provide 50% (150G) and 100% (200G) more capacity than standard quadrature phase shift keying (QPSK) albeit at the expense of reducing transparent reach to around half and one-third, respectively. This catered to the need of extremely high equipment density and maximizing optical transport capacity per fiber and per transceiver. However, there was another need growing up from the continuously shifting traffic pattern. And one answer to all these needs mentioned above was the flexible-rate interface modules, which grant software-switchable modulation (QPSK, 8-QAM, and 16-QAM supported in the same device), flexible channel spectral width, and flexible frequency tunability to provide the ideal balance between performance, capacity, and reliability across the most challenging networks [4].

A key aspect of transport networks is their capability to withstand failure scenarios, given the very large amount of traffic they carry. This requirement is usually met via protection and restoration techniques [5]. Protection is a static mechanism to protect against failures, where the resources for both the primary and the backup paths are reserved prior to the data communication. Restoration is a dynamic mechanism where the backup path is not set up until the failure occurs. Survivability using these techniques is usually provided to handle single/ multiple link or node failures in the network with each scheme, claiming a different stake in terms of network resources and recovery time. Moreover, protection/restoration can be supported at different layers of the OTN and the selection of which layer to utilize, either the ODU or the optical channel (OCh) layer, also involves similar trade-offs [6].

This chapter is organized as follows. Section 2 overviews the current role of OTN switching in transport networks and introduces the concept of the universal OTN switch, highlighting the motivation behind it and the key benefits of adopting it. In order to support the case studies presented in the remaining of the chapter, a routing and grooming framework is detailed in Section 3. Section 4 addresses the relevance of mechanisms for failure survivability in transport networks, presenting a case study comparing the cost-effectiveness of supporting restoration at different layers of the transport network. Moreover, the benefits of combining universal traffic switching and flexible-rate line interfaces are investigated in Section 5 and are quantitatively assessed via a case study using reference transport networks. Section 6 elaborates on the prospects of further cost savings in metropolitan and core networks as a result of adopting coherent-detection technology in both network segments. Finally, Section 7 presents the concluding remarks.

2. The role of OTN switching in transport networks

Technological advancements have seen different node architectures being proposed, each having their pros and cons. Transponders and muxponders still provide the simplest approach to getting traffic on and off a 100G transmission interface, by multiplexing one or more client interface to a single high-speed line interface. On one hand, this approach incentivizes simplicity and relatively low CAPEX, while on the other hand, penalizes the operator because of its limited ability to groom traffic, inability to perform remote configuration, and being inefficient to combine add/drop traffic with pass-through traffic from other line interfaces.

2.1. First-generation OTN switch

The present day traffic represents a blend of packet traffic and extensively installed legacy TDM traffic. To address this varied traffic pattern and mix, the modern day network architectures requires an OTN switch to cater service-agnostic switching where the multiple client service types can be mapped into ODU frames and the same can be switched at the ODU level. This will not only allow subwavelength services to be aggregated at their source and destination nodes but also allow them to be groomed at intermediated nodes and thus finally contribute to a reduction in the number of expensive WDM line interfaces in use. **Figure 1** depicts the transponder/muxponder and OTN switch architectures. The later architecture introduces a digital OTN switch, which is separated from the WDM box using short-reach optics or integrated with the WDM box to reduce footprint and power consumption [7]. According to Infonetics, the majority of service providers (86%) are choosing OTN switching as the technology best suited to fill 100G optical channels, because it enables efficient aggregation of diverse services and protocols over a single optical link. Noteworthy, it is being embraced and deployed by network providers throughout Asia, Europe, and North America [8].

Another key feature enabled by OTN switching is fast shared protection and restoration schemes with fine granularity, which cannot be achieved with a transponder or muxponder solution. Other than contributing to a reduction of CAPEX, OTN switching does include several other benefits. Among them, high scalability, fast end-to-end service provisioning, multiple traffic type support, subwavelength switching, router port offloading, and client service level mapping are a few to name [7]. But then, certain specific traffic patterns and network topology can result in the switchless architecture to have a CAPEX advantage over

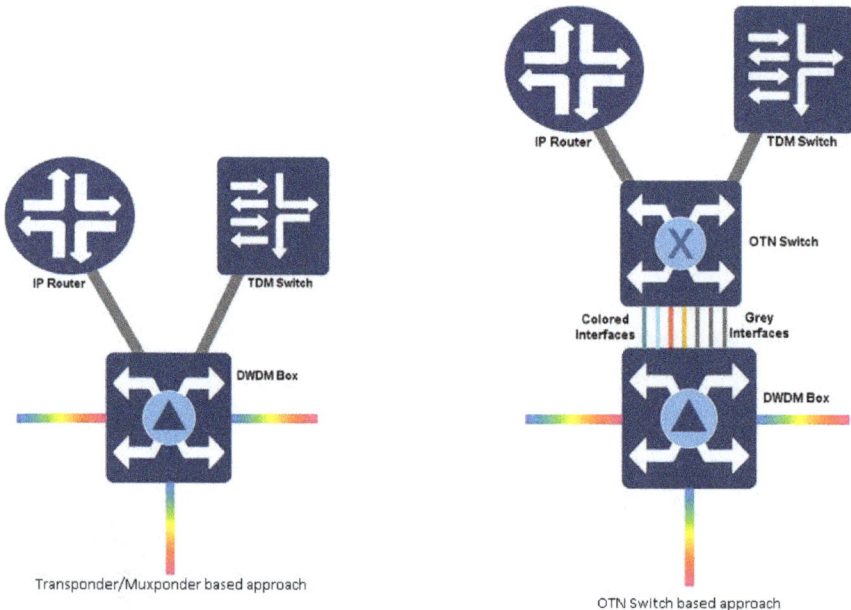

Figure 1. Transponder/muxponder-based approach vs. OTN-based approach.

the switched architecture. The same is witnessed when client services data rates match with the WDM channel data rates or even when the traffic type is not a mix and only comprises of packet data, and in these cases, the overall node architecture can be simplified to a switchless one with simplified equipment spanning L3-L0. Another perspective also challenges network operators while including OTN switch and that is contributed by the additional power and space consumption increasing the OPEX.

The traditional or first-generation OTN switch focuses only on switching ODUs and involves dedicated cards for circuit-based switching and dedicated cards for packet-switching, as illustrated in **Figure 2**. Consequently, switching is restricted within the same switch-type domain (packet/circuit). Furthermore, within the same switching domain, interworking between different technologies, such as the OTN and the synchronous digital hierarchy (SDH)/synchronous optical network (SONET), may be also blocked. Essentially, traditional OTN switches fail to deliver "universal client port" functionalities, which would allow to support any mix of client interfaces in a single switching domain, further improving the ability to efficiently utilize the transport network resources.

2.2. The universal OTN switch

Conversely to the traditional or first-generation OTN switch, next-generation universal OTN switch as shown in **Figure 3** can aggregate different protocols on the client side and enable transparent multiplexing of packet and TDM traffic, allowing a single device to be used in multiple applications efficiently.

The universal OTN switch is backed by a universal transport platform which is capable of switching traffic flows based on any L1-L2.5 protocol and on every port, including multiple protocols on the same port simultaneously. Hence, it can offer network operators the best of both worlds by dynamically controlling every flow on every port as a circuit or packet and providing the most efficient, future-proof solution for virtually all applications [9].

These next-generation OTN switches employ universal cards to handle TDM, OTN, carrier Ethernet (CE), and multiprotocol label switching-transport protocol (MPLS-TP) traffic and

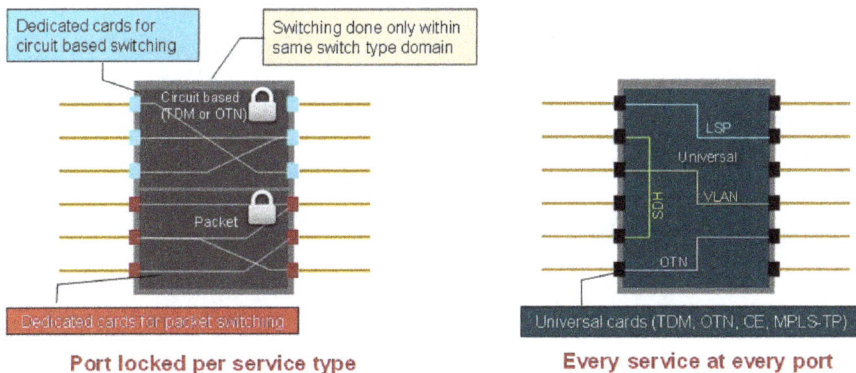

Figure 2. Traditional vs. universal OTN switch.

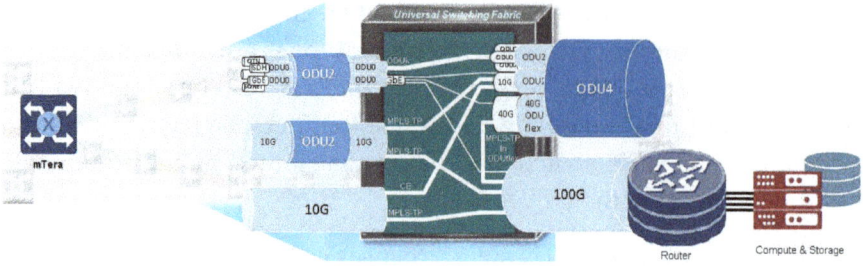

Figure 3. Universal OTN switch.

provide grooming efficiency, granularity, and service classification of a packet switch along with scalability, operational efficiency, and performance of an OTN switch, as illustrated in **Figure 3**. Furthermore, the universal OTN switch can also assist in reducing router ports by distributing services from aggregated router hand-offs with virtual local area network (VLAN) to ODU mapping, as shown in **Figure 4**.

Naturally, several other benefits, common to first-generation OTN switches, are also present and include fast end-to-end service provisioning, rapid restoration, high scalability, sub-wavelength level switching, and easy support of new/multiple traffic types. Likewise, if the service data rates are the same as the data rates of the WDM wavelength channels or when only packet traffic is present, then the universal switch might not have a CAPEX advantage over traditional OTN switch or even the conventional transponder/muxponder with OTN encapsulation.

Depending on the location in the network and the traffic matrix, switching at the packet or STS-1/VC-4 level can have efficiency benefits over switching OTN at the ODU0 (1.25G) level or above, as long as one of two conditions are met. Either there must be a significant number of client interfaces below 1G or there must be the potential for large statistical gains from

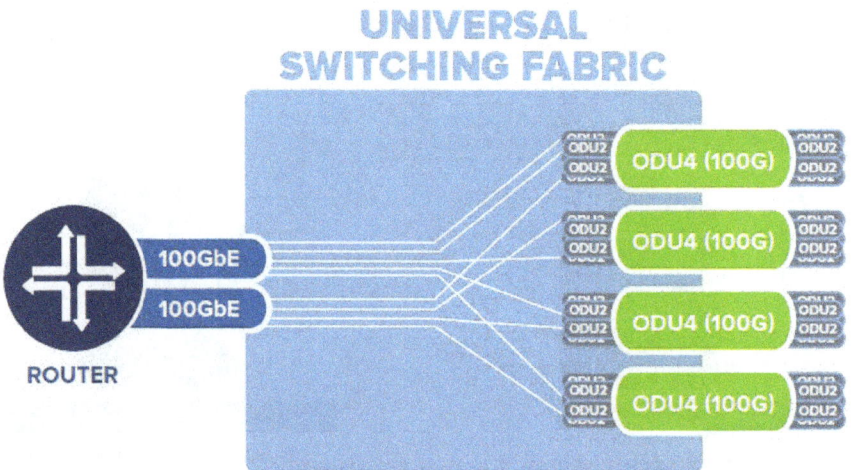

Figure 4. Universal switching fabric [1].

multiplexing uncorrelated bursty traffic flows. These conditions are more likely to occur in a metro network versus the long haul network where statistical gains have often already been achieved at the IP/MPLS layer and client interface speeds are likely to be above 1G.

Importantly, universal switching platforms enable OTN, SONET/SDH, and packet switched traffic to share the same high-speed interface, with SONET/SDH mapped to ODU2, packet traffic mapped to ODU flex, and the remaining capacity available for OTN switched traffic, thus making the most efficient use of each high-speed interface and the optical spectrum it consumes. In addition, universal switching provides investment protection against changes in traffic patterns and client types. With universal fabrics, and the ability to define in software interfaces and virtual interfaces for OTN, CE (Bridging, VLAN cross-connect), or MPLS-TP/virtual private LAN service (VPLS), and without impacting the capacity of the switch, operators can easily evolve from first-generation to universal OTN switching.

3. Routing and grooming framework

The planning of a transport network requires appropriate dimensioning algorithms to guarantee that all traffic demands can be successfully supported and at the expense of minimum CAPEX, therefore giving the network operator the best positioning to run a profitable business. Moreover, the multilayer nature of transport networks exploiting both OTN switches and reconfigurable optical add/drop multiplexers (ROADMs) and the additional requirements entailed by protection and restoration mechanisms have to be taken into account by the network design framework.

The multilayer routing and grooming framework developed and implemented to support the different network scenarios analyzed in the remaining of this chapter can accommodate the different constraints imposed by the specific node architectures and available line rates and consists of the following main steps:

Step 1. For each node pair sd, compute a set of K routing options (K shortest paths for unprotected demands, K shortest disjoint cycles for demands with protection/restoration) over the network graph $G(V, L)$, where V denotes the set of nodes and L denotes the set of links, and store them in set Π.

Step 2. Order all traffic demands from the same planning period according to a given criteria (e.g., largest first, longest first).

Step 3. For the next ordered traffic demand t_d perform the following steps:

    a. Set $k = 1$. Create an auxiliary graph, $G(V', L')$, where each network node $v \in V$ belonging to routing solution $\pi_k \in \Pi$ is mapped as node $v' \in V'$ and where each existing light path overlapping with π_k and with available capacity to support t_d is mapped as link $l' \in L'$. For unconnected nodes in V', determine the feasibility of a new overlapping light path with data rates of 100G and 200G. If feasible, map links $l_{100}' \in L'$ and $l_{200}' \in L'$.

  b. Set the cost of all links in L', such that their cost verifies $c(l') \ll c(l_{200}') < c(l_{100}')$, that is, reusing an existing light path is always preferred over creating new light paths and creating a new 200G light path is slightly more economic than creating a 100G light path (as to give preference to more spectral efficient 200G light paths in case of a tie).

  c. Route over the auxiliary graph to determine the least cost path/cycle and let the routing and grooming solution cost be $c(\pi_k^*)$. Increment k. If $k = K$, go to **(Step 3d)**. Otherwise, repeat the steps from **(Step 3a)**.

  d. Return the routing and grooming solution for $t_{d'}$ over all K solutions, with smallest cost. If none is found, the traffic demand is blocked.

Step 4. If all traffic demands have been considered, end the algorithm. Otherwise, repeat from **(Step 3)** for the next traffic demand.

Importantly, the framework is ready to support flexible-rate line interfaces, namely, capable of operating at 100G (QPSK) and 200G (16-QAM), and it can easily incorporate more line rates (e.g., 150G and 300G via 8-QAM and 64-QAM modulation formats, respectively) or can be executed for a single line rate by disabling all others. Moreover, in the case of utilizing shared restoration mechanisms to recover from failure scenarios, the fact that network resources can be shared by the backup paths of ODUs/OChs whose working paths are link- or node-disjoint also needs to be modeled. In order to control the degree of resource sharing, the number of ODUs/OChs that can share the same resource is limited by a maximum resource sharing value S.

4. Protection/restoration in transport networks

Due to several failure issues, one of the key aspects to be considered while planning and operating telecommunication networks is resiliency. With the evolution toward 5G, every-day network operators and equipment vendors struggle to come up with innovative network resiliency mechanisms which are robust, faster, and cost-effective. Transport networks are no alien to this behavior and the resiliency schemes for the same dates back to the time when SDH/SONET-based networks were at helm. Currently, OTN is the predominant transport technology. Given the multiple layers defined within the OTN, resilience mechanisms acting at the electrical and optical domains are available, the former being enforced at the ODU layer via ODU switching network elements, whereas the latter are supported at the OCh layer through ROADMs [6].

As referred, in addition to the layer at which failure survivability mechanisms are enforced, these mechanisms can be further classified as protection or restoration. Usually, protection relies on dedicated backup resources determined in advanced, reserved, and preconfigured for a particular service and which can be quickly triggered to replace the working resources. On the other hand, restoration does not require dedicated resources, i.e., backup resources can be shared, and the backup path and its resources are only assigned upon recovery from a failure. Moreover, restoration is typically triggered by the control plane, i.e., via the generalized multiprotocol label switching (GMPLS). GMPLS-based restoration can by dynamic, designated as dynamic source rerouting (DSR) and where backup resources are determined once a failure is detected, or preplanned, named preplanned shared restoration (PSR), in which case the backup resources are known in advance of the failure. The main advantage of DSR is that it can grant (best effort) survivability against a larger number of failure scenarios, namely, when multiple failures take place. On the other hand, PSR provides faster recovery of the failed service/demand since it avoids the time required to compute the backup path and associated resources [6].

The economics and qualitative behavior of PSR in transport networks are evaluated in the following, with particular emphasis on comparing the use of such scheme at the ODU and OCh layers and identifying in which network and traffic scenario is each one of them expected to be a more suitable option for a transport network operator.

Based on the resiliency schemes and the quality-of-service (QoS) they cater to, the working and protection/restoration paths are either link disjoint or node disjoint (the later forces the links to be disjoint though). Resources on the protection/restoration path (backup path) can be shared among multiple demands contrary to the same being dedicated to a single demand and this is enabled by the shared restoration scheme as illustrated in **Figure 5**. The sharing of resources is a novelty when compared to 1:1 dedicated protection schemes, which increase CAPEX due to a higher number of line interfaces required. Solid red and yellow highlights the working path of both the demands, respectively, while both these demands are using the same backup path. This is only possible when the working paths of both the demands are link disjoint and both the demands simultaneously stay unaffected by a single link failure and thus the same resource can be used for restoring either of the demands.

When resource sharing is enforced, savings are expected with respect to the amount of required additional resources, when compared to dedicated schemes (e.g., 1 + 1 protection), although restoration schemes usually provide slower recovery to failures. Furthermore, a

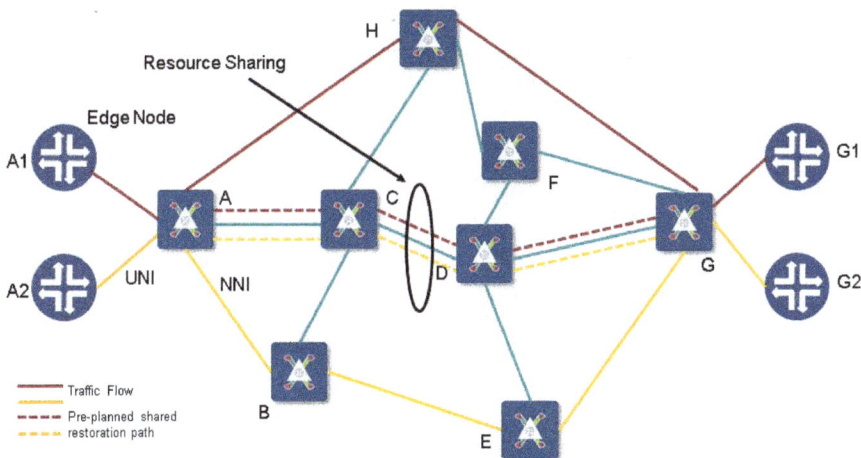

Figure 5. Illustration of resource sharing for shared restoration schemes.

qualitative and quantitative comparison of both schemes (ODU and OCh based restoration) can be performed, addressing key aspects such as restoration time, network element complexity, switching granularity, planning complexity, and line interface count. **Table 1** intends to summarize the qualitative comparison of both schemes.

In order to gain further insight on how PSR at the ODU and OCh level compare in terms of their resource requirements, above all in number of line interfaces, a planning case study is presented in this section. The study is realized over the 31-node backbone network covering Italy that is already used in previous studies [10] and illustrated in **Figure 6**. It is assumed that each network link supports up to 96 wavelength channels and each wavelength is operated at 100 Gb/s (i.e., carries one ODU4). A very simplified performance model is used to determine the transmission reach between regeneration sites: it consists of a maximum transmission reach of 1500 km and a reach penalty of 60 km per node traversed. With respect to the traffic pattern, 30% of the node pairs were randomly selected to exchange traffic demands. Each client traffic demand is mapped into an appropriate ODU k, with $k = \{0, 1, 2, 3\}$, which correspond to data rates of $\{1.25, 2.5, 10, 40\}$ Gb/s, respectively. The traffic load offered to the network is evenly distributed over the different ODU rates, meaning that each ODU k accounts for around 25% of all offered traffic load [6]. As to understand the impact of the network traffic load on the effectiveness of both restoration schemes, the total traffic load is varied from 2 up to 20 Tb/s, with load increments of 2 Tb/s.

A comprehensive comparison requires not only to compute the resource requirements of enforcing ODU or OCh restoration for every traffic demand, but also to benchmark these results against the resources needed when the traffic demands are either unprotected or require 1 + 1 protection. The routing and grooming framework of Section 3 is used to plan the network. An upper bound on the number of demands sharing a common restoration resource is set to $S = 10$. Importantly, in the case of PSR at the OCh layer, the traffic is first groomed into ODU 4s and afterward the resulting ODU 4 are routed considering also the protection/restoration mechanism being enforced.

	Restoration layer	
	ODU	**OCh**
Restoration time	• Faster: few hundreds of milliseconds	• Slower: hundreds of milliseconds up to seconds
ROADM complexity	• Low: can use simple ROADMs	• High: colorless and directionless
ODU switching	• Mandatory	• Not required
Switching granularity	• Finer: starting from 1.25 Gb/s	• Coarser: e.g., 40 or 100 Gb/s
Planning complexity	• Multilayer: with intermediate grooming	• Single-layer: without intermediate grooming
Line IF count	• Expected to be higher	• Expected to be lower

Table 1. Qualitative comparison of ODU and OCh restoration.

Figure 6. Network topology – Telecom Italia National Backbone 31 Node network.

For both ODU and OCh protection/restoration, disjointness is applied at the link level and the number of candidate routing paths/cycles K is set to 5 working and backup paths. **Figures 7** and **8** present the line interface count and wavelength channel utilization as a function of the offered traffic load when using resilience mechanisms acting at the ODU layer. The wavelength channel utilization is defined as the fraction of wavelengths being used overall network links.

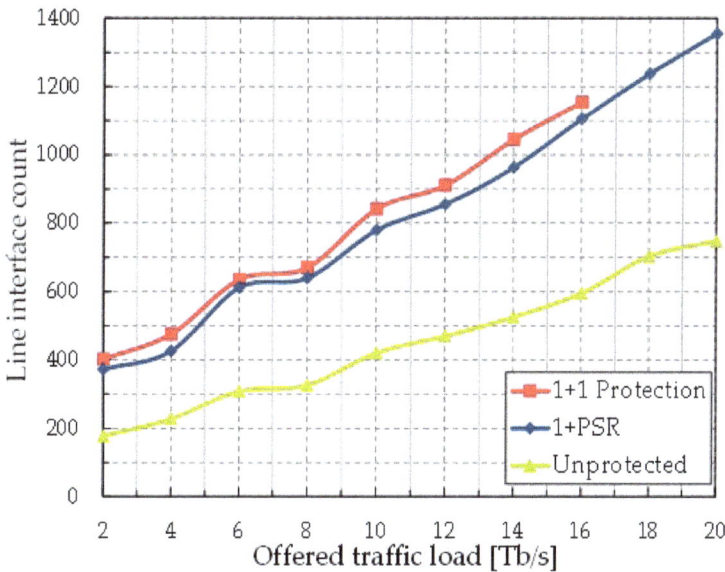

Figure 7. Line interface count for ODU restoration and protection.

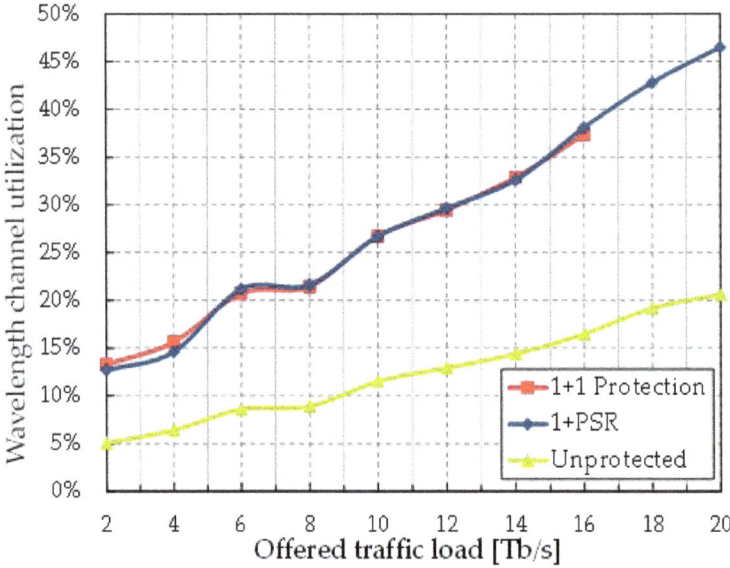

Figure 8. Wavelength channel utilization for ODU restoration and protection.

The equivalent pair of plots when supporting resilience mechanisms that operate in the OCh layer is depicted in **Figures 9** and **10**.

The main outcome of this comparison is that OCh PSR is a more cost-effective scheme than ODU PSR in a network of this size and supporting the traffic pattern defined as input [6]. In addition, it is also clear that PSR enables the network to withstand a single link failure with less resource overprovisioning than that of 1 + 1 protection.

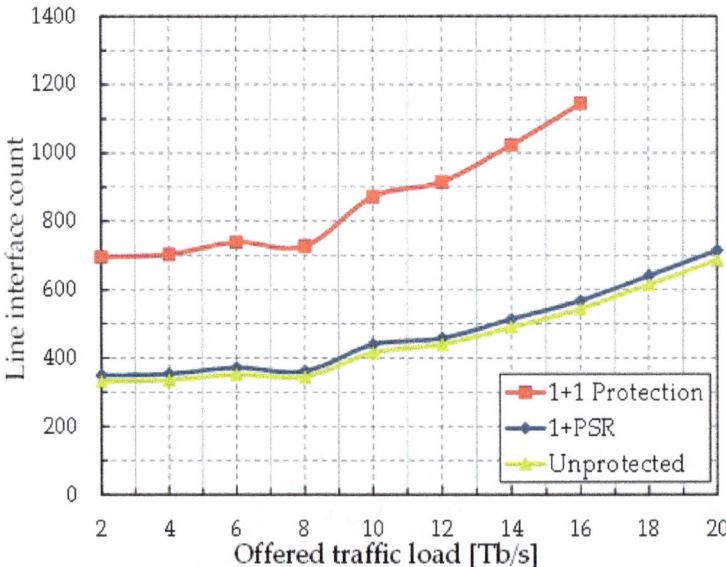

Figure 9. Line interface count for OCh restoration and protection.

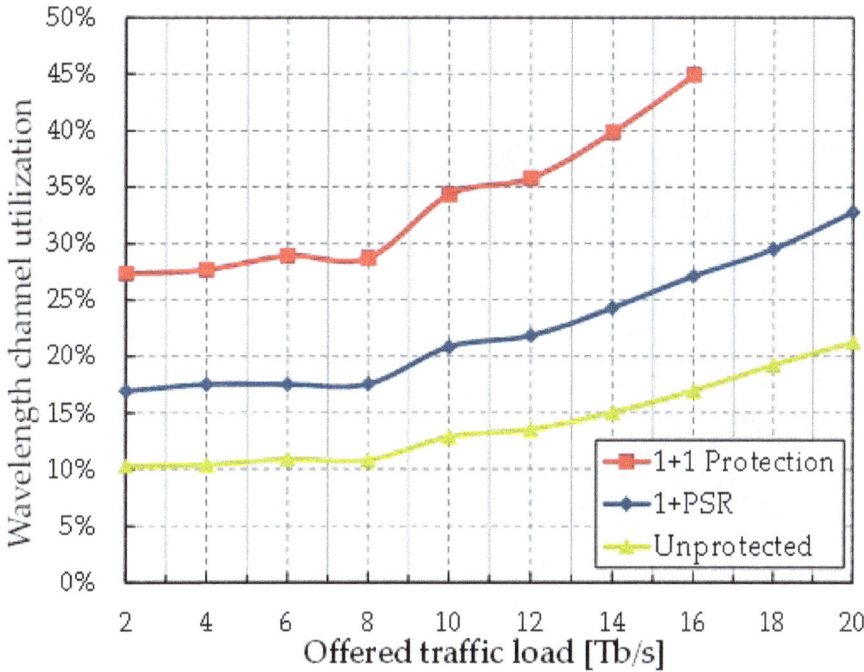

Figure 10. Wavelength channel utilization for OCh restoration and protection.

5. Combining universal OTN switching with flexible-rate DWDM networks

The wide deployment of 100G in core networks began with QPSK modulation format, where binary electrical signals are converted to a format with four constellation points, which is transmitted over two orthogonal polarizations. The applied coherent detection and advanced digital signal processing technologies enable detecting arbitrary multilevel schemes, which can be used to transmit more bits per time slot (e.g., 16-QAM, 64-QAM).

To cater network operators with the capability to address continuously shifting traffic patterns and increased capacity demands, flexible-rate interfaces are being introduced to deliver optimal transmission reach, performance, and agility. **Figure 11** exemplifies the demand reach distribution in a variety of long-haul and ultra-long-haul networks along with the expected cumulative distribution of the utilization of BPSK, QPSK, 8-QAM, and 16-QAM modulation formats for a fixed symbol rate providing data rates of 50, 100, 150, and 200 Gb/s, respectively.

There are several benefits obtained from the flexible-rate interfaces, but to highlight a few will be cost-optimized coverage from intracity data center interconnection (DCI) to ultra-long-haul demands, single sparing blade for all modulation schemes, and restoration of higher reach modulation schemes using higher reach modulation schemes. A joint design considering flexible-rate and universal OTN switching technology can unfold crucial benefits that go beyond the separate claims of improved capacity on the line side (flexible-rate) and more efficient traffic aggregation (universal OTN switching).

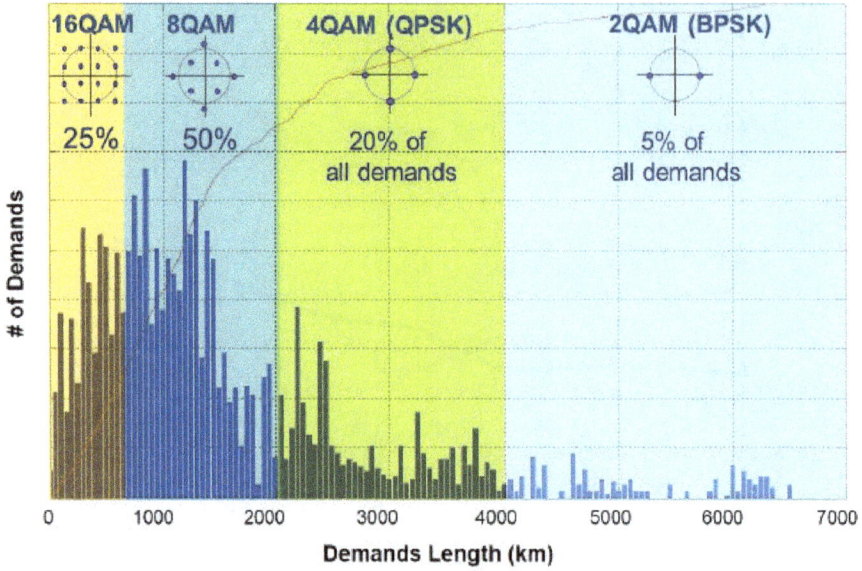

Figure 11. Example of demand length distribution and expected cumulative distribution of the modulation format utilization in core networks with Raman amplification [11].

Recent work shows evidence that if both features are combined in an optimal way, a considerable reduction of up to 30% could be achieved in terms of the number of deployed light paths, with proportional savings in a number of line interfaces and spectrum used [12]. In the scope of this work, two long distance optical transport networks, as depicted in **Figure 12** and inspired in networks operated by Telecom Italia/TIM, were considered. One is a more recent Italian national backbone, which has 44 nodes and 71 fiber links and has already been used in other studies [13, 14].

It meets the needs of circuits at a national level, mainly for IP router interconnection and for connectivity of big clients. With its shortest paths under 2200 km and backup paths (disjoint

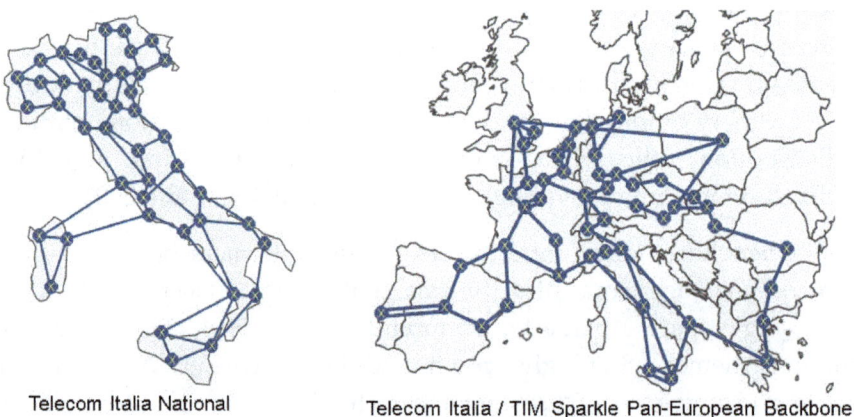

Telecom Italia National   Telecom Italia / TIM Sparkle Pan-European Backbone

Figure 12. Network topologies – Telecom Italia National Backbone 49 Node network and Sparkle.

from the shortest one) under 2600 km, it is between a regional and long-haul network. The other network is TIM Sparkle Pan-European backbone [15], a geographically expanding network that currently covers Central, Southern, and Eastern Europe with 49 nodes and 72 fiber links. It classifies as an ultra-long-haul network (shortest paths under 5500 km and backup paths under 7000 km). With respect to the client rates to be serviced by the networks during their entire lifecycle, the following data rates are considered: 10G, 40G, and 100G for the Italian National Backbone and 1.25G, 2.5G, 10G for the European Backbone. Traffic comprises both Internet Packet traffic and SONET/SDH TDM traffic. Two traffic periods are considered wherein the later phase, traffic was extrapolated to be at the end of 4 years of the current period, with 25% growth for each year. Total traffic for the two periods under analysis and the partitioning of bandwidth among different client rates for the Italian national backbone and European backbone network is as described in Ref. [14]. Using the multilayer optimization algorithm described in Section 3, five routing options were calculated for each node pair ($K = 5$) and the algorithm was run considering all four combinations given by traditional versus universal OTN switching and using only 100G (QPSK) line rates versus optimizing the line rate between 100G (QPSK) and 200G (16-QAM) according to the properties of the routing path [16].

The required number of light paths, which impacts the number of required expensive line interfaces as well as the amount of spectrum used, is shown in **Figure 13**.

As expected, for TIM national backbone network, the benefit of supporting higher data rates with the same line interface is more effective in reducing the number of light paths when compared to the Sparkle network, mostly because the former topology benefits from average shorter routing paths. Noticeably, universal switching always grants savings when compared to its traditional counterpart due to the packet-level aggregation and universal nature of the switch. Moreover, the combined effect of using both universal switching and flexible-rate leads to the highest savings in light path count and, consequently, the most cost-effective solution. Although not shown here due to the lack of space, savings in router port count are also achieved, which further contributes to decrease TCO when leveraging universal OTN switching and flexible-rate line interfaces. Thus, a joint network design considering both flexible-rate and universal switching technology can unfold crucial benefits that go beyond the separate claims of improved capacity on the line side (flexible-rate) and more efficient traffic aggregation (universal switching).

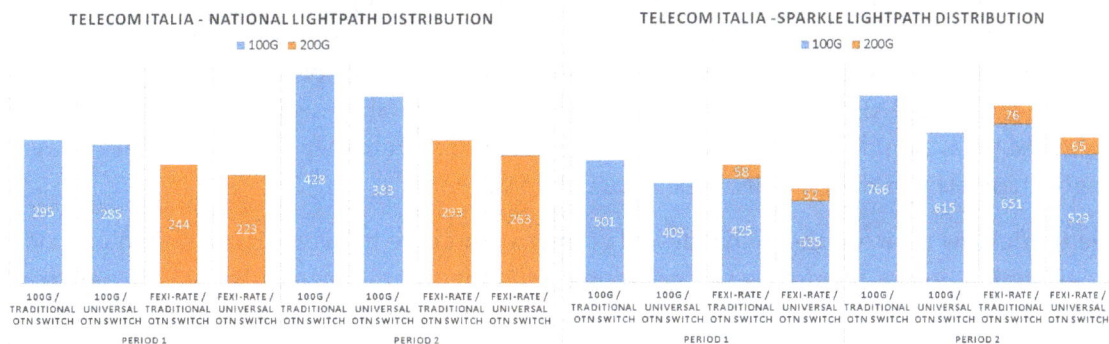

Figure 13. Number of light paths as function of the network and node architecture.

6. Transparent handover between metro and core segments of next-generation transport networks

Transport networks can hierarchically be broadly classified into access, metro, and core domain. End-user connectivity is catered to by the access network and a few dozen kilometers could be covered based on the specific technology requirements and the density of user population. In between, the core and the access part, lies the metro network which on one hand aggregates traffic coming from the access networks and on the other hand transports intrametro traffic (e.g., intrametro data center interconnection) and covers often up to a few hundred kilometers. Following the level of traffic aggregation, metro networks could further be segmented into metro aggregation and core. At the end, covering a bigger geographical area (e.g., state or country) is what is called the core network interconnecting the metro networks and typically spans over several hundreds to thousands of kilometers.

In the metro transport networks, the predominant technology and topology are in the form of SDH/SONET rings. But while these SDH/SONET rings had intrinsic scalability limitations, there was always an increasing need for capacity. And this gave rise to the introduction of coarse wavelength division multiplexing (CWDM) and DWDM in these networks catering to wavelength switching support at intermediate nodes thus reducing optical-electrical-optical (OEO) conversions which was initially a cost burden to the operator. In present days, the metro network is mostly infested with 2.5G or 10G wavelengths using direct-detection modulation formats and as always the increasing demand of data will be catered to using higher data rate channels (e.g., 40G and 100G) in the coming days. Compared to these, core network has a different behavior and is severely dependent on DWDM and ROADMs exhibiting asymmetrical meshed topology. The wavelength channels have a good share of 10G, 40G, and 100G. Moreover, QPSK is now the dominant modulation format for the higher data rates of 40G and 100G with coherent-detection replacing direct-detection. The utilization of higher-order modulation formats, such as 16-QAM, enables to increase channel capacity. But, the poor reach performance of these higher modulation formats penalizes a widespread deployment of the same. To mitigate this negative effect, adoption of super channels is being looked upon as an alternative to core networks where multiple carriers are grouped together to realize these higher bit-rates (e.g., 2 × 100G QPSK to create a 200G super channel on a 100GHz spectrum) [17]. Another important aspect to look upon is the optical add-drop multiplexers (OADMs) being used in core and metro networks. Metro networks are comparatively simpler with lesser nodal degrees and thus simpler OADMs of the likes of broadcast-and-select or fixed add/drop ROADMs or to the extent of nonreconfigurable fixed OADMs (ROADMs) are in use today. On the other hand, core networks use much higher capacity and also wavelength channels and thus require expensive OADMs.

The traffic flow between the above described metro and core network at the boundary node is depicted in **Figure 14** and the same is represented by metro-to-core (M2C) in the next part of this chapter.

To be more detailed, let us presume one data channel running from one metro network to another metro network via a core network and is designated as metro-to-core-to-metro

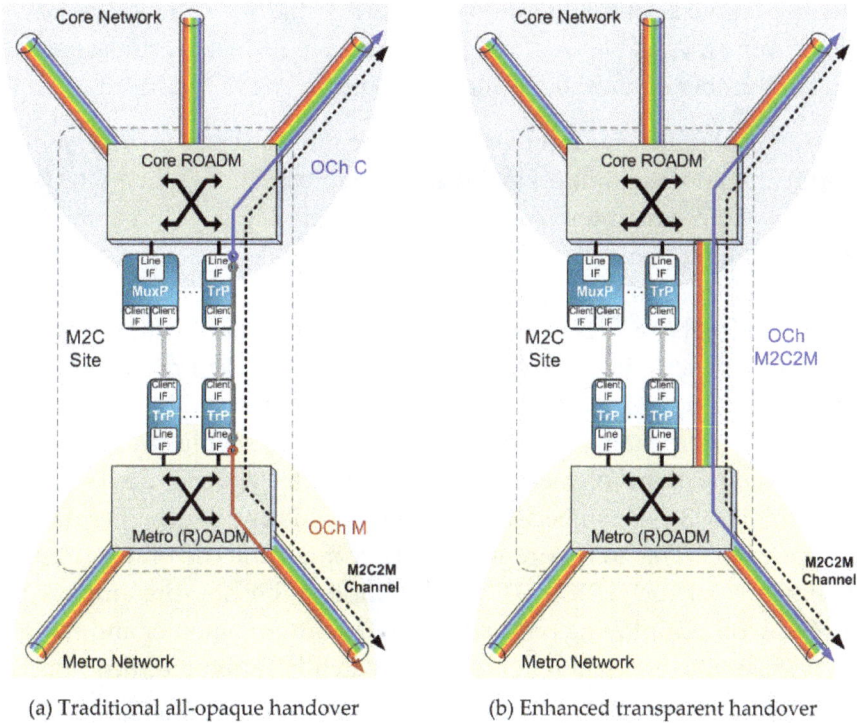

(a) Traditional all-opaque handover   (b) Enhanced transparent handover

Figure 14. Architectures for interconnecting metro and core networks (TrP – transponder, MuxP – muxponder). (a) Traditional all-opaque handover; (b) enhanced transparent handover.

(M2C2M) channel. The current mode of operation forces this M2C2M channel to ride on an optical channel in the metro domain (OCh M) till the metro (R)OADMs. At this (R)OADM node, the same OCh M gets terminated at a line interface only to be transferred to a transponder or muxponder which maps the same into the optical channel (OChC) ferrying inside the core network. At the boundary node in between the core and metro node, a similar process makes sure that the channel is handed over from the core to the metro network. It is to be noted that, these handover sites could be a single site at the same physical location (housing the (R)OADMs) or could be physically separated sites within a comparatively shorter distance hosting each (R)OADM looking toward the core and metro network.

This architecture represents an all-opaque architecture depicting a clear enough demarcation between the metro and core network segments where the hardware (client interfaces) performs the handover. As it already undertakes the OEO operation, these sites are used for grooming subrate data signals into higher data-rate pipes/channels. For example, the M2C2M channel with a data rate of 2.5G can be multiplexed/groomed into 10G and 100G OCh M and OCh C, respectively. Therefore, the M2C2M channel can be multiplexed/groomed in OCh M along with other subrate data channels of equal or smaller capacity starting at the same node and destined to another metro network, while in the core network the M2C2M channel can be multiplexed/groomed with other data channels that start and end in the same metro networks pair. This process directly influences the fill ratio of the core network light paths and thus contributing to the reduction of CAPEX as a result of lesser number of expensive line interfaces and also the network occupied bandwidth. It is noteworthy that the bandwidth usage in the core network is much more a valuable asset when compared to its metro counterpart due to longer fiber links, higher number of amplifiers, and advanced, larger ROADMs.

In this approach, the M2C2M channel crosses one line interface pair in the first metro network where they originate and client interface pair in the first M2C site, line interface pair in the following core network, client interface pair in the second M2C site, and finally a line interface pair in the terminating metro network, thus spending in total 6 and 4 line and client interfaces, respectively. But still these are being shared with other smaller data rate M2C2M channels. But this is not the entire scenario because soon we start encountering higher data rate M2C2M channels (e.g., 10G or even higher) which cannot be multiplexed/groomed into the already existing OCh C due to nonterminating OChs in the same metro networks or capacity non-availability in the existing channels. Such use-cases foresee savings by the transparent handling of M2C2M channel between metro and core networks, which is possible if an additional fiber link exists in the M2C site in between the (R)OADMs and this higher data rate M2C2M channel is mapped on it. This transparent M2C interconnection can only be deployed by augmenting nodal degree of each (R)OADMs and additional booster/pre-amplifier deployment for each direction while employing one extra port of splitter/combiner and wavelength selective switches (WSSs) [18, 19]. Adding an optical switch between add/drop ports subset and using them directly for the optical bypass will be a good alternative [20].

In order to gain insight on expected design, implementation, and operational differences between a metro core network with traditional all-opaque M2C handover versus the same network with enhanced transparent handover, a subset of these differences is highlighted in **Table 2**, to highlight the different aspects to be assessed with care before opting for either of

	All-opaque M2C handover	Enhanced transparent M2C handover
Interface requirements	• Separate (local) optimization of line IFs required in the metro and core networks • Intensive usage of client IFs at M2C sites	• Joint (global) optimization of line IFs used at metro and core networks • Savings in client IF count at M2C sites
(R)OADMs at M2C sites	• All traffic handover done via the add/drop ports of both (R)OADMs	• Both (R)OADMs have to be augmented with one additional degree
Spectral efficiency	• Independent channel format selection at the metro and core networks • Best suited (cost-wise, performance-wise) format can be selected in each network: e.g., direct-detect 10 Gb/s in metro, more spectral efficient but expensive 100 Gb/s in core	• Format used in OCh M2C2M impacts spectral efficiency of both networks • Additional spectral inefficiencies associated to coexistence of direct detect and coherent formats in the core network: extra spacing between neighboring channels [20]
Planning complexity	• Lower: due to sequential planning of metro networks first and core network afterward	• Higher: requires integrated planning of metro and core networks

Resource provisioning	• Separate provisioning of resources at each network segment	• Requires end-to-end provisioning of the OCh M2C2M
	• Easy support for deploying different vendors hardware (e.g., line interfaces, common equipment) and software (e.g., network management systems) in the metro and core networks	• Complex support for deploying different vendors equipment: interworking between optical and network management systems, limited interoperability between control plane instances, use of alien wavelength concept, shortening transparent reach

Table 2. Qualitative comparison of all-opaque metro-to-core handover with enhanced transparent handover.

these strategies for a M2C traffic handover. Further, the analysis targets to capture technology landscape influenced differences in today's metro and core networks [21].

7. Conclusions

The overriding purpose of this chapter is to depict the relative importance of the universal OTN switch and flexible-rate line interfaces in the context of different network scenarios and traffic conditions. To accomplish the same, results from different network studies were presented. It is evident that, when the traffic pattern involves multiple subrates coming from varied data sources (TDM, Packet), universal OTN switch yields the maximum savings from CAPEX point of view. This is contributed by the reduced number of light paths and router port savings. Needless to mention here that, the more is the mismatch between the client traffic rates and the line-side rate(s), more savings could be achieved by exploiting a multilayer optimization approach. Furthermore, exploiting the higher capacity light paths enabled by flexible-rate line interfaces and combining the same with universal OTN switching, the better of both worlds can be achieved. But it is important to mention that flexible-rate line interfaces supporting 16-QAM, for example, will be less important when the network links are very long due to the limited transparent reach with this modulation format. On the other hand, the role of these state-of-the-art node architectures in the context of protection and restoration was also visited and different alternatives were presented. It could be inferred that OCh restoration would be preferable from a CAPEX point of view, with lesser resources being required, while opting for a better QoS (e.g., in terms of recovery time), ODU restoration can be a better choice. At the end, this chapter overviews the present day metro-to-core network scenario which enforces all-opaque traffic engineering and also prospects the economic and technological feasibility of transparent handovers. Recent technological developments, for example, metro core spectral efficiency gap narrowing, alien wavelength/black link standardization aided by state-of-the-art multivendor and multilayer resource provisioning platforms will overcome the hurdles of transparent handovers, and thus exploiting the big saving potentials in terms of interfaces at these boundary nodes.

Author details

Bodhisattwa Gangopadhyay[1*], João Pedro[1,2] and Stefan Spälter[3]

*Address all correspondence to: bodhisattwa.gangopadhyay@coriant.com

1 Coriant Portugal, Amadora, Portugal

2 Instituto de Telecomunicações, Lisboa, Portugal

3 Coriant GmbH, Munich, Germany

References

[1] Coriant. The Role of OTN Switching in 100G & Beyond Transport Networks [Internet]. 2016. Available from: http://www.lightwaveonline.com/content/lw/en/whitepapers/2016/03/the-role-of-otn-switching-in-100g-beyond-transport-networks.whitepaperpdf.render.pdf

[2] ITU. G.709: Interfaces for the Optical Transport Network [Internet]. Available from: http://www.itu.int/rec/T-REC-G.709/en, 2009

[3] Cinkler T, Coudert D, Flammini M, Monaco G, Moscardelli L, Munoz X, Sau I, Shalom M, Zaks S. Traffic grooming: Combinatorial results and practical resolutions. In: Koster A, Muñoz X, editors. Graphs and Algorithms in Communication Networks: Studies in Broadband, Optical, Wireless and Ad Hoc Networks. 28th ed. Springer, Heidelberg; 2010. p. 426, p.97. DOI: 10.1007/978-3-642-02250-0_2

[4] Eira A, Pedro J, Pires J. Optimized design of fixed/flex-rate line-cards and transceivers over multiple planning cycles. In: Optical Society of America, editor. Optical Fiber Communication Conference, 2014. San Francisco: Optical Society of America; 2014. p. W2A-47. DOI: 10.1364/OFC.2014.W2A.47

[5] Gerstel O. Opportunities for optical protection and restoration. In: IEEE, editor. Optical Fiber Communication Conference and Exhibit. IEEE; 22–27 February 1998, San Jose, California, USA. DOI: 10.1109/OFC.1998.657388

[6] Pedro J, Gangopadhyay B. On the trade-offs between ODU and OCh preplanned shared restoration in transport networks. In: IEEE, editor. Telecommunications Network Strategy and Planning Symposium (Networks), 2014 16th International. Madeira: IEEE; 2014. pp. 1–6. DOI: 10.1109/NETWKS.2014.6959252

[7] Schmitt A. Infonetics Research. Integrated OTN Switching Virtualizes Optical Networks [Internet]. 2012. Available from: https://www.infinera.com/wp-content/uploads/2015/07/Infinera-WP-Infonetics-Integrated-OTN-Switching.pdf

[8] Schmitt A. Infonetics Research Excerpt. Global Service Provider Survey Excerpts: OTN, MPLS, and Control Plane Strategies [Internet]. 2013. Available from: https://www.

infinera.com/wp-content/uploads/2015/07/Infonetics-Global_Service_Provider_Survey_
Excerpts-OTN_MPLS_Control_Plane_Strategies.pdf

[9] Coriant. Universal Switching and Transport [Internet]. Available from: http://www.cori-
ant.com/solutions/technologies/universal-switching-and-transport.asp, 2015

[10] Pedro J, Santos J, Morais R. Dynamic setup of multi-granular services over next-gener-
ation OTN/DWDM networks: Blocking versus add/drop port usage. In: IEEE, editor.
Proceedings of IEEE ICTON 2012. IEEE; 2012, Coventry, United Kingdom. DOI: 10.1109/
ICTON.2012.6253705

[11] Coriant. Flexing Next-Generation Optical Muscles [Internet]. Available from: http://
www.coriant.com/spotlights/flexi-rate-white-paper.asp, 2016

[12] Gangopadhyay B, Pedro J, Spälter S. Cost-effective next-generation information high-
ways leveraging universal OTN switching and flexible-rate. In: Optical Society of
America, editor. Optical Fiber Communication Conference; 19–23 March, 2017. Los
Angeles: Optical Society of America; 2017

[13] Costa N, Pedro J, Quagliotti M, Serra L. Preplanning framework to evaluate the potential of different modulation formats in DWDM networks. In: 10th Conference on
Telecommunications (ConfTele).; TECHDAYS, Aveiro. 2015

[14] Gangopadhyay B, Pedro J, Spälter S. Comparative assessment of network architectures
for transporting Packet and TDM traffic. In: IEEE, editor. 21st European Conference on
Networks and Optical Communications (NOC), 1–3 June, 2016. Lisbon: IEEE; 2016. DOI:
10.1109/NOC.2016.7507000

[15] Telecom Italia. Sparkle [Internet]. Available from: http://www.tisparkle.com/

[16] Cantono M, Gaudino R, Curri V. A statistical analysis of transparent optical networks
comparing merit of fiber types and elastic transceivers. In: IEEE, editor. 18th International
Conference on Transparent Optical Networks (ICTON), 10–14 July, 2016, Trento, Italy.
IEEE; 2016. DOI: 10.1109/ICTON.2016.7550511

[17] Eira A, Pedro J, Pires J. On the impact of optimized guard-band assignment for super-
channels in flexible-grid optical networks. In: Optical Society of America, editor. Optical
Fiber Communication Conference/National Fiber Optic Engineers Conference 2013.
Optical Society of America; 2013, Anaheim, California, USA. DOI: 10.1364/OFC.2013.
OTu2A.5

[18] Rambach F, Konrad B, Dembeck L, Gebhard U, Gunkel M, Quagliotti M, Serra L, López V.
A multilayer cost model for metro/core networks. IEEE/OSA Journal of Optical
Communications and Networking. 2013;5(2013):210–225. DOI: 10.1364/JOCN.5.000210

[19] Pedro J, Pato S. Impact of add/drop port utilization flexibility in DWDM networks [Invited].
IEEE/OSA Journal of Optical Communications and Networking. 2012;4(11):B142–B150.
DOI: 10.1364/JOCN.4.00B142

[20] Woodward S, Feuer M, Kim I, Palacharla P, Wang X, Bihon D. Service velocity: Rapid provisioning strategies in optical ROADM networks. IEEE/OSA Journal of Optical Communications and Networking. 2012;4:92–98. DOI: 10.1364/JOCN.4.000092

[21] Pedro J, Gangopadhyay B. Prospects for transparent handover between the metro and core segments of next-generation transport networks. In: IEEE, editor. 17th International Conference on Transparent Optical Networks (ICTON), 2015; 5–9 July, 2015; Budapest, Hungary: IEEE; 2015. pp. 1–5. DOI: 10.1109/ICTON.2015.7193475

5

Evaluation of Parametric and Hybrid Amplifier Applications in WDM Transmission Systems

Vjaceslavs Bobrovs, Sergejs Olonkins, Sandis Spolitis,
Jurgis Porins and Girts Ivanovs

Additional information is available at the end of the chapter

Abstract

Over the past two decades, a rapid expansion of the amount of information to be trans-ferred has been observed. This tendency is explained by the rapid increase of Internet and other service users, as well as with the increasing availability of these services. This rapid growth in the amount of globally transmitted data is also associated with the expansion of the range of services offered, including such resource-consuming ser-vices as high-resolution video transmission, videoconferencing, and cloud computing, as well as with increasing popularity of such services. To satisfy this constantly increasing demand for higher network capacity, fiber optical transmission systems have been stud-ied and applied with a growing intensity. Currently, optical transmission systems with wavelength-division multiplexing (WDM) have attracted much attention, as this technol-ogy allows using the available optical fiber resources more effectively than alternative technologies.

Keywords: optical amplifiers, parametric amplifiers, hybrid amplifiers, fiber optics, EDFA, WDM

1. Introduction

According to the latest Cisco forecast, the total amount of global IP traffic in 2016 reached 1.1 zettabytes, whereas in 2018 it will reach 1.6 zettabytes. The forecasted increase in the monthly transferrable IP traffic over the period from 2013 to 2018 is shown in **Figure 1a**. Studies per-formed by Cisco show that in comparison with 2012 the amount of Internet traffic transferred in the peak hours in 2013 increased by 32%, whereas the average daily volume of transfer-rable Internet traffic increased by 25% [1]. If this tendency remains, then in 2018 the volume of

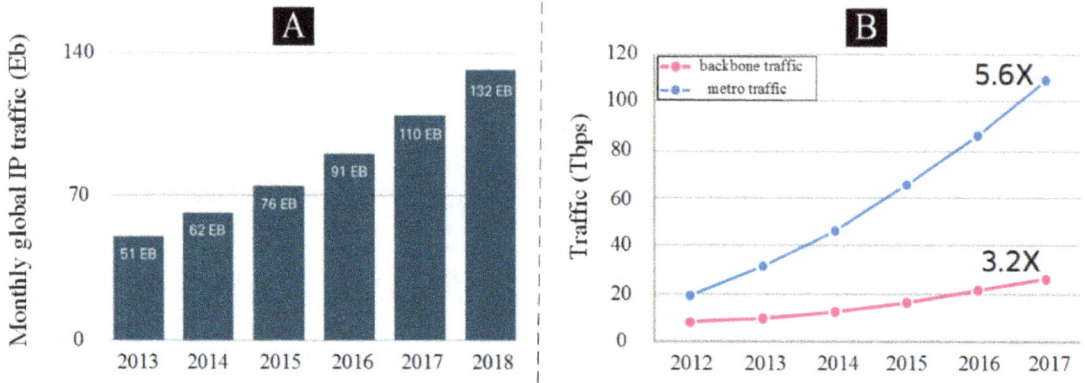

Figure 1. Cisco forecast of the monthly transferrable IP traffic (A) and Bell Labs forecast of the transferrable data amount in backbone and metro networks (B) [1, 2].

transferrable Internet traffic during the peak hours will reach 1 petabit per second, whereas the daily average will reach 311 terabits per second [1, 3]. According to the Bell Labs forecast, results of which are shown in **Figure 1b**, during the period from 2012 to 2017, the increase of traffic in backbone networks will reach 320%, whereas in metro networks, it will reach by 560% [2].

It is possible to increase the wavelength-division multiplexing (WDM) system throughput capacity either by increasing the data transmission speed in channels or the number of channels. The wavelength band that is used for transmission in WDM systems is limited due to the wavelength dependence of optical signal attenuation in optical fibers [4, 5]. In modern transmission systems, the minimum attenuation of single-mode optical fiber is 0.2 dB km^{-1}, and it is observed in the "C" wavelength band, which corresponds to wavelengths from 1530 to 1565 nm. Regardless of the fact that the attenuation value is so low, its impact accumulates with every next kilometer. In long-haul transmission systems, where transmission lines are several hundreds and even thousand kilometers long, the attenuation substantially degrades the quality of the received signal, as the photodetector sensitivity is limited [6–8]. As the number of channels increases, the attenuation caused by the optical signal division also increases, especially in cases where power splitters are used [9]. However, by increasing the speed of data transmission, it becomes necessary to reduce the optical noise produced by optical components (light sources, modulators, amplifiers, receivers, etc.), as higher transmission speed signals have lower noise immunity.

Therefore, solutions are needed for compensating the ever-increasing accumulated signal attenuation in an ever-broader wavelength range. Currently, erbium-doped fiber amplifiers (EDFAs) are most commonly used around the globe for compensation of optical signal attenuation. The amplification bandwidth of EDFAs is strictly limited (for conventional EDFA solutions, it is only 35 nm), which restricts the wavelength range used for the transmission in existing systems [10–12]. It is, thus, necessary to seek for new solutions to amplifying optical signals and for opportunities of expanding the range of amplified wavelengths and increasing the attainable amplification level for the already-existing optical signal amplification solutions. This can be achieved by combining amplifiers of various types. In such a way, it is possible to combine the positive properties and partly compensate the drawbacks of different types of amplifiers.

During recent years, the need to increase transmission capacity of existing optical networks together with requirements for reducing the total cost of construction and maintenance of optical networks has induced increasing interest in all-optical signal processing [13–16]. In contrast to solutions with optical-electrical-optical (O/E/O) signal conversion, which induces the so-called bottlenecks in optical transmission systems, all-optical signal processing is performed in real time, whereas the signal is transmitted through a nonlinear medium [17]. Therefore, all-optical signal processing allows avoiding the part of transmission capacity limitation that is caused by O/E/O signal conversion.

The progress in nonlinear material research has resulted in commercial production of optical fibers and other components with high values of the nonlinear coefficient. Therefore, the optical power, required to initiate fiber nonlinearities, has become lower [15]. Fiber nonlinearity is the main mechanism that is used for all-optical signal processing. Optical amplifiers are the only optical devices capable of rising the power of optical signal high enough to induce manifestation of nonlinear effects during transmission. That is why the usage of optical amplifiers for all-optical signal processing purposes has been intensively studied all over the world during recent years, and various applications of optical amplifiers have been demonstrated [13–16, 18–20].

2. Main principles of optical amplification

Amplification of optical signals is based on the energy transfer from pumping optical radiation or another type of energy to the amplifiable optical signal. This process is implemented differently in various types of optical amplifiers. In general, the amplification process uses the stimulated emission phenomenon in the amplification environment, such as, for instance, semiconductor optical amplifiers or doped fiber optical amplifiers. Furthermore, such nonlinear optical effects such as Raman, Brillouin, and four-wave mixing (FWM) are used to amplify optical signals in cases of Raman, Brillouin, and parametric optical amplifiers, respectively [21].

The mechanism of amplifying optical signals is based on occurrence of stimulated light emission in the gain medium. The light emission phenomenon can be explained using the Rutherford-Bohr atomic model. Bohr has stated that atoms may jump from one energy state to another, by performing what is known as the quantum jumps, corresponding to a change of orbit. This orbit change requires a change in the energy level; therefore, if the atom jumps from the higher energy state to the lower energy state, it will produce a photon. A photon contains energy, which corresponds to the difference between the initial higher energy level and lower occupied level energy, as the overall energy of the process must remain unchanged. This assumption derives from the law of conservation of energy [22]. Thus, photon energy can be determined according to the following equation [23]:

$$E_{photon} = E_2 - E_1 = h\,v_{photon} \qquad (1)$$

where E_{photon} is the generated photon energy, E_1 and E_2 are the high and low energy level, h is the Planck constant, and v_{photon} is the generated photon frequency.

Optical amplifiers can be classified according to the nature of the amplification process [23]:

a. Amplifiers, in which amplification is obtained, using linear properties of the material (semiconductor optical amplifiers (SOAs) and amplifiers on rare-earth element-doped fiber basis (xDFA))

b. Amplifiers, for which the principle of operations is based on nonlinear properties of the material (Raman optical amplifiers, Brillouin optical amplifiers, and fiber optical parametric amplifier (FOPA))

A second way of classifying optical amplifiers is according to the medium, in which amplification takes place:

• Amplifiers, in which semiconductor material is used (SOA)

• Amplifiers, which are produced on the basis of optical fibers

The main parameters that are used to characterize optical amplifiers are the level of amplification, the gain bandwidth, the saturation power of the amplifier, the polarization sensitivity of the produced gain, and the amount of signal impairments produced by the amplifier.

The achievable level of amplification is determined as the relation of the output signal power to the power of the same signal in the input of the amplifier. Amplifiers are sometimes also described with amplification efficiency, which describes the amplification as a function of the pumping power. The unit of measurement of efficiency of amplification is dB/mW [24].

The bandwidth of the amplifier produced gain is applied to the wavelength or frequency range, in which the use of the amplifier is effective, namely, where it can ensure an increase in signal power. This value is especially important in WDM transmission systems, as it limits the number of channels in such systems [23].

The saturation point for an optical amplifier is the maximum attainable output power value, namely, when the optical signal power in the amplifier output no longer increases while raising the signal power at the amplifier input. When the input power is increased above the saturation point, all carriers in the gain medium are already in a saturated status, and a higher level of energy transfer to the amplified signal is no longer possible. The saturation power is defined as the output power, at which 3 dB decrease in amplification is observed, in respect to the maximum possible level of amplification [23].

The dominating source of noise in optical signal amplifiers is the amplified spontaneous emission (ASE), which originates in the gain medium [25]. The amount of noise generated by amplifiers depends on various factors. The most important of these are the gain medium material parameters (e.g., the spontaneous lifetime of the energy level), gain spectrum, noise bandwidth, amplifier saturation, and population inversion parameters. The problem of noise generated by an amplifier is most explicit in systems, where it is required to use multiple amplification stages, therefore placing the amplifiers in a cascade, such as backbone optical networks. Each amplifier in such cascades not only amplifies the transmitted signal but also the noise generated by the amplifier from the previous amplification stage and additionally adds ASE noise of its own [23]. To assess the amount of ASE noise generated by the amplifier, the noise figure (NF) parameter is normally used. This value describes the optical signal-to-noise ratio (OSNR) changes, as the signal passes through the amplifier [23, 26].

In the studies conducted by the authors, using simulation software OptSim, the performance of SOA, EDFA, lumped Raman amplifier (LRA), and the distributed Raman amplifiers (DRA) under equal operating conditions has been compared. The simulation scheme introduced for this purpose is displayed in **Figure 2**. Such a structure of the WDM transmission system simulation model will also be used further in the research, when the operations of an amplifier are analyzed.

The performance of different types of amplifiers has been compared in a 16-channel dense wavelength division multiplexing (DWDM) transmission system with 10 Gbps transmission speed per channel, 50 GHz channel spacing, and non-return-to-zero on-off keying (NRZ-OOK) (on-off keying) modulation format. In each case, also the length of the dispersion-compensating fiber (DCF) has been determined. Optical amplifiers have been used as in-line amplifiers. The comparison of SOA, EDFA, LRA, and DRA performance is available in **Table 1**.

The largest transmission distance has been achieved in a system with the DRA. Here, just like in the case of LRA, the attainable amplification is limited by the impact of fiber nonlinearity on the quality of the amplified signal. A 1150 mW co-propagating pumping radiation is used for DRA pumping. The amplification process occurs in the transmission line section between the DRA pumping source and the receiver block. Thus, the single-mode fiber (SMF) attenuation reduces the signal amplification rate in the direction from the amplifier to the receiver block, which allows achieving much larger amplification than in the case of LRA, and accordingly increases the attainable transmission distance. Irrespective of the

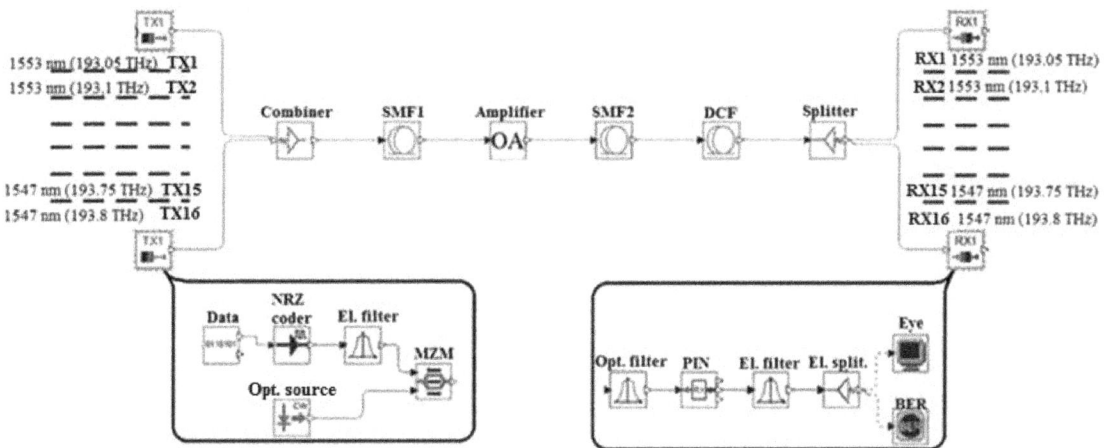

Figure 2. Simulation model of the 16-channel 10 Gbps DWDM transmission system used for comparison of optical amplifier performance.

Amplifier type	-	SOA	EDFA	LRA	DRA
Transmission distance (km)	69	112	135	119	146
DCF length (km)	5	15	20	17	20
Gain in wavelength range from 1546 to 1553 nm (dB)	-	17.4	23.4–25.1	19.9–20	24.9–25
NF in wavelength range from 1546 to 1553 nm (dB)	-	-	4.5–4.6	3–3.1	−8.6
Level of interchannel cross talk in the channel with the highest bit error rate (BER) (dBm)	−55.5	−50	−47.9	−48.3	−49.3

Table 1. Summary of the results obtained in the 16 channel 10 Gbps DWDM transmission system depending on the type of amplifier used (Column 2—without using an amplifier).

fact that the average amplification in the case of DRA is larger just only by 0.7 dB than in the case of the EDFA amplifier, the achieved transmission distance is larger by 11 km than in the system with EDFA. This can be explained by the low amplification efficiency of the Raman amplifiers at low powers of the amplified optical radiation. Thus, the signal, the power of which is much larger than the noise power, will be amplified more effectively than the noise generated by the amplifier. Nevertheless, such characteristic of the amplifier should also be interpreted as a serious drawback of the distributed Raman amplifiers, as the need arises to use powerful pumping lasers (1150 mW strong pumping radiation is necessary to achieve amplification of 25 dB). EDFA pumping source power is equal to 316 mW. EDFA is able to ensure a high level of signal amplification; however, this could be achieved only in a 35 nm wavelength region in the "C" optical band. The typical noise figure of EDFAs is higher than in the case of LRA and DRA. The main deficiency of SOAs is a very high number of produced signal impairments; therefore, this type of amplifiers is rarely used in WDM systems, even though their gain spectrum is much broader in comparison with EDFAs.

Taking into account the excessive number of SOA produced signal impairments, the strong wavelength and unevenness of the EDFA produced gain, and the low amplification effectivity of Raman amplifiers, it is clear that, if Cisco and Bell Labs forecasts are correct, then it will be necessary to find another optical signal amplification solution that could ensure a higher level of amplification over a broader wavelength band and at the same time that would amplify signal impairments as little as possible.

The first possible solution is to combine the aforementioned optical amplifiers into a hybrid optical amplifier, which would allow compensating for the negative properties of various amplifier types, for instance, to expand and equalize the EDFA gain spectrum, or would reduce the SOA-generated noise proportion in the amplifier output.

Another possible solution is the use of fiber optical parametric amplifiers (FOPAs). This type of amplifiers can ensure a high level of amplification over a broad wavelength band, and, if compared to other lumped amplifier types, given an optimized configuration, they produce very small number of signal impairments. Moreover, parametric amplifiers can also be used for all-optical signal processing purposes, for example, for wavelength conversion [27, 28], dispersion compensation [29], time-division-multiplexed signal demultiplexing [20], and 2R and 3R all-optical signal regeneration (2R—signal power and form regeneration; 3R—signal power, form, and phase regeneration) [30, 31].

3. Hybrid optical signal amplification

This chapter is dedicated to studies of hybrid optical amplifiers, which were obtained by applying the combinations of currently commercially used optical amplifiers (SOA, EDFA, and Raman amplifiers). The possibilities of applying hybrid Raman-EDFA and Raman-SOA solutions in WDM transmission systems for improving the operations of existing lumped in-line amplifiers have been studied and demonstrated. Due to the excessive number of SOA produced signal distortions and the strong wavelength dependency of EDFA produced gain, the implementation of EDFA-SOA hybrid solution has not been considered.

The unevenness of the EDFA gain spectrum and signal distortions caused by ASE noise significantly affect the performance of the whole transmission system, especially in systems with several amplification spans. To demonstrate the impact of the unevenness of EDFA gain spectrum and of the generated signal distortions, a 16-channel 10 Gbps DWDM transmission system simulation model has been introduced with four amplification spans. Equal power of the optical flow has been ensured at each amplifier input.

The obtained results are shown in **Figure 3**. After each amplification span, BER value of the detected signal increases by 2–3 orders (given the same input signal power). Upon comparing the EDFA gain spectra after the first and fourth amplification span, it is found that amplification decreases on average by 11.6 dB, whereas the amplification difference between the channels increases from 1.3 to 4.3 dB. The following conclusions are drawn:

- Every additional EDFA not only generates the amplified spontaneous emission noise but also amplifies the noise produced by the previous amplification spans. This significantly degrades the quality of amplifiable signal.

- The ASE power level after each amplifier is gradually increasing. Accordingly, part of the erbium ion population inversion is used to amplify the noise generated in the previous amplification spans. As a result, the part of the obtained population inversion, which was used for signal amplification, has decreased.

The slope of the gain spectrum increases after each amplification span. Uneven amplification is undesirable in multichannel WDM systems, especially in systems with several cascaded EDFA in-line amplifiers, as it leads to difference between power levels of various channels, which, accordingly, will lead to a signal quality degradation in channels with a lower amplification level.

Summing up all the aforementioned results, it has been concluded that it is necessary to configure the EDFA amplifier in a way to obtain the overall amplification spectrum that is as even as possible in the frequency range used for transmission, as well as to reduce the number of EDFA produced signal distortions.

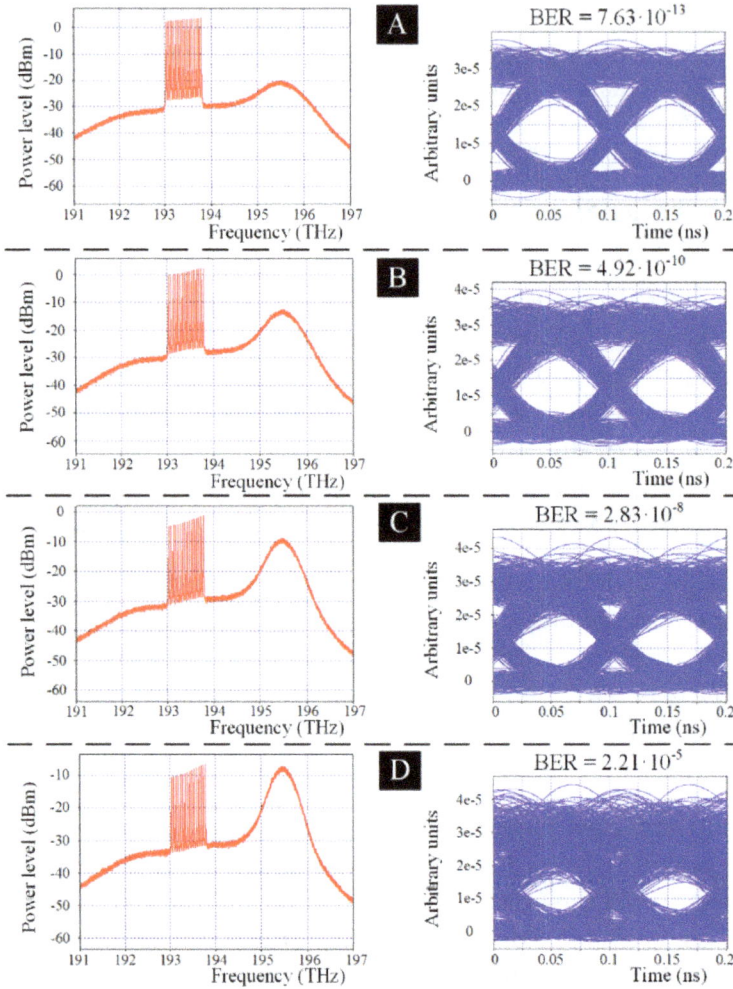

Figure 3. Optical spectra (the power level depending on frequency) at the output of the EDFAs (to the left) and eye diagrams of the signal detected in the ninth channel (to the right) after first (A), second (B), third (C), and fourth (D) stages of amplification.

3.1. Raman-EDFA hybrid amplifier

In the Raman-EDFA optical amplifier combination, most noise is generated by the EDFA amplifier. Therefore, in most cases, the Raman amplifier is used as a preamplifier in such cascades. EDFA amplifiers provide lower noise figures when functioning closer to the saturation point. Therefore, in hybrid amplifiers, EDFA with a relatively short doped fiber should be used (the longer the doped fiber, the higher level of amplification is obtained by the photons generated by spontaneous emissions). For further analysis of the hybrid Raman-EDFA solution, a simulation model is used, which is shown in **Figure 4**.

In the simulation model, the optical flows that are produced by the 16 transmitters are combined and transferred through a 150 km long standard single-mode fiber (SMF1). The signal power level at the SMF1 fiber output in all 16 cannels has reached −37.1 ± 0.1 dBm. The overall optical flow has been amplified by the EDFA in-line amplifier or by the hybrid Raman-EDFA

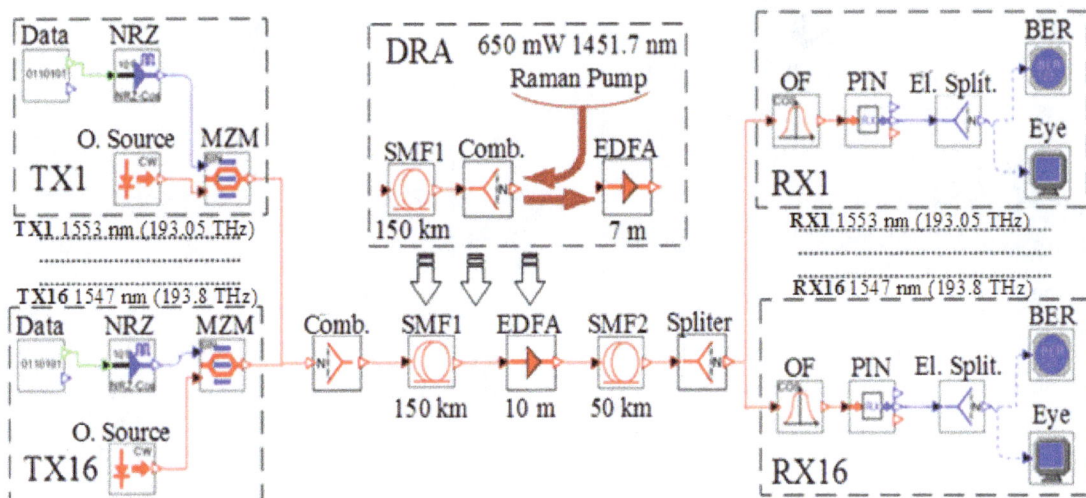

Figure 4. Simulation model of the 16-channel 10 Gbps DWDM transmission system with an EDFA in-line amplifier or with a hybrid Raman-EDFA amplifier.

amplifier (arrows in **Figure 4** show the layout of the hybrid amplifier) and afterward transferred through a 50 km long SMF (SMF2). Dispersion compensation has been performed using a fiber Bragg grating (FBG), and then the optical flow has been divided among 16 receivers, using an optical power splitter.

After comparing the gain spectra produced by the EDFA in-line amplifier and the hybrid Raman-EDFA amplifier (see **Figure 5**), it has been found that implementation of the hybrid solution allows reducing the gain difference among all 16 channels from 1.5 dB (in the case of the EDFA) to 0.1 dB (in the case of the hybrid amplifier).

As can be seen in **Figure 6**, implementation of the hybrid solution has ensured OSNR improvement in all 16 channels from 1.7 up to 2.6 dB, that is, an average increase of ~2 dB. Such OSNR improvement can be explained by the following facts:

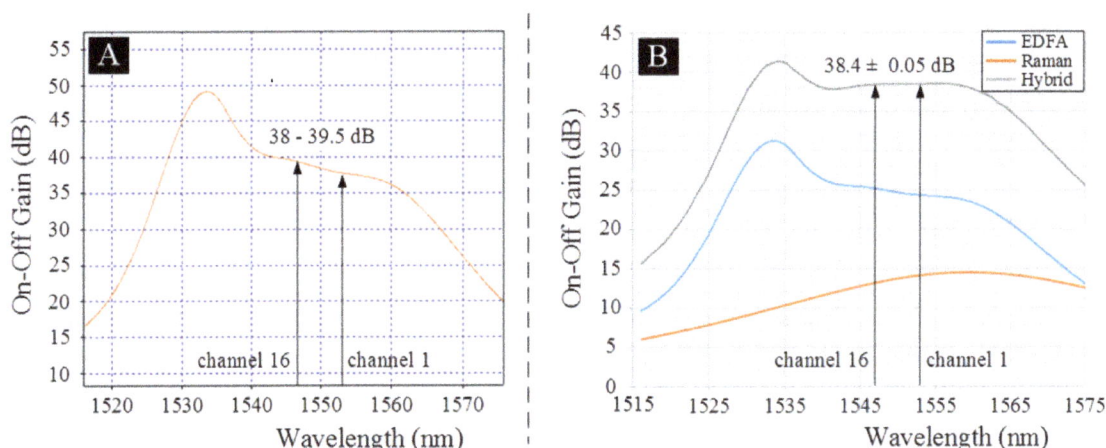

Figure 5. Gain spectra of the EDFA in-line amplifier (A) and of the hybrid Raman-EDFA amplifier (B).

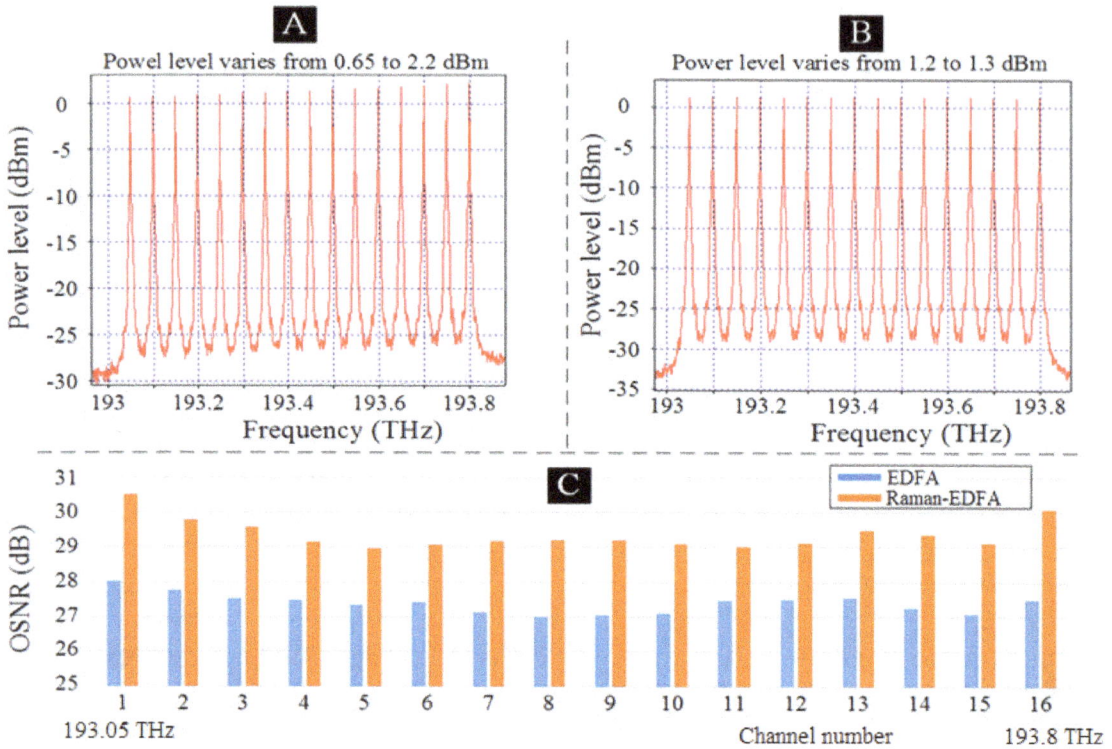

Figure 6. Signal spectra at the output of the EDFA (A) and at the output of the hybrid Raman-EDFA amplifier (B) and OSNR comparison among all 16 channels in the system with the EDFA in-line amplifier and the hybrid Raman-EDFA amplifier (C).

- The usage of the distributed Raman amplifier has raised signal power at the input of the EDFA by 13.1–14.1 dB; therefore, the EDFA functions closer to the saturation point.

- The EDFA fiber length has decreased by 3 meters, which allows reducing the required input signal power for saturation of the EDFA.

- The coherent nature of stimulated Raman scattering (SRS) ensures that in SMF1 optical fiber, the signal is amplified more effectively than the low power optical noise, which allows obtaining negative noise figure values (from −0.4 to −0.6 dB in the wavelength region used for transmission), and accordingly improved OSNR.

In the case of the hybrid amplifier, it has been found that raising the signal power at the input of the EDFA and reducing the length of the erbium-doped fiber (EDF) allow obtaining lower noise figure values by 0.3–0.4 dB for the EDFA.

Upon performing a comparison of operations of the aforementioned EDFA and Raman-EDFA solutions, it can be concluded that the hybrid amplifier can ensure more even amplification over a broader wavelength region and higher OSNR values. However, more powerful lasers are necessary for implementing such solutions, which increases the costs of developing this solution. For the EDFA in-line amplifier, 316 mW of pumping power is required to amplify the −37.1 dBm input signal by more than 38 dB. In the case of the hybrid solution, the Raman amplifier required 650 mW of pumping power to ensure that gain is high enough and that its slope can compensate the slope of the EDFA with 200 mW pump gain spectrum, but the total pumping power of the hybrid amplifier has reached 850 mW. However, the hybrid solution ensured gain difference below 1 dB over a 23 nm wavelength range (from 1538 to 1561 nm, by 17 nm more than that used for transmission of all 16 channels), which allows significantly increasing the number of channels in WDM transmission systems.

3.2. Raman-SOA hybrid amplifier

The Raman-SOA hybrid solution is configured in a way to reduce the number of signal distortions produced by the semiconductor optical amplifier and also to increase the attainable transmission distance. The introduced simulation model of the transmission system used for studying this amplifier combination is similar to the one used previously (see **Figure 7**). Wavelength grid is chosen based on ITU-T G.694.1 recommendation where the central frequency is 193.1 THz.

The transmission line span length between the transmitter block and SOA is specifically selected to ensure optimum signal power at the input of the semiconductor amplifier. Inserting the distributed Raman amplifier in a cascade before the semiconductor amplifier would increase the signal power in SOA input, which would lead to a more explicit manifestation of nonlinear optical effects in the semiconductor material and would, accordingly, deteriorate the quality of the amplifiable signal. Therefore, it is the semiconductor amplifier that is used as the first in the cascade.

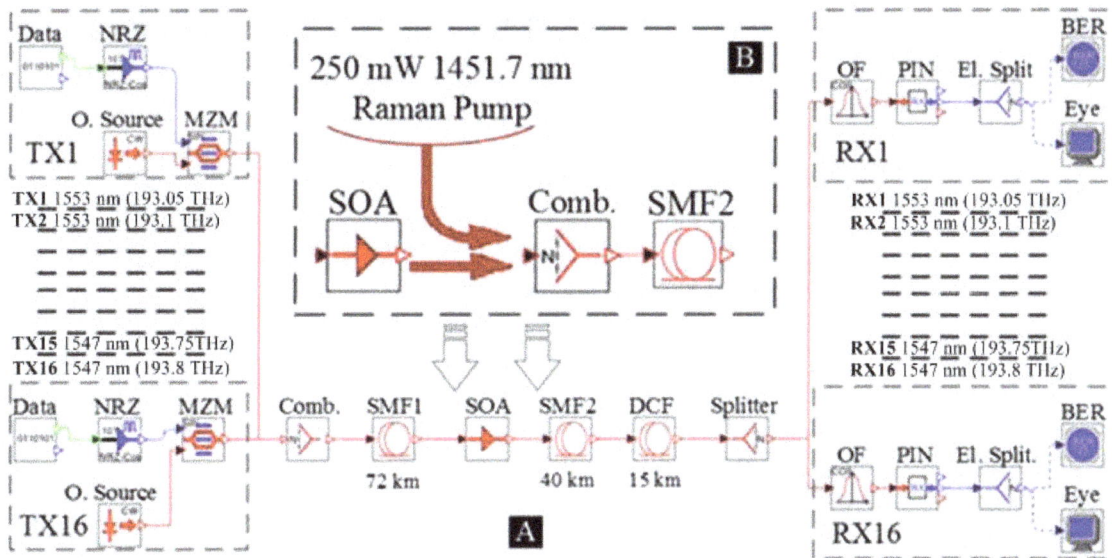

Figure 7. Simulation model of the 16-channel 10 Gbps DWDM transmission system with the SOA in-line amplifier (A) or with a hybrid Raman-SOA amplifier (B).

The implementation of the hybrid Raman-SOA solution allows using such mode of the semiconductor amplifier, in which it produces minimum distortions of the amplified signal, whereas the amplification deficit, which occurs after reducing the pumping current value by 43 mA, is compensated by the DRA with a 250 mW 1451.7 nm co-propagating pump. The implementation of the Raman-SOA hybrid solution allows increasing the attainable transmission distance by 12 km. The gain spectrum of the DRA is shown in **Figure 8a**. Eye diagrams for channels with the highest BER value in a system with the SOA amplifier (ninth channel f = 193.45 THz) and in a system with the Raman-SOA hybrid amplifier (tenth channel f = 193.5 THz) are shown accordingly in **Figure 8b** and **c**. From **Figure 8a**, it can be seen that the DRA produced amplification is large enough to compensate the amplification deficit of 5.3 dB that occurs after reducing the SOA pumping current by 43 mA. SOA amplifier (9th channel) and Raman-SOA hybrid amplifier (10th channel) be selected, because they are the worst channels.

After comparing **Figure 8b** and **c**, it has been found that implementation of the Raman-SOA hybrid solution allows obtaining approximately the same BER level as in the case of SOA in-line amplifier, but at signal power lower by 1.5 times. This shows that, by using SOA together with the distributed Raman amplifier and introducing relevant SOA pumping current adjustments, it is possible to substantially lower the amount of SOA produced noise and, therefore, to improve the quality of the amplified signal.

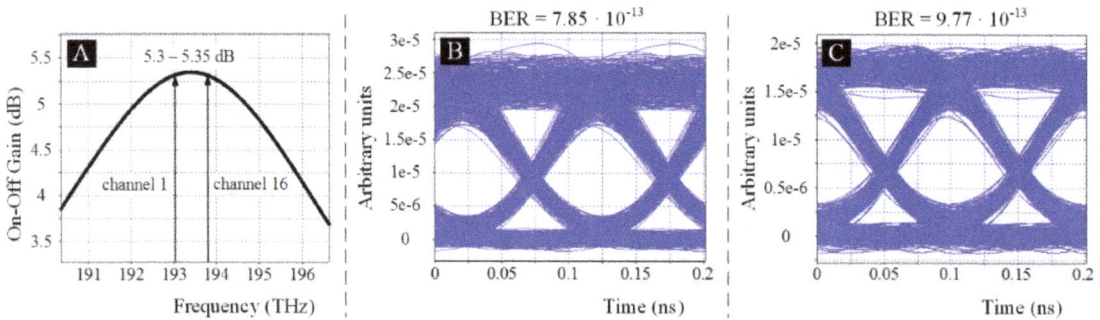

Figure 8. DRA produced gain spectrum (A) and eye diagrams of the ninth channel in the system with the SOA in-line amplifier (B) and the tenth channel in the system with the Raman-SOA hybrid amplifier (c).

4. Evaluation of parametric amplifiers and its application

Parametric amplifiers can be based on degenerate FWM (in the single-pump case) and on nondegenerate FWM (in the dual-pump case). FOPA produced gain will reach its maximum, if the phase-matching condition is met or if the phase-mismatch parameter k is equal to zero. In the case of a single-pump FOPA, irrespective of the broad amplification range, the amplification spectrum is not even. In the experimental transmission system, described before, the gain −3dB bandwidth has reached 2.2 THz (see **Figure 9**). It has been found that for ensuring an optimum operation mode of a single-pump FOPA, it is necessary to maintain a small negative linear phase deviation in respect to the zero-dispersion frequency, which would compensate the nonlinear phase mismatch. That is why the pumping radiation wavelength must be slightly larger than the fiber zero-dispersion wavelength (ZDWL).

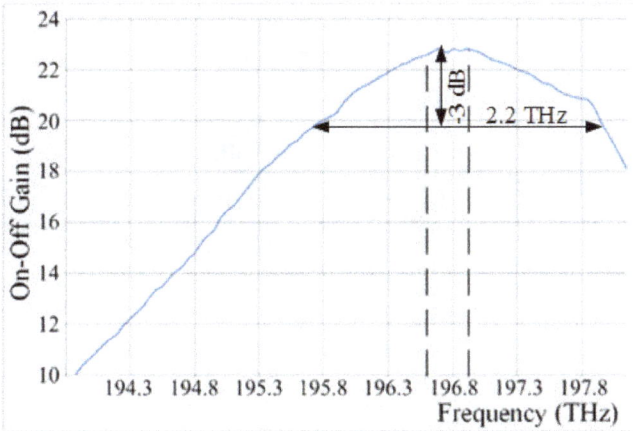

Figure 9. Gain spectrum of the single-pump FOPA with 660 mW 1553.9 nm pumping radiation.

The gain spectrum bandwidth is very dependent on the nonlinearity parameter of the medium and on the pumping radiation power and the length of the gain medium highly non-linear fibers (HNLFs). Thus, by increasing the fiber length, it is possible to achieve a higher level of amplification, but in this case, the gain spectrum width will be reduced accordingly (the longer the fiber, the larger the accumulated phase mismatch is). Due to this reason, when constructing FOPA amplifiers, it is not recommended to use HNLFs that are longer than 1 km. If they are configured in a way to achieve as wide gain spectrum as possible, it is required to use as short HNLF as possible, but to maintain the achievable amplification level, the pump power must be increased, or a fiber with a higher nonlinearity coefficient must be used. When selecting the pumping radiation parameters, one must keep in mind that, by changing the pumping radiation power, also the nonlinear phase mismatch is changed. Therefore, along with adjusting the pump power, its wavelength also needs to be reconfigured.

The performance of single-pump parametric amplifiers is affected by various factors, which must be taken into account when designing a specific FOPA. It is necessary to selectively choose the pumping radiation parameters to ensure as high amplification efficiency as possible and to avoid occurrence of excessive channel-channel four-wave mixing (CC-FWM) and pump-channel four-wave mixing (PC-FWM) produced interchannel cross talk, which in its turn is produced due to excessive pumping. The SBS threshold increase is also very important; otherwise, the amplification effectiveness will decrease, and the amplified signal will be distorted.

One of the most effective solutions of increasing the SBS threshold is phase modulation of the pumping radiation. However, as can be seen from the results shown in **Figure 10**, if the choice of frequencies modulating the pumping radiation phase is not thoroughly considered, a substantial expansion of spectrum of the idler spectral components will occur. Therefore, in systems, in which idlers are used for all-optical signal processing, it must be ensured that the chosen pumping radiation phase-modulation does not produce excessive spectral broadening of idlers (in the results shown in **Figure 10**, idler spectral broadening has reached 54% at the level of −15 dB from the maximum power spectrum). For initiating the FWM process, it is also necessary to preserve the angular momentum among the four photons involved in the parametric interactions, as the parametric gain has explicit polarization dependence.

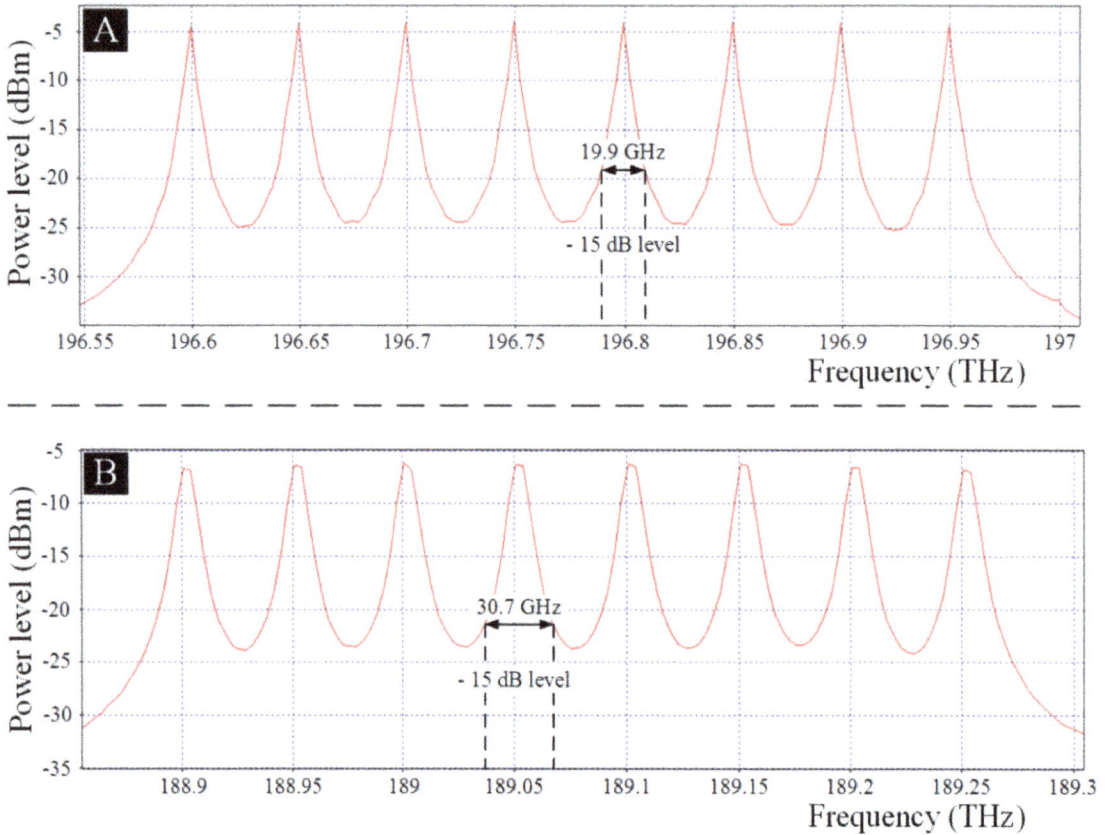

Figure 10. Spectra of the amplified signal (A) and the generated idlers (B) at the output of the single-pump FOPA.

Unlike single-pump FOPAs, dual-pump FOPAs can ensure even amplification over a very broad wavelength band. To achieve even amplification in a broad wavelength band, it is necessary to ensure that the wavelengths of the pumps are placed symmetrically in respect to the gain medium ZDWL, whereas the frequency distance between ZDWL and pumping radiations must be large enough (depending on the specific FOPA configuration), to avoid the impact of PC-FWM-generated components on the quality of the amplified signal.

Since dual-pump FOPAs both degenerate and nondegenerate FWM, for one input signal, it is possible to obtain at least five idlers (see **Figure 11**, where ω_1 and ω_2 are the pumping source frequencies, but ω_3 is the signal), which are directly related to the amplifiable signal frequency. This leads to amplification spectrum depressions at frequencies near the frequencies of the pumps. It has been concluded from the results obtained in this study that at 0.5 mW pump power it is recommended that the amplified signal frequency is at a distance of at least 1.2 THz from pump frequencies.

Just like in the case of single-pump FOPAs, dual-pump FOPAs also require the usage of one of the methods for mitigation of the negative impact of SBS. However, in the dual-pump case, it is important to note that by manipulating with the phase of both pumping radiations it can

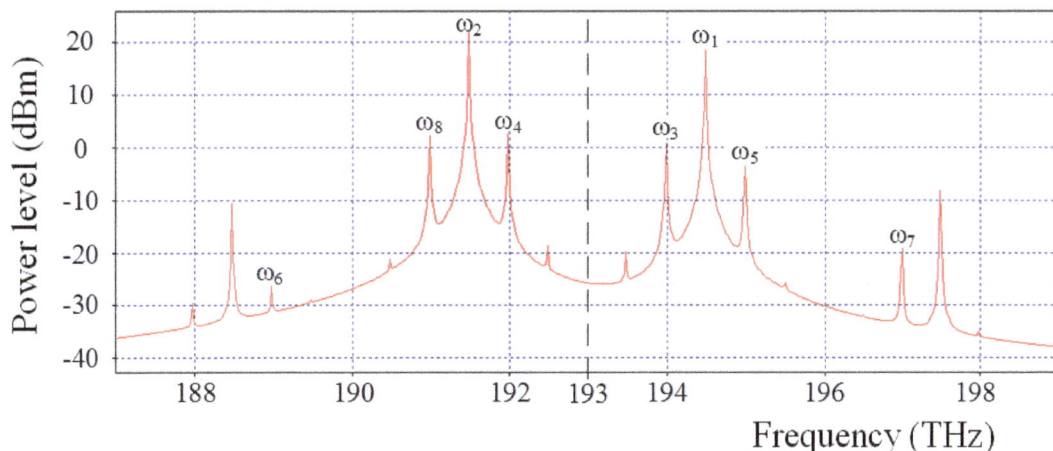

Figure 11. Optical spectrum at the output of a dual-pump FOPA with 200 mW 191.5 THz pumps and 500 meter long HNLF.

be achieved that the relevant idler will not experience spectral broadening. The amplification efficiency in the case of dual-pump FOPAs is highly dependent on the SRS triggered energy transfer between the pumps. It is not possible to fully avoid this effect. To reduce its impact, normally higher power is used for the pump with the higher frequency than for the other pumping radiation, thus achieving that the average power difference between the pumps is minimum over the entire gain medium.

In traditional WDM transmission system architecture, one optical source is required to produce a single-channel carrier. It is not the most cost-effective solution, as, by increasing the number of transmission channels, the number of required light sources increases accordingly. Due to this reason, an increasing number of studies are conducted to find such transmission system architecture, which would be able to ensure a higher number of signal carriers using fewer optical sources [32–34]. FOPAs during the process of parametric amplification generate idler spectral components, which, in essence, are phase-conjugated copies of the amplified signal. These idlers could be used not only for wavelength conversion or 2R and 3R signal regeneration but also for increasing the number of carriers on the transmitter side of a WDM transmission system.

Therefore, a model of a dual-pump FOPA has been introduced for doubling the number of existing carriers in a WDM transmission system. For this reason, a simulation model of a 32-channel DWDM transmission system has been created with 10 Gbps transmission speed per channel, 100 GHz channel spacing, and NRZ-OOK modulation format. This system simulation model is displayed in **Figure 12**. The authors chose SMF length of 20 km, because it is the typical line length of optical access networks. EDFA preamplifier is used here for insertion loss compensation of transmission line and other transmission elements.

The main feature in the simulation model, which is presented in **Figure 12**, is that the FOPA is placed before the transmitter block or at the 32-channel modulator inputs. The optical

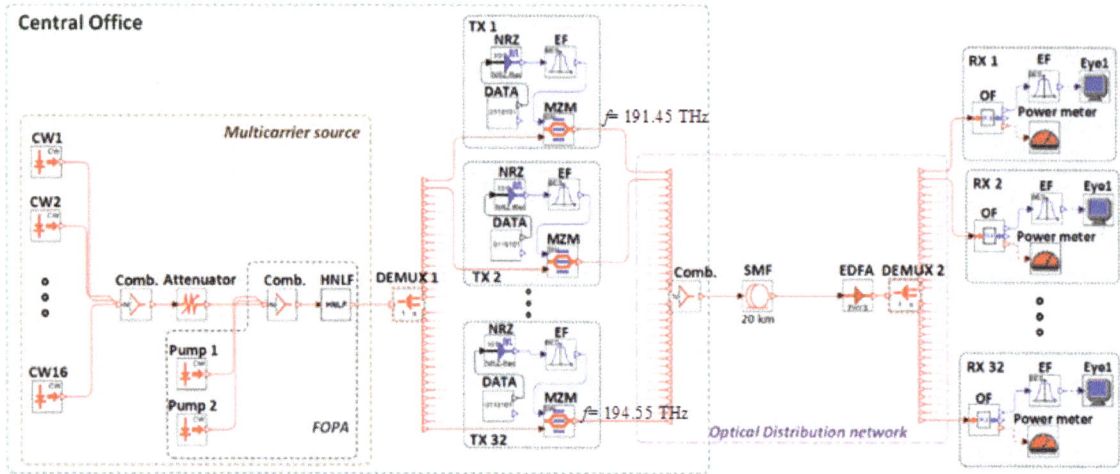

Figure 12. Simulation model of the 32 channel 10 Gbps WDM transmission system with the provided multicarrier source solution, which is based on wavelength conversion using a dual-pump FOPA.

multicarrier source consists of continuous radiation lasers (CW1–CW16), an optical attenuator, two powerful pumping sources, two optical splitters, and a 500 m long HNLF. One of the main goals of this experiment is to obtain 32 carriers with even frequency distribution (equal channel spacing), which can be achieved by using idlers ω_4. Taking into account that the distribution of idlers ω_4 and the initial light source frequencies are symmetrical in respect to the gain medium ZDWL, it has been decided to place the carrier with the lowest frequency higher by 50 GHz than the HNLF ZDWL (193 THz). Therefore, the frequencies of the 16 initial carriers are distributed in a range from 193.05 to 194.55 THz with 100 GHz channel spacing (see **Figure 13a**). The optical flow sent through the parametric amplifier is not modulated and basically represents a continuous radiation set. At the output of the HNLF, a combination is obtained consisting of 16 initial carriers, 16 idlers ω_4 (generated as a result of parametric processes), 2 pumps, and other third-order spectral components (see **Figure 13b** and **c**). The pump power for both pumps is set to 400 mW each (26 dBm), and 190 and 196 THz frequencies are temporarily chosen.

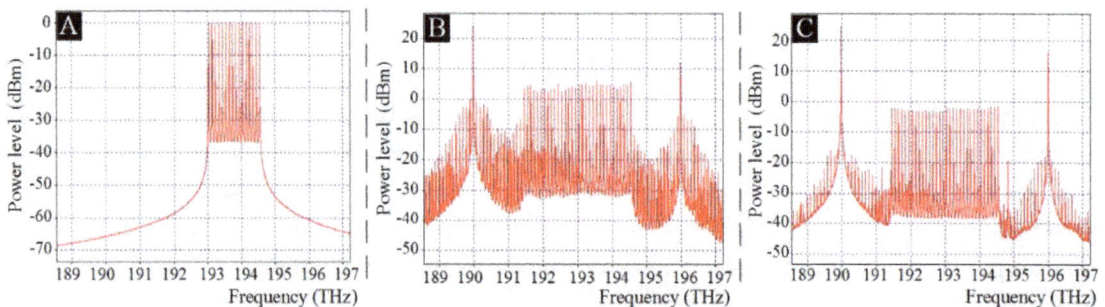

Figure 13. Optical spectrum at the input of the FOPA when initial carrier power is set to 0 dBm (A) and optical spectra at the output of the HNLF when the power of the initial carriers is set to 0 dBm (B) and to −10 dBm (C).

It has been found that when given an excessively high level of input signal power, CC-FWM processes will trigger explicit interchannel cross talk (see **Figure 13b**). Due to this reason, when the idlers obtained as a result of parametric processes are used for increasing the number of carriers, it is necessary to limit the power of the carriers at the amplifier input. Based on the obtained results, the power level of the initial carriers at the input of the amplifier is reduced to −10 dBm. The alignment of idlers in respect to the central frequencies of the throughput band of demultiplexer filters is achieved by changing the frequency of the first pumping radiation (frequency obtained in simulations −196.01 THz). With the aforementioned amplifier configuration, the maximum power level difference among all the 32 channels has been reduced to 1.9 dB.

To assess the performance of the proposed system architecture solution, BER value dependence on the received signal power in the channel with the poorest signal quality (the highest BER) is obtained. These results are compared to the corresponding results obtained in a system with traditional architecture (with 32 laser sources, which function in a continuous radiation mode). The obtained results are shown in **Figure 14**. It has been found that power penalty of 1.8 dB exists between the system with the proposed multicarrier source and the conventional 32-channel solution. It is important to note that part of the obtained power penalty is directly related to the large amount of ASE produced by the EDFA used as a preamplifier for ensuring the necessary signal power at the input of the received block.

There are at least two alternatives to the proposed system architecture, which can produce more than one carrier per optical source: spectrum-sliced systems [33, 35] and systems, which are based on FWM use for producing third-order spectral components (without the initial carriers) [34, 36]. Nevertheless, the use of idlers ω_4 produced by FOPA for doubling the number of carriers in WDM systems ensures the best carrier signal stability and, therefore, the highest quality of the transmittable signal.

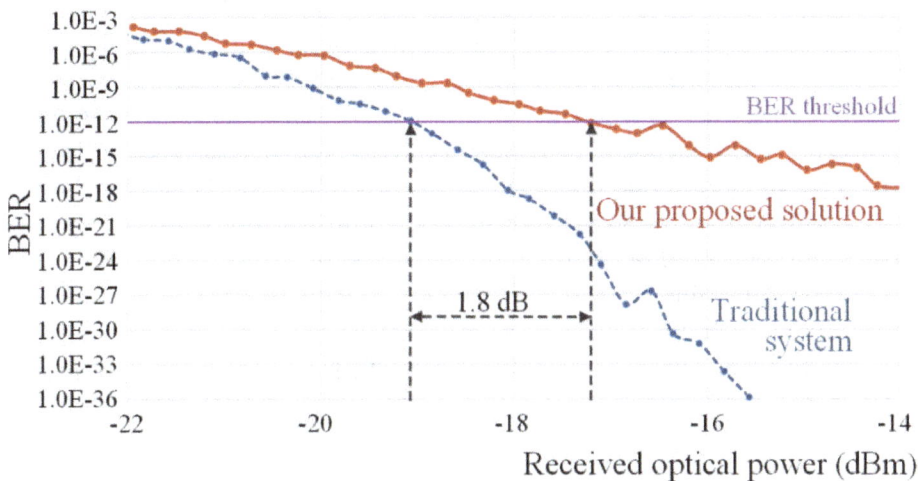

Figure 14. BER value dependence on the power of the detected signal for the 12th channel (f =192.55 THz; λ = 1557 nm) in the system with the proposed multicarrier source solution (solid line) and in the system with traditional architecture (dashed line).

It has been found that for dual-pump FOPAs the maximum amplification efficiency is achieved when both pumps are linearly polarized with the same state of polarization (SOP) and their SOP corresponds to the SOP of the amplified signal. However, when the SOP of both linearly polarized pumps is orthogonal to the SOP of the amplified signal, amplification decreases to its minimum value, and in a broad frequency region, it is equal to zero. The results obtained in this paper have shown that the same situation is observed also in the case of single-pump FOPAs.

This property of parametric amplifiers can be used for emphasizing one state of polarization from a combination of two orthogonally polarized optical components. The key problem, which is observed when the FOPA is used for emphasizing a specific state of polarization, is ensuring the conservation of the relative positioning of signal and pump SOP throughout the entire length of HNLF. This problem occurs due to the following reasons:

- Due to the effect of fiber birefringence, SOP of optical radiation changes along the fiber, and, as result, random SOP rotation is observed. It is very difficult to compensate such a random SOP change, as the rotation rate is affected by various factors, such as temperature, the frequency of the transmitted radiation, internal and external mechanical loads, etc. It is possible to avoid rotation of SOP of the pumps and the amplified signal by using polarization-maintaining HNLF as the gain medium.

- When the amplified signal and the pumps are propagating in the gain medium, additionally to fiber birefringence, their states of polarization are also affected by self-phase modulation (SPM) and cross-phase modulation (XPM) nonlinear effects. Therefore, when configuring the parametric amplifier, it is first necessary to avoid excessing pumping; otherwise, it can lead to a more explicit occurrence of SPM and XPM, which decreases the efficiency of the FWM process in the gain medium.

It is not possible to completely avoid changes in relative positioning of the SOP of the signal and the SOP of the pump. To demonstrate this, a simulation model is introduced, where a single-pump FOPA (500 mW, 1533.9 nm) amplifies a signal with −31 dBm total optical power at the input of the HNLF. At first, both the signal and the pump are linearly co-polarized. During the simulation, the SOP of the pump is rotated in respect to the SOP of the amplified signal, and the power of the signal is observed at the output of the FOPA. It has been found that by using polarization-maintaining fibers, under the influence of SPM and XPM, a change in the relative positioning of the signal SOP and the pump SOP is observed. As a result of this change, even when the signal is orthogonally polarized in respect to the SOP of the pump at the input of the HNLF, the signal obtains 1.5–1.6 dB gain. When the SOP of the amplified signal is co-polarized with the SOP of the pump, the obtained amplification reaches 18.3 dB, which is by 16.7 dB higher than in case of orthogonal relative positioning of the SOP at the input of the HNLF.

As it has already been previously mentioned, the polarization dependence of the parametric gain can be used for emphasizing radiation with a specific SOP from the flow of orthogonally polarized optical components, which in its turn can be used for emphasizing polarization-multiplexed signals and 2PolSK to NRZ-OOK modulation format conversion.

For conversion of 2PolSK signal to NRZ-OOK modulation format, cases of single-channel and multichannel systems are considered. To avoid changes in relative positioning of the states of polarization between pumping radiations, single-pump parametric amplifiers are used in both cases. In both cases, the FOPA is placed at the receiver (or receiver block) input.

At first, a single-channel transmission system simulation model is introduced with 2PolSK modulation format, 150 km long optical fiber, a FOPA preamplifier (which simultaneously performs modulation format and wavelength conversion functionality), and two receivers for detecting the converted NRZ-OOK signal at signal and idler frequencies. The introduced simulation model is displayed in **Figure 15**.

In case of a single-channel system, the primary task is to assess the new modulation format conversion solution created within the scope of this paper, by obtaining a power penalty introduced specifically by the process conversion of the modulation format. Based on the obtained results, 535 mW, 1554.1 nm pumping radiation is chosen, the phase of which is modulated with the following frequency tones: 180 MHz, 420 MHz, 1.087 GHz, and 2.133 GHz. Such configuration ensures 14.8 dB gain for the logical "1" component of 2PolSK signal, which is sufficient for ensuring BER value below the 10^{-12} mark for the obtained NRZ-OOK signal.

Based on the obtained results, it has been concluded that the idler requires lower pump power to ensure BER values below the 10^{-12} mark, even though the gain for idler is lower by 0.8 dB than the signal (see **Figure 16**). Therefore, in the case of a single-channel system, it is recommended to process the idler spectral component as the informative signal. These results can be explained by the fact that the signal at its initial frequency contains the orthogonally polarized logical "0" component, which for the obtained NRZ-OOK signal is interpreted as noise.

It has been found that the power of the obtained NRZ-OOK signal that is necessary to ensure BER value below 10^{-12} is −23.65 dBm, whereas the necessary idler power is −23.8 dBm (see **Figure 17**). In a standard single-channel system with NRZ-OOK modulation format, the signal power required to ensure BER value below the 10^{-12} threshold has reached −24 dBm. Thus, there is a power penalty of 0.4 dB between the NRZ-OOK signal from the standard

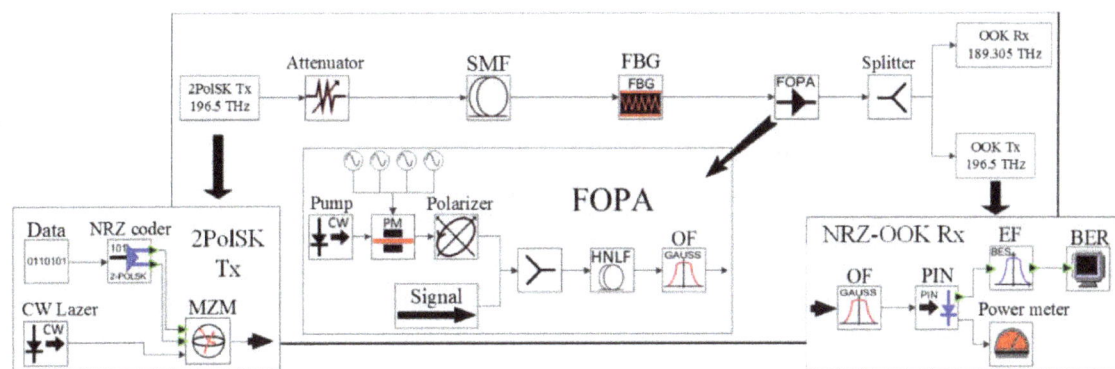

Figure 15. Simulation model of the single-channel transmission system, where a single-pump FOPA with linearly polarized pumping radiation is used for 2PolSK to NRZ-OOK modulation format conversion.

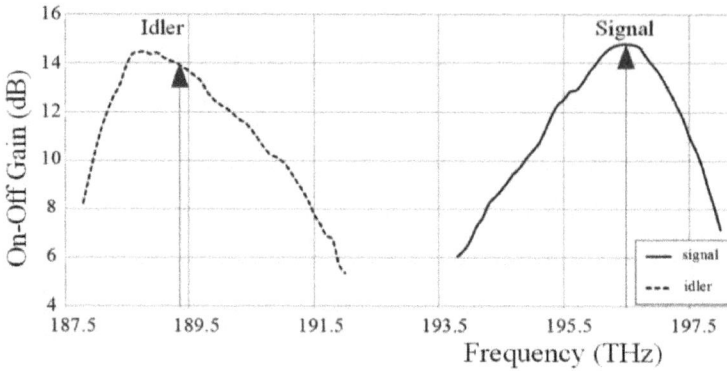

Figure 16. Gain spectrum produced by the single-pump FOPA with linearly polarized pumping radiation at signal frequencies (solid line) and at idler frequencies (dotted line).

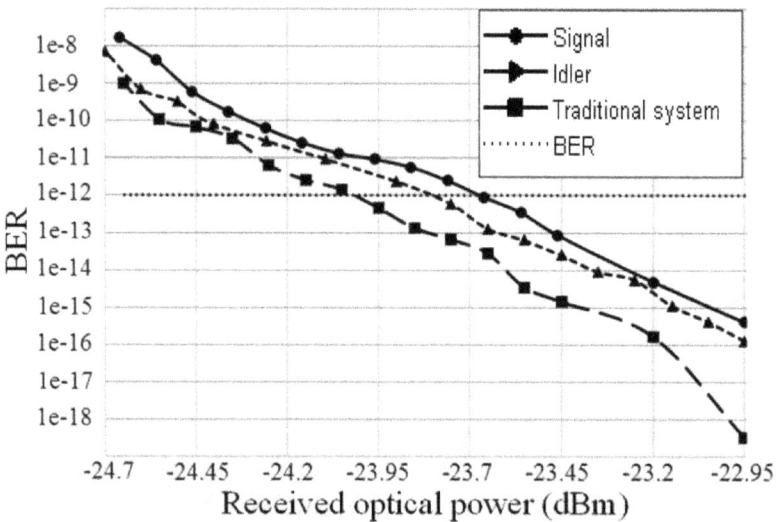

Figure 17. BER value dependence on the detected signal power in the standard single-channel transmission system (dashed line) and in the system with modulation format conversion at the initial signal frequency (solid line) and at idler frequency (dotted line).

single-channel system and the converted signal. It is important to note that the power penalty between the NRZ-OOK signal from the standard single-channel system and the generated idler is lower by 0.1 dB (only 0.2 dB). These results are explained by the fact that the idler produced during the FWM process does not contain the logical "0" component of the initial 2PolSK signal, which in this case is interpreted as noise for the converted NRZ signal. The obtained power penalty values are also attributable to the relative intensity noise, which is transferred from the pumping radiation to the amplifiable signal, as well as to the phase SOP mismatch between the pump and the amplified signal that occurs due to SPM and XPM.

In the case of the multichannel system, the goal is to assess the performance of the developed modulation format conversion solution in the presence of interchannel cross talk. When converting 2PolSK signal to NRZ-OOK modulation format, using FOPA with linearly polarized pumping radiation, one must pay special attention to the control of the level of interchannel cross talk produced by the CC-FWM interactions because if the SOP pumping radiation and SOP signal coincidence, the FWM process takes place with its maximum efficiency, including also production of the CC-FWM interchannel cross talk.

To assess the performance of the proposed solution in the presence of interchannel cross talk, a 16-channel 10 Gbps DWDM transmission system is introduced with 2PolSK initial modulation format and 100 GHz channel spacing (see **Figure 18**). In this system, the access network is divided into two branches, eight channels in each. The first branch consists of eight channels, occupying frequency range from 194.5 THz (1541 nm in wavelength) to 195.2 THz (1536 nm), whereas the second branch is occupying frequency range from 196.2 THz (1528 nm) to 196.9 THz (1523 nm). Only those results are included in this paper, which are obtained in the second access network branch, where the signal is divided among eight receivers using an optical splitter with 10.5 dB attenuation.

Based on the obtained results, in the second access network branch, 790 mW 1554.15 nm pumping radiation is used, the phase of which is modulated with the same frequency tones as in the case of a single-channel system: 180 MHz, 420 MHz, 1.087 GHz, and 2.133 GHz. It has been found that to ensure BER values below the 10^{-12} threshold, all eight idlers require pump power that is by 35 mW higher than in the case of the signals at their initial frequency. The obtained gain for the idler spectral components is lower by at least 2.2 dB, whereas the gain spectrum slope near its maximum is higher than in the initial signal frequency band (see **Figure 19**). The obtained level of amplification in the initial signal frequency band changes in the range from 30.3 to 30.9 dB among all eight channels; thus, the amplification difference between the channels reaches 0.6 dB. Between the idler spectral components, such difference has reached 2.6 dB (from 26.1 to 28.7 dB), but the biggest amplification difference between the signal at its initial frequency and the corresponding idler has reached 4.2 dB. This explains the need for pump power exceeding 35 mW to ensure BER values below the 10^{-12} mark in the idler frequency band.

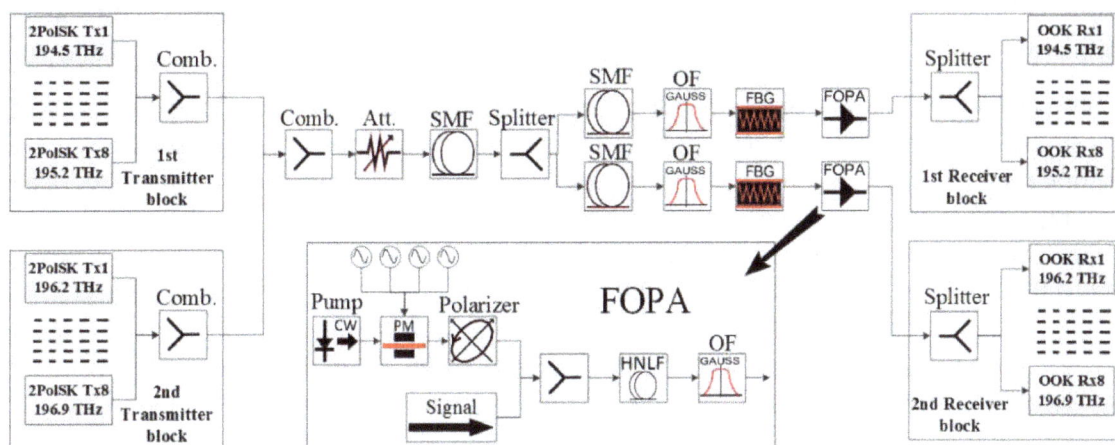

Figure 18. Simulation model of the 16-channel WDM transmission system with two access network branches, where in each branch the FOPA preamplifier is used for 2PolSK to NRZ-OOK (on-off keying) modulation format conversion.

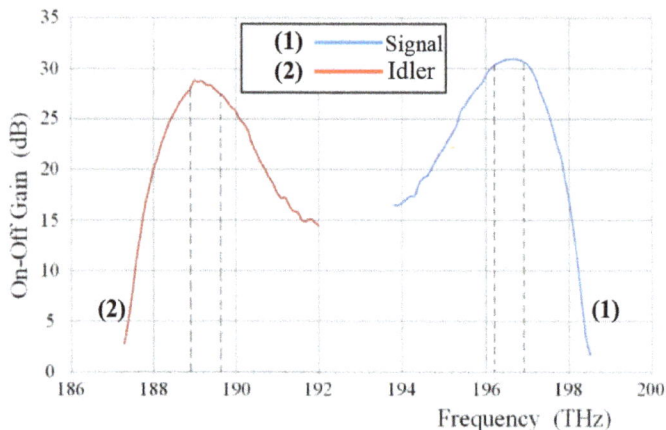

Figure 19. Gain spectrum ensured by the FOPA with 790 mW 1554.15 nm linearly polarized pumping radiation at initial signal frequencies (1) and idler frequencies (2).

To assess the performance of the proposed solution, the results obtained in the second branch of the access network are compared to the standard eight-channel DWDM system without signal amplification. The detected signal power required to ensure a certain BER value in the fifth channel of the second access network branch both at the initial signal and idler frequencies is compared with the same results obtained in the standard eight-channel DWDM system. As seen in **Figure 20**, in a system with modulation format conversion, to ensure BER values below the 10^{-12} mark, the required signal power is at least −23.5 dBm. In the standard eight-channel system, the corresponding required power level is −23.9 dBm; therefore, in this case, the power penalty between the signal with the converted modulation format and the signal from the standard eight-channel solution is 0.4 dB. It must be particularly emphasized that, contrary to the single-channel system, in the multichannel system, the idler BER values are higher than those of signals at their initial frequencies—when receiving the idler corresponding to the fifth channel, at least −23.15 dBm is required to ensure a BER value below the 10^{-12} threshold, which is more by 0.3 dB than receiving the signal of the fifth channel at its initial frequency.

It has been found that in case of the idler, larger amplitude fluctuations are observed, which densify the logical "0" and "1" component levels of the eye diagrams. The cause behind generating noise of such range is CC-FWM produced interchannel cross talk, which is produced as a result of parametric amplification and creates third-order spectral components, the frequencies of which correspond to the frequencies of the amplified signals. As mentioned previously, the gain difference between the idlers is much larger (by 2 dB) than between the signals at their initial frequencies. Therefore, more explicit manifestation of CC-FWM processes is observed, which also produces additional interchannel cross talk. Moreover, the cross talk caused by CC-FWM is not only transferred from signals at their initial frequencies to the idlers but is also generated between the idlers.

It has also been found that the pumping radiation-phase modulation leads to spectral expansion of the idlers by approximately 40%, which, accordingly, results in additional interchannel

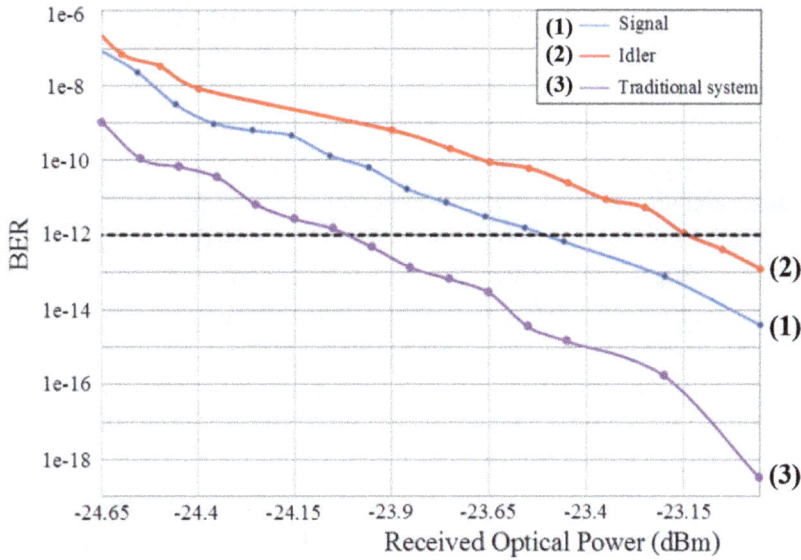

Figure 20. BER value dependence on the detected signal power in the standard eight-channel transmission system (3) and in the fifth channel on the second branch of the multichannel system with modulation format conversion at the initial signal frequency (1) and idler frequency (2).

cross talk. Interchannel cross talk caused by the CC-FWM interactions and idler spectral broadening is the main reason why, in the case of idlers, the power penalty in relation to the standard system is larger by 0.4 dB than for signals at their initial frequencies.

The second studied application of parametric gain polarization dependence is the emphasizing of a signal with a specific SOP from a combination of two polarization-multiplexed NRZ-OOK signals. For this purpose, a two-channel 10 Gbps transmission system with NRZ-OOK modulation format and polarization multiplexing is introduced (see **Figure 21**). Both signals, the SOP of which are mutually orthogonally allocated, are transmitted using the same frequency −196.5 THz.

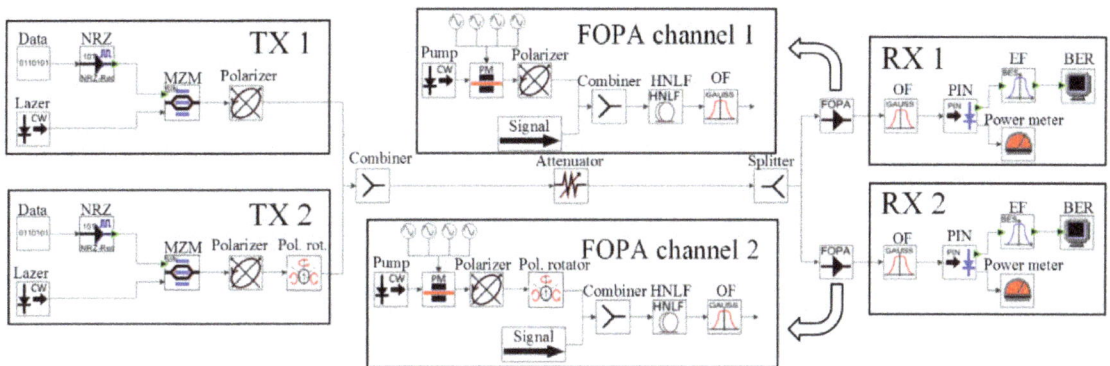

Figure 21. Simulation model of the two-channel optical transmission system with FOPA for division and amplification of polarization-multiplexed signals.

Based on the obtained results, a decision has been made to use 530 mW 1553.9 nm pumping radiation in the case of both FOPA, because this is the lowest pump power that can ensure BER values below the 10^{-12} threshold in both the first and the second channel. Previously described results have shown that phase modulation of the pump can cause spectral broadening of idlers. In a situation, when the probability that both orthogonally polarized optical radiations are observed simultaneously in the logical "1" level is high, the mutual deviations of the SOP of optical components have a bigger impact on the quality of the amplified signal than in the case when such simultaneous transmission of logical "1" is not performed (e.g., in the case of 2PolSK signal). Therefore, it is important to minimize the phase mismatch between the pumping radiation and the amplified signals, which can also cause a change of the relative positioning of SOP between the signal and the pump. Bearing in mind this fact and having observed the FOPA produced gain spectrum and in the OSNR at the output of the amplifier, the following frequency tones have been selected for pumping-radiation phase modulation: 0.13 GHz, 0.42 GHz, 1.087 GHz, and 1.94 GHz. The obtained signal gain in the first channel has reached 20 dB, whereas in the second channel, it has reached 20.1 dB. Idler component gain maximum is lower by 0.7 dB (see **Figure 22**).

To assess the performance of the proposed solution, BER value dependence on the received signal power is obtained, and these results are compared with the same results obtained in a standard single-channel transmission system with NRZ-OOK modulation format without optical signal amplification. It can be concluded from the results shown in **Figure 23** that there is a power penalty of 0.8 dB between the signal detected in the first channel and the signal from the standard single-channel NRZ-OOK system. The power penalty for the idler spectral component has reached 0.5 dB.

Unlike the system with modulation format conversion, in this case, a situation has been observed, when, given the same frequency, it is possible that the logical "1" of orthogonally polarized components is observed simultaneously in both channels. Thus, the effect of orthogonally polarized radiation on the divided signal quality is higher than in a system with modulation format conversion. This is the fact, which mainly explains a larger power penalty value than

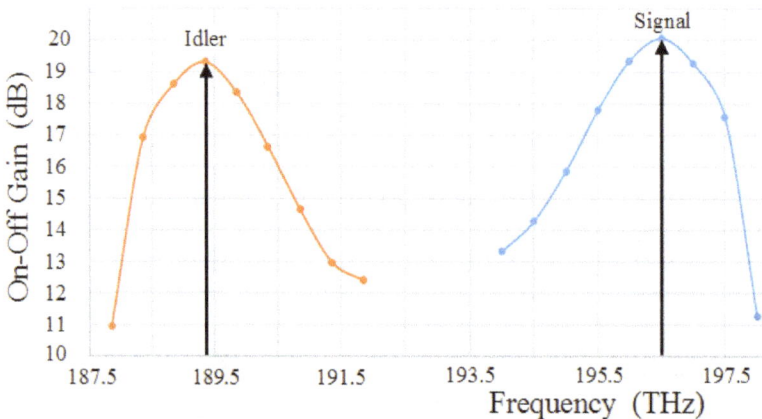

Figure 22. FOPA produced on-off gain for the first channel at idler frequencies (left side) and at the initial signal frequencies (right side).

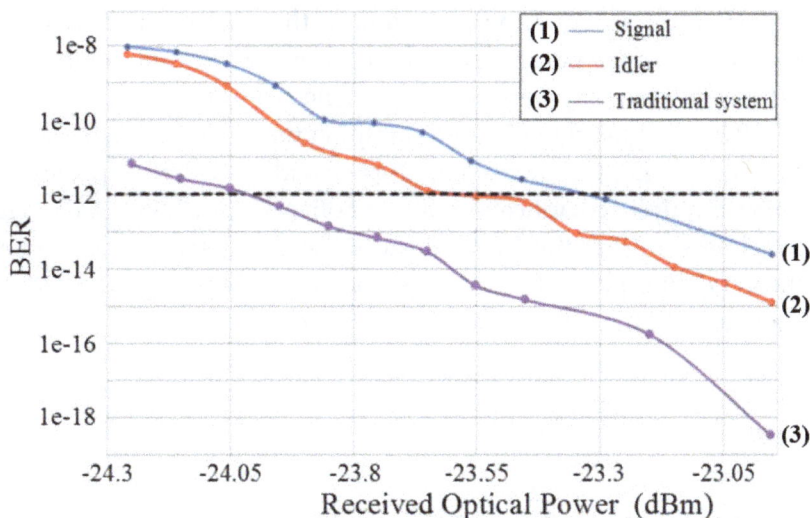

Figure 23. BER value dependence on the power of the detected signal in the standard single-channel system (3) and in the first channel of the system with the chosen FOPA configuration of the signal at its initial frequency (1) and of the idler (2).

in a single-channel system with modulation format conversion. A lower power penalty value in the case of idler component is explained by the fact that the orthogonally polarized radiation of the second channel is not included in the parametric amplification and idler generation process, and, therefore, it is not reflected in the idler itself. Irrespective of the fact that the amplification of orthogonally polarized (second channel) radiation is much lower (1–2 dB), its power level is still sufficiently high to affect the BER value of the divided signal, which in our case produces an additional power penalty of 0.3 dB. The idler use has allowed achieving a lower power penalty in respect to the standard single-channel system with the NRZ-OOK modulation format, whereas to achieve a BER value below the 10^{-12} threshold, it is necessary to use pump power that is larger by 10 mW than that when receiving the initial signal with 196.5 THz frequency.

Upon comparing the amplification spectra in a single-channel system with modulation format conversion and in a system with polarization-multiplexed signal division, it has been concluded that the amplification obtained in the latter case is larger by 5.2 dB (14.8 and 20 dB, respectively), irrespective of the fact that the pumping radiation power differs only by 5 mW. This is explained by the following two factors:

- In a system with signal division from orthogonally polarized signal combination, the signal power level at the input of the HNLF fiber is lower by 3.4 dB (−44.1 dBm). Therefore, the amplifier requires a lower pumping radiation power for ensuring a specific amplification level.

- Secondly, to minimize the idler spectral broadening in a system with polarization-multiplexed signal division, the frequency tones used for pumping radiation-phase modulation are reconfigured, and the achieved amplification difference clearly shows that with the given configuration SBS mitigation is more effective than in a system with modulation format conversion.

Upon summing up all the information presented in this chapter, it can be concluded that polarization dependence of the parametric amplification can be used for emphasizing optical radiation with a specific state of polarization from a flow of two orthogonally polarized optical radiations. FOPA with linearly polarized pumping radiation can be used both for 2PolSK signal conversion into NRZ-OOK modulation format and for signal emphasizing from a flow of two polarization-multiplexed optical signals. In both cases, such FOPA configurations have been found, which ensure that the BER values of the processed signal are below the 10^{-12} mark, and at the same time, none of the proposed solutions cause power penalty that exceeds 1.8 dB in comparison with the relevant standard solutions.

5. Conclusions

The implementation of the hybrid Raman-EDFA amplifier has allowed not only equalizing the gain spectrum but also increasing OSNR in all channels by 1.7–2.6 dB. The usage of the distributed Raman amplifier in cascade as a preamplifier has allowed the EDFA to operate closer to the saturation point, and, therefore, the EDFA noise figure decreased by 0.3–0.4 dB. The OSNR increase is also related to the fact that due to the coherent nature of Raman scattering, DRA amplifies the signal more effectively than the noise, which allows obtaining negative noise figure values (from −0.4 to −0.6 dB).

The implementation of the hybrid Raman-SOA amplifier has enabled the use of such SOA configuration, at which SOA produced the lowest amount of amplified signal distortions. As a result, by using the Raman-SOA hybrid amplifier, it is possible to obtain approximately the same BER level as in the case when the SOA is used as a single in-line amplifier for a signal, which is weaker by 1.5 times. This has allowed increasing the attainable transmission distance by 12 km or by 11%.

While changing the power of FOPA pumping radiation, the nonlinear phase mismatch of the parametric process also changes. Therefore, while configuring the pump power, the wavelength must also be changed accordingly.

Modulation of the FOPA pumping radiation phase, which has been used for increasing the SBS threshold, has caused spectral expansion of idlers by 54%. Therefore, the frequency tones used in systems with wavelength conversion for pumping radiation-phase modulation must be selected in a way that ensures that the spectral expansion of idlers remains as low as possible.

By manipulating with the parameters of dual-pump FOPA, it is possible to achieve an increase in the number of carrier signals from 16 to 32, simultaneously ensuring equal channel spacing of 100 GHz and maximum difference in power levels of 1.9 dB among all channels. It has been found that, in case when each carrier signal power at the input of the FOPA is equal to 0 dBm, the CC-FWM interactions produce considerable interchannel cross talk. Due to this reason, when using the idlers to double the number of carriers, it is necessary to control the power of carrier signals in the amplifier input.

It has been found that even when using polarization-maintaining HNLF fibers, due to the influence of self-phase modulation and cross-phase modulation, a change in interposition of SOP of the signal and FOPA pumping radiation has been observed, as they are transmitted through HNLF. As a result, the signal, the SOP of which is orthogonal in respect to the SOP of the pumping radiation at HNLF input, has been amplified by 1.5–1.6 dB.

In a single-channel system with 2PolSK signal conversion to NRZ-OOK modulation format, power penalty of 0.4 dB has been observed between the NRZ-OOK signal from the standard single-channel system and the converted signal, whereas in the case of the idler spectral component, the power penalty is lower by 0.2 dB. These results can be explained by the fact that the idler produced by the parametric FWM process does not contain the logical "0" component radiation from initial 2PolSK signal, which for the converted NRZ signal is interpreted as noise.

In the multichannel system with modulation format conversion, a more explicit manifestation of CC-FWM processes has been observed among the idlers rather than among channels at initial frequencies. This can be explained by the fact that pump power exceeding 35 mW is required to obtain BER values below the 10^{-12} threshold in all eight channels in the idler frequency band than for signals at their initial frequencies. As a result of such amplification difference, more explicit CC-FWM manifestation has been observed, which accordingly has led to additional interchannel cross talk. Additionally, cross talk generated by CC-FWM has not only been transferred to the idlers from the signals at their initial frequencies but also generated among the idlers.

Unlike the system with modulation format conversion, in the system with signal emphasizing from the flow of two orthogonally polarized signals, a situation is observed, when at the same frequency it is possible that the logical "1" components are observed in both channels simultaneously. Therefore, the influence of orthogonally polarized radiation on the quality of the emphasized signal is larger by 0.4 dB (in the case of idlers, by 0.3 dB) than in a system with modulation format conversion.

Acknowledgements

This work has been supported by the National Research Program in Latvia within the project Nr. 10-4/VPP-4/11.

Author details

Vjaceslavs Bobrovs*, Sergejs Olonkins, Sandis Spolitis, Jurgis Porins and Girts Ivanovs

*Address all correspondence to: vjaceslavs.bobrovs@rtu.lv

Riga Technical University, Institute of Telecommunications, Latvia

References

[1] Cisco Systems Inc., "The Zettabyte Era: Trends and Analysis," white paper, June 2014, pp. 1-24.

[2] BELL LABS Metro Network Traffic Growth, An Architecture Impact Study Strategic White Paper, 2013.

[3] Cisco Systems Inc., "Cisco VNI: Forecast and Methodology, 2013-2018," white paper, June 2014, pp. 1-14.

[4] Agrawal G.P., "Fiber Optics Communications Systems", John Wiley & Sons, USA, 2002, 561 p.

[5] Thyagarajan K. and Ghatak A., "Fiber Optic Essentials", John Wiley & Sons Inc., Canada, 2007, 259 p.

[6] Kikuchi K., "Ultra-long-haul optical transmission characteristics of wavelength-division multiplexed dual-polarization 16-quadrature-amplitude-modulation signals," Electronics Letters, Vol. 46, No. 6, March 18, 2010, pp. 433-434, DOI: 10.1049/el.2010.3533.

[7] Downie J.D. "High-capacity long-haul transmission using ultra-low loss optical fiber," 17th Opto-Electronics and Communications Conference (OECC), July 2-6 2012, pp. 172-173, DOI: 10.1109/OECC.2012.6276426.

[8] Šalik P., Čertik F. and Roka R., "Duobinary modulation format in optical communication systems," Advances in Signal Processing, Vol. 3, No. 1, 2015, pp. 1-7.

[9] 3MTM technical specifications, PLC Optical Splitters," USA, 2008, pp. 1-2.

[10] Baney D.M. and Stimple J., "WDM EDFA gain characterization with a reduced set of saturating channels," IEEE Photonics Technology Letters, Vol. 8, No. 12, Dec. 1996, pp. 1615-1617. DOI: 10.1109/68.544695.

[11] Sharma M. and Sharma V.R., "Gain flattening of EDFA in C-band using RFA for WDM application," 2nd International Conference on Signal Processing and Integrated Networks (SPIN 2015), 19-20 Feb. 2015, pp. 346-351, DOI: 10.1109/SPIN.2015.7095422.

[12] Becker P.C., Plsson N.A. and Simpson J.R., "Erbium-Doped Fiber Amplifiers", Academic Press, USA, 1999, 481 p.

[13] Houbavlis T., Zoiros K.E., Kalyvas M., Theophilopoulos G., Bintjas C., Yiannopoulos K., Pleros N., Vlachos K., Avramopoulos H., Schares L., Occhi L., Guekos G., Taylor J.R., Hansmann S. and Miller W., "All-optical signal processing and applications within the esprit project DO_ALL," Journal of Lightwave Technology, Vol. 23, No. 2, Feb. 2005, pp. 781-801, DOI: 10.1109/JLT.2004.838854.

[14] Oxenlowe L.K., Pu M., Ding Y., Hu H., Da Ros F., Vukovic D., Jensen A.S., Ji H., Galili M., Peucheret C. and Yvind K., "All-optical signal processing using silicon devices," European Conference on Optical Communication (ECOC 2014), 21-25 Sept. 2014, pp. 1-3, DOI: 10.1109/ECOC.2014.6964241.

[15] Willner A.E., Khaleghi S., Chitgarha M. R. and Yilmaz O. F. "All-optical signal process-ing", Journal of Lightwave Technology, Vol. 32, No. 4, February 15th, 2014, DOI: 10.1109/JLT.2013.2287219.

[16] Chu C.-H., Lin S.-L., Chan S.-C. and Hwang S.-K. "All-optical modulation format con-version using nonlinear dynamics of semiconductor lasers," IEEE Journal of Quantum Electronics, Vol. 48, No. 11, Nov. 2012, pp. 1389-1396, DOI: 10.1109/JQE.2012.2212877.

[17] Ozolins O., Parts R., Bobrovs V., "Impact of cascaded MRRs on all-optical clock recov-ery from 40 Gbit/s RZ-OOK signal," 9th International Symposium on Communication Systems, Networks & Digital Signal Processing (CSNDSP 2014), 23-25 July 2014, pp. 541-545, DOI: 10.1109/CSNDSP.2014.6923888.

[18] Hisano D., Maruta A. and Kitayama K.-I., "Timing detector using cross gain modula-tion in semiconductor optical amplifier for adaptive all-optical signal processing," 18th Microoptics Conference (MOC 2013), 27-30 Oct. 2013, pp. 1-2.

[19] Hu H., Jopson R.M., Gnauck A.H., Dinu M., Chandrasekhar S., Xie C. and Randel S. "Parametric amplification, wavelength conversion, and phase conjugation of a 2.048-Tbit/s WDM PDM 16-QAM signal," Journal of Lightwave Technology, Vol. 33, No. 7, April 1, 2015, pp. 1286-1291, DOI: 10.1109/JLT.2014.2370038.

[20] Hansryd J. and Andrekson P.A., "O-TDM demultiplexer with 40-dB gain based on a fiber optical parametric amplifier," IEEE Photonics Technology Letters, Vol. 13, No. 7, July 2001, pp. 732-734, DOI: 10.1109/68.930430.

[21] Shama H.B., Gulati T. and Rawat B., "Evaluation of optical amplifiers," International Journal of Engineering Research and Applications, Vol. 2, No. 1, 2012, pp. 663-667.

[22] Csele M., "Fundamentals of Light Sources and Lasers," John Wiley & Sons Inc., Canada, 2004, 343 p.

[23] Thyagarajan K. and Ghatak A., "Fiber Optics Essentials," John Wiley & Sons Inc., Canada, 2007, 259 p, DOI: 10.1002/9780470152560.

[24] Forestieri E. "Optical Communication Theory and Techniques," Springer Science + Business Media Inc., USA, 2005, 216 p, DOI: 10.1007/b100765.

[25] Kaminow I.P., Li T. and Willner A. E., "Optical Fiber Telecommunications: Components and Subsystems," Academic Press, USA, 2008, 945 p.

[26] Agrawal G.P., "Applications of Nonlinear Fiber Optics," Academic Press, USA, 2001, 473 p.

[27] Karasek M., Honzatko P., Vojtech J., Radil J., "Multi-wavelength conversion at 10 Gb/s and 40 Gb/s based on 2 pumps FOPA," 13th International Conference on Transparent Optical Networks (ICTON 2011), 2011, pp. 1-4, DOI: 10.1109/ICTON.2011.5970951.

[28] Temprana E., Ataie V., Peric A., Alic N., Radic S., "Wavelength conversion of QPSK signals in single-pump FOPA with 20 dB conversion efficiency," Optical Fiber Communications Conference and Exhibition (OFC 2014), 9-13, Mar. 2014, pp.1-3. DOI: 10.1364/OFC.2014.Th1H.2.

[29] Ruo-Ding L., Kumar P., Kath W.L., "Dispersion compensation with phase-sensitive optical amplifiers," IEEE Journal of Lightwave Technology, Vol.12, No. 12, 1994, pp. 541-549, DOI: 10.1109/50.285338.

[30] Wang J., Ji H., Hu H., Mulvad H.C.H., Galili M., Palushani E., Jeppesen P., Yu J.L., Oxenlowe L.K., "All-optical 2R regeneration of a 160-Gbit/s RZOOK serial data signal using a FOPA," IEEE Photonics Conference (IPC 2012), 23-27 Sept. 2012, pp. 108-109, DOI: 10.1109/IPCon.2012.6358512.

[31] Wang J., Yu J., Meng T., Miao W., Sun B., Wang W., Yang E., "Simultaneous 3R regeneration of 4*40-Gbit/s WDM signals in a single fiber," IEEE Photonics Journal, Vol. 4, No. 5, pp. 1816-1822, 2012, DOI: 10.1109/JPHOT.2012.2215955.

[32] Lee H.H., Cho S.H., Lee J.H., Lee S.S., "Excess intensity noise suppressed 100-GHz spectrum-sliced WDM-PON with pre-spectrum-sliced seed light source," 36th European Conference Exhibition on Optical Communication (ECOC 2010), 19-23 Sept. 2010, pp.1-3, DOI: 10.1109/ECOC.2010.5621291.

[33] Spolītis S., Olonkins S., Poriņš J., "Realization of dense bidirectional spectrum sliced WDM-PON access system," 9th International Symposium on Communications Systems, Networks and Digital Signal Processing (CSNDSP 2014), Conference Proceedings, 2014, pp.1-6, DOI: 10.1109/CSNDSP.2014.6923890.

[34] Ivanovs Ģ., Ļašuks I., Ščemeļevs A., "A hybrid TDM/WDM-PON system with Fwm-generated source of multiwavelength optical signals," Latvian Journal of Physics and Technical Sciences, Vol. 47, No. 5, 2010, pp.3-14, DOI: 10.2478/v10047-009-0020-3.

[35] Mathlouthi W., Vacondio F., Rusch L.A., "High-bit-rate dense SS-WDM PON using SOA-based noise reduction with a novel balanced detection," Journal of Lightwave Technology, Vol. 27, No. 22, Nov. 15 2009, pp.5045-5055, DOI: 10.1109/JLT.2009.2026062.

[36] Fok M.P., Shu C., "Bandwidth enhanced multi-wavelength source from an SBS-assisted fiber ring laser using four-wave mixing in a highly nonlinear bismuth oxide fiber," Conference on Lasers and Electro-Optics / Quantum, DOI: 10.1109/CLEO.2006.4628216.

6

Holograms in Optical Wireless Communications

Mohammed T. Alresheedi,

Ahmed Taha Hussein and Jaafar M.H. Elmirghani

Additional information is available at the end of the chapter

Abstract

Adaptive beam steering in optical wireless communication (OWC) system has been shown to offer performance enhancements over traditional OWC systems. However, an increase in the computational cost is incurred. In this chapter, we introduce a fast hologram selection technique to speed up the adaptation process. We propose a fast delay, angle and power adaptive holograms (FDAPA-Holograms) approach based on a divide and conquer methodology and evaluate it with angle diversity receivers in a mobile optical wireless (OW) system. The fast and efficient fully adaptive FDAPA-Holograms system can improve the receiver signal to noise ratio (SNR) and reduce the required time to estimate the position of the receiver. The adaptation techniques (angle, power and delay) offer a degree of freedom in the system design. The proposed system FDAPA-Holograms is able to achieve high data rate of 5 Gb/s with full mobility. Simulation results show that the proposed 5 Gb/s FDAPA-Holograms achieves around 13 dB SNR under mobility and under eye safety regulations. Furthermore, a fast divide and conquer search algorithm is introduced to find the optimum hologram as well as to reduce the computation time. The proposed system (FDAPA-Holograms) reduces the computation time required to find the best hologram location from 64 ms using conventional adaptive system to around 14 ms.

Keywords: adaptive hologram, delay spread, SNR

1. Introduction

In general, holography is the storage of the phase and amplitude information of a wavefront. It is usually used as an approach to create three-dimensional (3D) images of objects through interference between a wavefront diffracted and a coherent reference beam. In optical wireless communications (OWC), the hologram is a transparent or reflective device that is used to

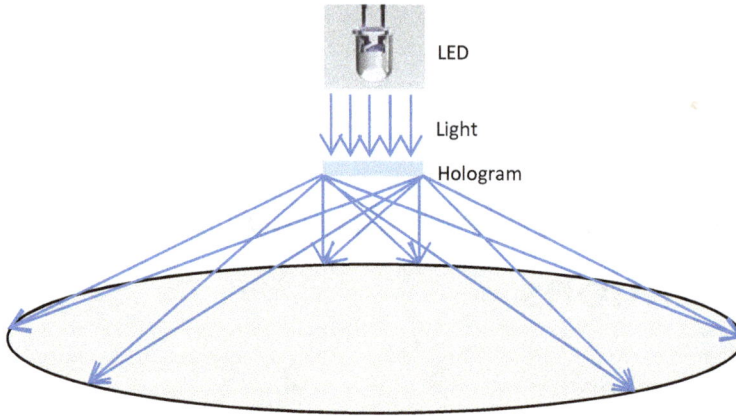

Figure 1. Holographic diffuser with uniform intensities that cover desired area.

spatially modulate the phase or amplitude of the energy passing through it. **Figure 1** illustrates the effect of a diffusing hologram on a set of rays from a light source. Here, the light source (light emitting diode (LED)) is split into a number of beams that cover the desired area.

Holograms can be produced from mathematical description or physical object. In mathematical approach, any wavefront can be generated. If mathematical method is implemented with a computer, the hologram is called a computer generated hologram (CGH). A ground glass diffuser can be given as a simple example of physical object diffuser, where ground glass can be placed at the output of a laser to change it from a point source to a large area source. However, this type of hologram cannot be controlled.

Beam steering has been widely studied in wireless communication systems to maximise the signal to noise (SNR) at the receiver [1, 2]. It is also considered as an attractive option in optical wireless communication (OWC) systems to enhance the system performance [3, 4]. New adaptive technique using beam steering is introduced in visible light communication (VLC) links in Ref. [5]. The goal is to maximise the SNR at the receiver in all possible locations within an indoor environment. Simulations results have shown that high data rate up to 20 Gb/s can be achieved by partially steering some of the beams towards the receiver location. Multi-input-multi-output (MIMO) infrared (IR) links employing beam steering method has been introduced in Ref. [6]. Furthermore, demonstration of IR-linked energy transmission using beam forming along with a spatial light modulator (SLM) is shown in Ref. [7]. An efficient power and angle adaptation technique is proposed in Refs. [1–4] in order to help the IR optical wireless (IROW) transmitter to optimise the diffusing spots distribution. These methods (power and angle adaptations) are able to enhance the received signal strength level, regardless of the receiver's location, the receiver's field of view (FOV) and transmitter's position. A significant performance improvement using beam angle and beam power adaptation in a line strip multi-beam system (APA-LSMS) is shown in Refs. [1, 2]. However, a cost has to be paid due to the complex adaptation requirements. The adaptive APA-LSMS transmitter needs to generate a single spot and scanning with all the possible locations (around 8000 locations) in the room in order to find the receiver and then generate the hologram with optimum powers and angles. This makes APA-LSMS system design very challenging.

In this chapter, we aim to point out the impairments of IROW links and propose new efficient solutions beyond those reported in Refs. [1–4]. We report an adaptive hologram selection method employing simulated annealing (SA) to generate diffusing spots (multi-beam). The proposed system is pre-calculated and stored all the holograms in memory. Each stored hologram is suited for a given transmitter and receiver location. Hence, it eliminates the need to calculate holograms real time at each transmitter-receiver location. We model fast angle and power adaptive holograms (FAPA-Holograms) and fast delay, angle and power adaptive holograms (FDAPA-Holograms) mobile OW systems, in conjugation with angle diversity receivers [8]. The conventional diffuse system (CDS) and line strip multi-beam systems (LSMS) are studied for comparison purposes. The ultimate goal of the proposed systems: FAPA-Holograms and FDAPA-Holograms is to reduce the time required to generate hologram at optimum transmitter and receiver location as well as to enhance the overall system performance such as SNR and channel bandwidth in a typical indoor environment. A significant improvement can be obtained by increasing the scanned stored holograms in our systems to approach the original power and angle adaptive methods proposed previously in Refs. [1, 2]. However, increasing the number of scanned stored holograms leads to an increase in the computation time needed to find the best hologram. To overcome this issue, we introduce a divide and conquer (D&C) algorithm to select the best hologram among a finite vocabulary of holograms, hence speed up the adaptation process associated with these adaptive systems [8]. High data rates of 2.5 and 5 Gb/s are considered for the FAPA-Holograms and FDAPA-Holograms systems.

The remainder of this chapter is organised into the following sections: Section 2 presents The IROW room setup and channel characteristics. Section 3 presents the proposed systems' configurations. Section 4 introduces the simulation results and discussion of the IROW systems. Finally, conclusions are drawn in Section 5.

2. The IROW room setup and channel characteristics

In order to study the impact of directive ambient light noise sources and multipath dispersion, as well as their effect on the received data flow, consideration was given to an unoccupied rectangular room that had no furnishings, with dimensions of 8 m × 4 m × 3 m (length × width × height). Researchers in Ref. [9] have studied and investigated the power reflected in indoor IROW system. The study found that the light reflected on either ceiling or wall is Lambertian in nature (mode $n = 1$). They also found that the wall reflected power by 80% whereas ceiling by only 30%. In this chapter, we consider that the rays reflected from door and windows are similar to those coming from walls. The reflecting elements from walls and ceiling can be modelled by dividing the surfaces into a small square shape, which can operate as secondary Lambertian transmitter with $n = 1$.

The accuracy of the impulse response profile is controlled by the size of the reflecting elements. Therefore, element sizes of 5 cm × 5 cm for the first order reflections and 20 cm × 20 cm for the second order reflections are employed for all arrangements. Previous work studied the received optical power within an indoor environment. They found that most of the received optical power is located within the two first order reflections (1st and 2nd). Third order and higher reflections are highly attenuated [10, 11]. Hence, two bounces are considered in our calculations. All the proposed systems use an upright transmitter with 1 W optical power. Furthermore, the significant signal to noise ratio (SNR) improvement of the hologram-proposed systems is used to reduce the transmit power to 80 mW reducing the power density on the adaptive hologram and helping eye safety.

In OW communication links, intensity modulation with direct detection (IM/DD) is considered the most viable approach. The indoor OW IM/DD channel can be fully specified by its impulse response $h(t)$, and it can be modelled as a baseband linear system given by

$$I(t, Az, El) = \sum_{m=1}^{M} Rx(t) \otimes h_m(t, Az, El) + \sum_{m=1}^{M} Rn(t, Az, El). \tag{1}$$

where $I(t, Az, El)$ is the current instantaneous due to m reflecting elements, El and Az are the directions of arrival in the elevation and azimuth angles, t is the absolute time, $x(t)$ is the optical power transmitted, $\otimes$ denotes convolution, M is the total number of receiving elements, R is the photodetector responsivity and $n(t, Az, El)$ is the background noise. The delay spread is a good tool to measure signal spread due to multipath propagation. The delay spread can be written as [12, 13]:

$$DS = \sqrt{\left(\sum_{\forall i} (t_i - \mu)^2 P_{r_i}^2\right) / \sum_{\forall i} P_{r_i}^2} \tag{2}$$

where the time delay t_i is associated with the received power P_{r_i} (P_{r_i} reflects the impulse response $h(t)$ behaviour) and μ is the mean delay given by

$$\mu = \sum_{\forall i} t_i P_{r_i}^2 / \sum_{\forall i} P_{r_i}^2. \tag{3}$$

The delay spared is deterministic for a given stationary transmitter-receiver and reflecting elements' positions. The delay spread can change for a given transmitter-receiver location when the reflecting elements moves or an object is entering and leaving the environment. However, the impact of such a change is not considered in this work and has not been investigated by other researchers.

The SNR of the received signal can be calculated by taking into account the powers associated with logic 0 and logic 1 (P_{s0} and P_{s1}), respectively. The SNR is given by [14]:

$$SNR = \left(\frac{R(P_{s1} - P_{s0})}{\sigma_0 + \sigma_1}\right)^2 \tag{4}$$

$$\sigma_0 = \sqrt{\sigma_{pr}^2 + \sigma_{bn}^2 + \sigma_{s0}^2} \text{ and } \sigma_1 = \sqrt{\sigma_{pr}^2 + \sigma_{bn}^2 + \sigma_{s1}^2} \tag{5}$$

where σ_{pr}^2 represents the receiver noise, which is a function of the design used for the pream-plifier; σ_{bn}^2 represents the background shot noise component and σ_{s0}^2 and σ_{s1}^2 represent the shot noise associated with the received signal (P_{S0} and P_{S1}), respectively. The signal-dependent noise (σ_{si}^2) is very small due to the weak received optical signal, see the experimental results reported in Ref. [15]. In this study, we used the PIN-FET transimpedance preamplifier proposed in Ref. [16]. The background shot noise calculations can be found in Ref. [3]. Nine branches angle diversity receiver is used to reduce the impact of multipath dispersion. In this work, we employed the non-imaging angle diversity receiver design proposed in Ref. [8]. We considered maximum ratio combining (MRC) scheme. Calculations of MRC method can be found in our previous work in Refs. [1, 19].

In order to consider the impact of background noise, we use eight light bulbs in the room. These lights are used for illuminations. However, in OW receiver, the signals arrived from each light are considered as undesired signals which can be modelled as background shot noise. In this study, we assumed that Philips PAR 38 Economic' (PAR38) was used as spotlight in which each unit of light radiates 65 W within a narrow beam width. The light from these units can be modelled as a Lambertian radiant intensity with order nl = 33.1 [11]. Additional simulation parameters are given in **Table 1**.

3. Proposed systems' configurations

A CDS and non-adaptive LSMS are two widely studied configurations in the literature; therefore, they are modelled and used for comparison purposes in order to evaluate the improvements offered through the proposed configurations. More information about CDS and LSMS system can be found in Refs. [9–11].

3.1. FAPA-hologram

Power adaptation and beam angle can be considered as an effective approach, which helps to have an optimum power allocation and distribution of the diffusing spots. A single spot is produced by the adaptive transmitter to scan the ceiling and walls at approximately 8000 possible locations (2.86° beam angle increment [3]) in order to identify the best location. A liquid crystal device (IR hologram) is used to change the spot location at each step. Power adaptation technique can be used with angle adaptation to further enhance signal to noise ratio. The transmitter switches spots one by one and the receiver calculates the SNR (weight) associated with each spot. Then the feedback signal is sent by receiver at a low data rate to inform the transmitter about the SNR associated with each spot. The transmitter re-distributes the power among the spots based on their SNR weights [1] .The transmitter generates the hologram after finding optimum angles and power levels of spots. Intensive calculations and time are required from a digital signal processor (DSP). An adaptation approach is proposed where a finite vocabulary of stored holograms is used in order to get rid of computing the holograms at each step to identify the best location. The ceiling is divided into 80 regions (0.4 m × 1m per region), see **Figure 2**. This large number of regions has been selected based on our recent optimisation in Ref. [17].

Parameter	Configuration								
Width	4 m								
Height	3 m								
ρ x z wall	0.8								
y-z wall	0.8								
x-z op wall	0.8								
y-z op wall	0.8								
Floor	0.3								
Transmitter									
Quantity			1						
Location (x,y,z)	(1 m, 1 m, 1 m)					(2 m, 7 m,1 m)			
Elevation			90°						
Azimuth			0°						
Receiver									
Quantity	9								
Photodetector's area	10 mm^z								
Acceptance semi-angle	12°								
Location (x, y, z)	(1,1,1),(1,2,1),(1,3,1,),(1,4,1),(1,5,1),(1,6,1),(1,7,1)								
Elevation	90°	65°	65°	65°	65°	65°	65°	65°	65°
Azimuth	0°	0°	45°	90°	135°	180°	225°	270°	315°
Resolution									
Time bin duration	0.5 ns				0.01 ns				
Bounces	1				2				
Surface elements	32,000				2000				
Number-of-spot lamps			8						
Locations			(1,1,1), (1,3,1), (1,5,1), (1,7,1) (3,1,1), (3,3,1), (3,5,1), (3,7,1)						
Wavelength			850 nm						
Preamplifier design	PIN-BJT				PIN-FET				
Bandwidth (BW)	50 MHz		2.5 GHz				5 GHz		
Bit rate	50 Mbit/s		2.5 Gbit/s				5 Gbit/s		

Table 1. Parameters used in simulation.

Holograms generated by means of a computer can produce spots with any prescribed amplitude and phase distribution. For FAPA-Holograms, all the spots have different weights (powers) and different phases. CGHs have many useful properties. Spot distributions can be computed on the

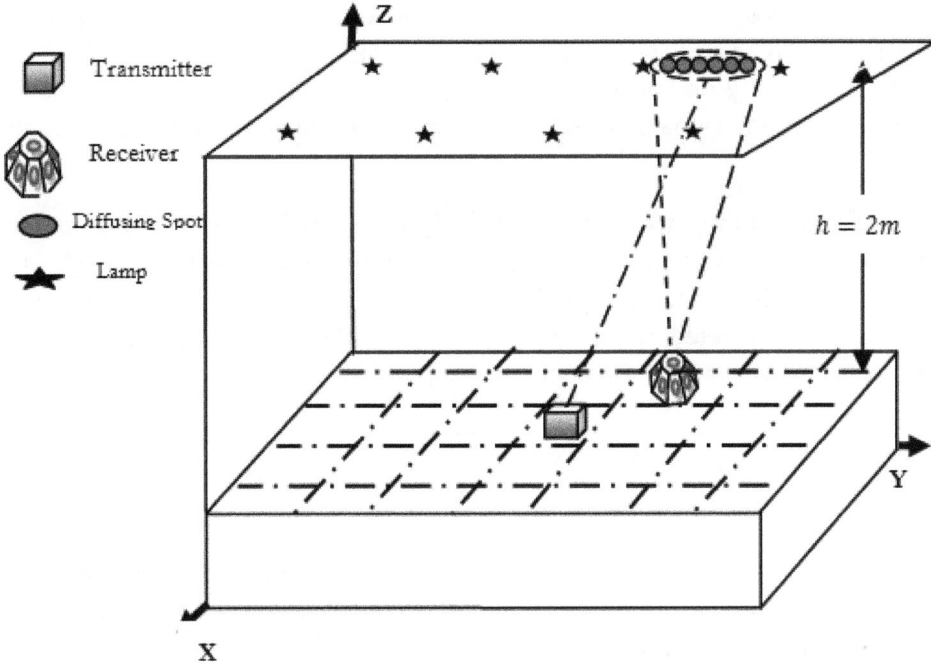

Figure 2. OW communication architecture of FAPA-Hologram system.

basis of diffraction theory and encoded into a hologram. Calculating a CGH means the calculation of its complex transmittance. The transmittance is expressed as follows:

$$H(u, v) = A(v, u).exp[j\phi(u, v)] \tag{6}$$

where $H(u, v)$ is complex transmittance function, $A(u, v)$ and $\phi(u, v)$ are amplitude and phase distribution, respectively. The parameters (u, v) are coordinates in the frequency space. The phase of incoming wavefront is modulated by hologram, whereas the transmittance amplitude is equal to unity. The analysis used in Refs. [18–20] has been employed for the design of the CGHs. The hologram $H(u, v)$ is considered to be in the frequency domain and the observed diffraction pattern $h(x, y)$ in the spatial domain. They are related by the continuous Fourier transform:

$$h(x, y) = \iint H(u, v)exp[-i2\pi(ux + vy)]dudv \tag{7}$$

The diffraction pattern of the hologram when it is placed in the frequency plane is given by

$$h(x, y) = RSsinc(Rx, Sy)\sum_{k=-\frac{M}{2}}^{\frac{M}{2}-1}\sum_{l=-\frac{N}{2}}^{\frac{N}{2}-1}H_{kl}exp[i2\pi(Rkx + Syl)] \tag{8}$$

where $sinc(a, b) = sin(\pi a) sin(\pi b)/\pi^2 ab$. The complex amplitude of the spots is proportional to some value of interest. But, the reconstruction will be in error because of the finite resolution of the output device and the complex transmittance of the resulting hologram. This error can be considered to be a cost function. Simulated annealing (SA) is used to minimise the cost function [21]. The phases and amplitudes of every spot are determined by the hologram pixel pattern and are given by its Fourier transform. The constraints considered in the hologram plane are to discretise the phase from 0 to 2π and a constant unit amplitude for the phase only CGH.

Let the desired spots in the far field be $f(x,y) = |f(x,y)|\exp(i\varphi(x,y))$. The main target is to find the CGH distribution $g(v, u)$ that produces optimum reconstruction $g(x, y)$ that is very close to the desired distribution $f(x, y)$. The cost function (CF) is defined as a mean squared error which can be interpreted as the difference between the normalised desired object energy $f''(x, y)$ and the scaled reconstruction energy $g''(x,y)$:

$$CF_k = \sqrt{\sum_{i=1}^{M}\sum_{j=1}^{N}\left(|f''(i,j)|^2 - |g''_k(i,j)|^2\right)^2}, \qquad (9)$$

where $f''(x, y)$ represents the normalised desired object energy and $g''_k(i,j)$ represents the scaled reconstruction energy of the kth iteration. Simulated annealing was used to optimise the phase of the holograms offline in order to minimise the cost function. The simulating annealing algorithm can help jump from local optima to close to a global optimum (minimising the cost function close to zero). The transition out of a local minima to global one is accomplished by accepting hologram phases that increase the mean squared error of the reconstruction with a given probability. The probability of accepting these phases is $exp(-\Delta CF/T)$, where ΔCF is the change in error and T is a control parameter (the temperature of the annealing process). First, we start with a high value of T so that all the change in the hologram phases are accepted and then slowly lower T at each iteration until the number of accepted changes is small. This method is similar to melting a metal at a high value of T and then reducing the T slowly until the metal crystals freezes at a minimum energy. The changes of hologram phases relate to a small perturbation of the physical system, and the resulting change in the mean squared error of the reconstruction corresponds to the resulting change in the energy of the system. Therefore, this technique finds a hologram configuration, which has a minimum mean squared error (CF). For phase only CGHs, the constraints are constant amplitude and a random phase distribution φ_0 ($M \times N$). In the first iteration, a random phase is applied to help in the convergence of the algorithm.

For a large room of 8 m × 4 m, the floor is divided into 80 regions. A library that contains 6400 holograms is optimised offline using SA. In order to accurately identify the receiver location, a large number of holograms are required [17]. The optimum diffusing spots were pre-calculated based on the power and angle techniques shown in Ref. [1]. A total of 80 holograms are stored in a library and allocated for each region, the transmitter should cover the 80 possible receiver positions in the room, which means 6400 holograms are required to cover the entire room. The total number of holograms required is N^2, where N represents the number of regions into which the floor/ceiling is divided. **Figure 2** illustrates one hologram when the receiver is present at (1 m, 6 m and 1 m) and the transmitter is placed in the middle of room at (2 m, 4 m and 1 m). SA is used to optimise the phase of the CGH. **Figure 3** shows

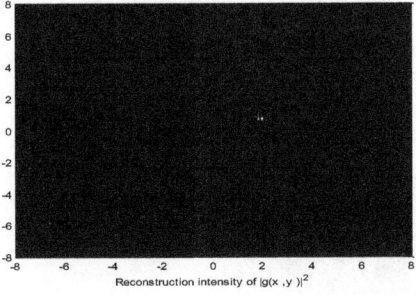
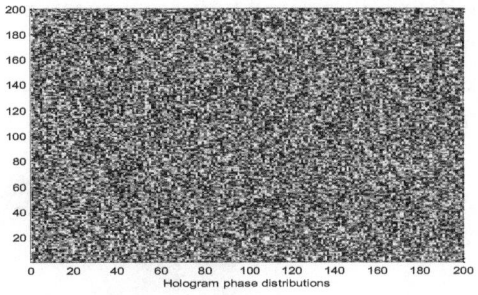

Iteration 1

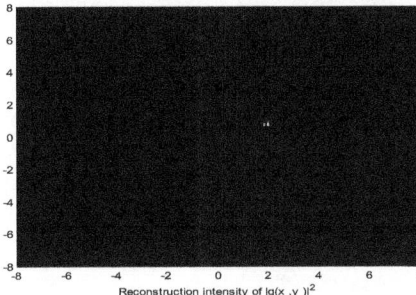
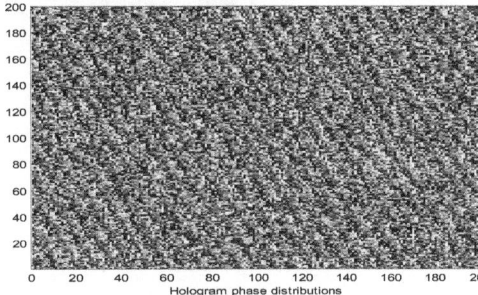

Iteration 5

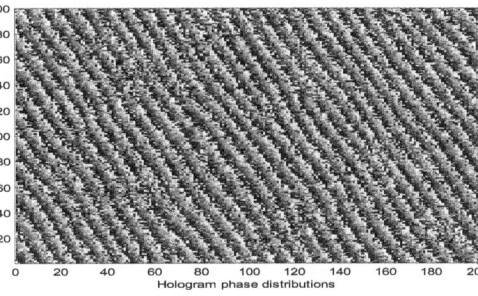

Iteration 15

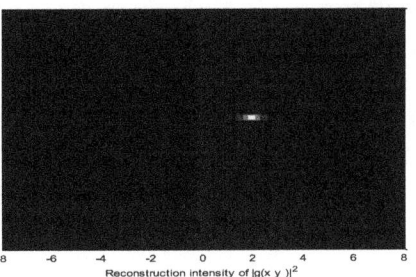
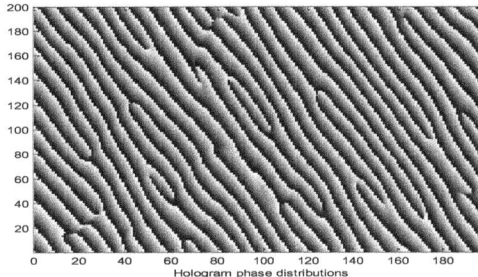

Iteration 100

Figure 3. The reconstruction intensity at the far field and hologram phase pattern at iterations 1, 5, 15, and 100 using simulated annealing optimisation. Different grey levels represent different phase levels ranging from 0 (black) to 2π (white).

phase distributions and hologram reconstruction intensity at the far field in four snapshots. When the number of iterations increases, the reconstruction intensities are improved. The desired spot intensity in the far field is shown in **Figure 4**. **Figure 5** shows the number of iterations versus cost function.

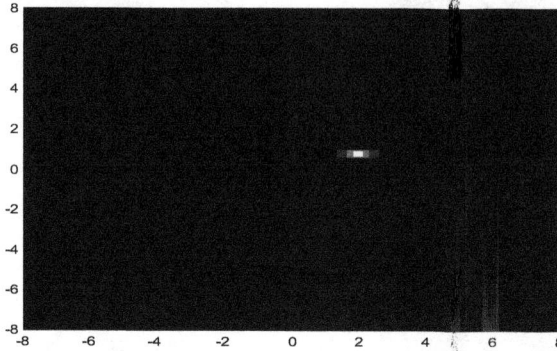

Figure 4. The desired spots intensity in the far field.

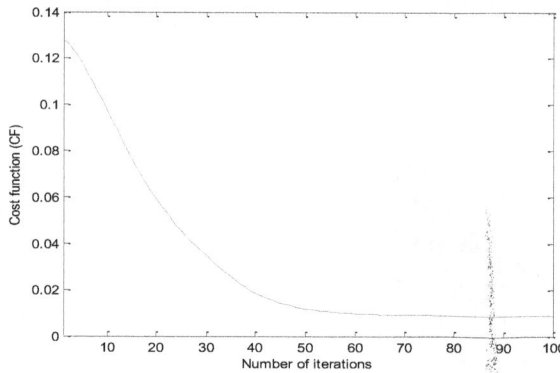

Figure 5. Cost function versus the number of iterations.

Scanning 6400 stored holograms in the system required a full search among all stored holograms, in order to select the best hologram. However, high complexity in term of the computation time is introduced. To solve this issue, a fast search technique is proposed to enhance the SNR via selecting the best hologram while reducing the computational time. The proposed search technique is based on a divide and conquer (D&C) method. Using the D&C algorithm, the transmitter is able to select the best hologram that can achieve the optimum receiver SNR. The fast search algorithm of our proposed system is applied for a single transmitter/receiver scenario as follows:

1. The stored holograms in the transmitter are first divided into four quadrants based on their transmission angles. The transmission angles associated with each quadrant are δ_{max-x} to δ_{min-x} and δ_{max-y} to δ_{min-y} in the x-axis and y-axis, respectively.

2. A single middle hologram is used in each group (quadrant) in order to find the first suboptimum hologram.

3. The receiver calculates the SNR for each hologram and sends a pilot signal at low rate (feedback channel) to inform the transmitter about the SNR associated with each scanned hologram.

4. The transmitter selects the hologram that achieves the best SNR and identifies the new optimal quadrant (first sub-optimum quadrant) for next iteration.

5. The transmitter divides the sub-optimum quadrant into four sub-quadrants and applies steps 2–4 in order to find the second sub-optimal quadrant.

6. The D&C process is carried out and steps 2–5 are repeated to find the best location that has the highest SNR level at the receiver.

The proposed system reduces the computation time from 64 ms (each hologram required 1 ms to scan) taken by the classic beam steering system to 13 ms (13 possible locations should be scanned in all iterations × 1 ms).

3.2. FDAPA-hologram

To additionally enhance the performance of the IROW system, we introduce beam delay adaptation technique coupled with beam angle and power adaptation using hologram selection approach. The delay spread in a multi-beam spot IROW system is influenced by the spots' numbers and locations seen by each detector's FOV [3]. Therefore, switching all the beams simultaneously can introduce a differential delay between the neighbouring signals received at the receiver, hence spreading the pulse and limiting the 3-dB channel bandwidth. Instead of transmitting all the beams simultaneously, the delay adaptation algorithm sends the signal that has the longest journey first, and then sends the other signals with different differential delays (Δt) so that all the signals reach the receiver at the same time. A total of 10 μs time delay computation is carried out for each beam [8]. Therefore, a total of 1ms delay adaptation time is required for a beam. Our FDAPA-Hologram (delay, power and angle methods) requires only 14 ms adaptation time in order to select the best hologram. Assume the receiver initiates the adaptation every 1 s. This time is associated with a pedestrian speed of 1 m/s. Therefore, employing FDAPA-Hologram with a total 14 ms adaptation time introduces only 1.4%. Array element delayed switching is used to implement the delay adaptation method. The delay adaptation algorithm is explained as follows:

1. A transmitter first switches only the first beam (spot).

2. The receiver estimates the mean delay (μ) associated with first beam in the branch.

3. The receiver calculates the mean delay for all other branches.

4. The transmitter repeats steps 1–3 for other beams.

5. The receiver sends a feedback channel to update the transmitter about the delay associated with each beam.

6. The transmitter calculates the delay difference (Δt) between the beams as follows:

$$\Delta t_i = \max(t_{i\_max}) - t_i \; 1 \leq i \leq N_{spot} \tag{10}$$

7. The transmitter sends the beams with different times associated with their differential delay estimated in Eq. (10) to help the beam arrive at the receiver at the same time

The differential delay depends on the distances among the beams. If all the beams touch each other on the ceiling, then the differential delay will be few nano seconds, hence requiring timing control to switch such a beam within the required time.

4. Simulation results and discussion

In this section, we evaluate the performance of the proposed support systems in an empty room in the presence of multipath dispersion, receiver noise, back ground noise (light units) and mobility. The results are presented in terms of delay spread and SNR.

4.1. Delay spread

The delay spread of our adaptive proposed systems (FAPA-Holograms and FDAPA-Holograms) compared with CDS and LSMS system is shown in **Figure 6**. The results are presented when the transmitter is located near the room corner while the receiver moves along x = 2 m. The conventional pure diffuse system (CDS) has the largest delay compared with other systems. This is due to the diffuse transmission along with wide FOV at the receiver. Moreover, the delay spread of non-adaptive system LSMS is increased as the distance between the transmitter and receiver increases. The delay spread is almost independent of the distance between the transmitter-receiver in our hologram configurations, FAPA-Holograms and FDAPA-Hologram. This is due to beam angle adaptation technique where the proposed systems choose the hologram that has the best SNR. A significant reduction in the delay spread to 0.04 ns is achieved in our FAPA-Hologram system. Moreover, our delay adaptation method in FDAPA-Hologram reduces the delay spread of FDPA-Hologram by factor of 8. This improvement enhances the 3 dB channel bandwidth and increases the SNR at high transmission rates.

4.2. SNR

The SNR results of our proposed systems are shown in **Figure 7**. The proposed systems are tested under the influence of background noise and transmitter/receiver mobility. The proposed adaptive holograms are compared with conventional CDS and multi-beam angle diversity LSMS system to facilitate the results with previous work published in Refs. [9–11]. The results are shown when transmitters operate at 50 Mb/s. High data rates of 2.5 and 5 Gb/s will be also considered in the next section. The transmitter is located near the room corner while the receiver moves 1 m step along x = 1 m and x = 2 m. The LSMS system provides better results than CDS with wide FOV receiver. This is due to providing direct link though spots and using non-imaging angle diversity receiver. Although the improvement has been achieved, there is

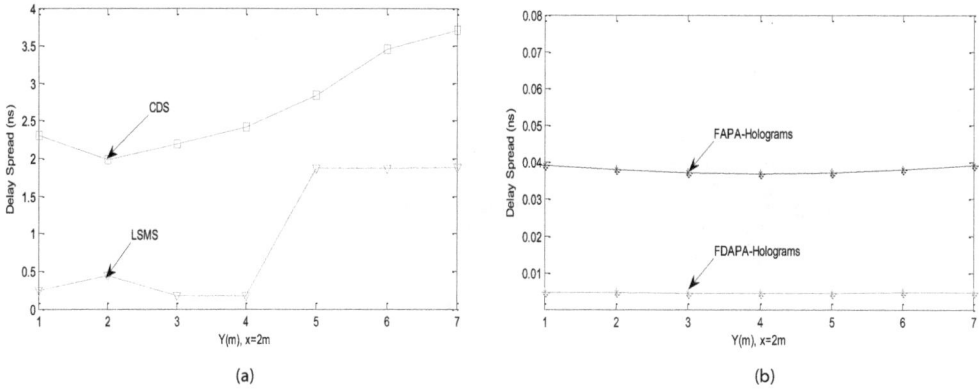

(a)   (b)

Figure 6. Delay spread of four configurations (a) CDS and LSMS, (b) FAPA-Holograms and APA-LSMS with angle diversity receiver, when the transmitter is placed at (1 m, 1 m, 1 m) and the receiver moves along $x = 2$ m line.

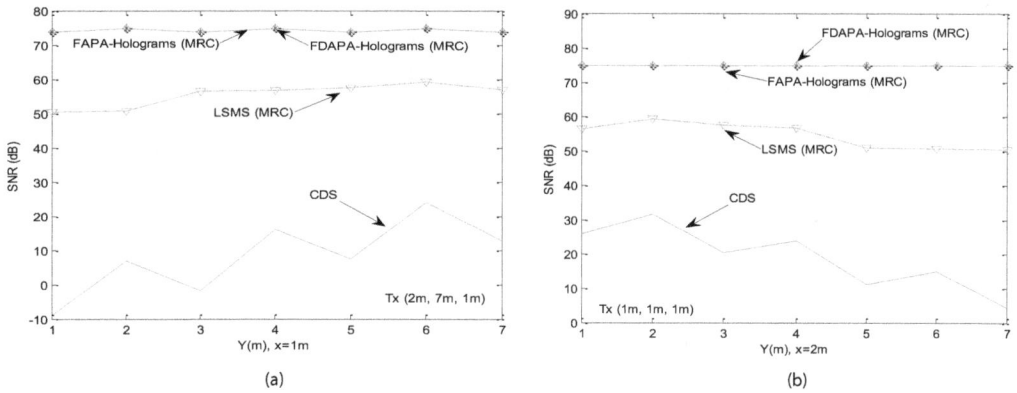

(a)   (b)

Figure 7. SNR of OW CDS, LSMS, FAPA-Holograms and FDAPA-Hologram at 50 Mbit/s, when the transmitter is located at (2 m, 7 m, 1 m) and (1 m, 1 m, 1 m) and the receiver mobiles along $x = 1$ m and $x = 2$ m lines.

degradation in the SNR results as LSMS transmitter move away from the receiver. For example, this is observed when the transmitter is moved towards the edge or the corner of the room at (2 m, 7 m and 1 m) and (1 m, 1 m and 1 m) while the receiver moves along $x = 1$ m and $x = 2$ m lines, respectively, as seen in **Figure 7(a)** and **(b)**. In order to overcome this significant reduction as well as improve the system performance, fast adaptive hologram (FAPA-Hologram and FDAPA-Hologram) systems are employed. Our proposed FAPA-Hologram achieves around 24 dB over the traditional LSMS system; see **Figure 7(b)**.

4.3. High data rate mobile IROW system

The results of the SNR achieved with our proposed FAPA-Hologram and FDAPA-Hologram allow the systems to reduce the total optical power transmit while operating at high data rates of 2.5 and 5 Gb/s. At high data rates, we considered a small photodetector area of 10 mm^2, in order to reduce the impact of high capacitance and improving receiver bandwidth. To the best of our knowledge, commercial photodetectors with a 10 mm^2 area and operating at high data rate are not common. However, researchers in Refs. [16, 22] have shown that the use of a small

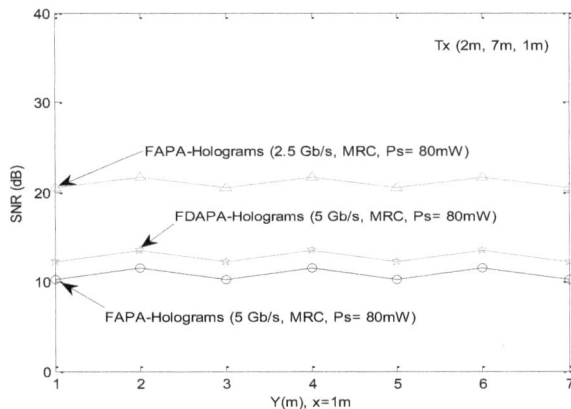

Figure 8. The SNR of proposed FDAPA-Holograms and FAPA-Hologram systems when operated at 2.5 and 5Gb/s, with a total transmit power of 80 mW.

detector area reduces the impact of high capacitance. In large commercial area, high speed detectors are starting to be used in free space optical systems. For example, Ref. [23] indicates that areas as large as 10 mm^2 and rise time as low as 10 ps are starting to emerge; however, the combination of large areas and fast response remains a challenge in photodetectors design. A 1 mW per beam is used to address the eye safety requirement in our proposed systems. Furthermore, we limit the power adaptation method where each beam cannot increase the power beyond 0.5 mW. The beams travel from the transmitter as group and spread until it reaches the object (reach to the ceiling in our case). At 10 cm distance each beam travel with different angle which can help in eye safety where the human eye cannot see more than one beam at time. Therefore, we propose that the transmitter is contained within a 10 cm deep enclosure to ensure that the human eye cannot be placed next to the transmitter. This can be achieved for example by placing the transmitter at the bottom of a laptop back cover (screen) and letting the beams emerge from the top of the back cover (screen). The proposed FAPA-Hologram achieves 20.5 dB at 2.5 Gb/s, see **Figure 8**. Moreover, at 5 Gb/s, the proposed FDAPA-Hologram offers around 2 dB SNR improvement over the FAPA-Hologram. This improvement is due to the use of beam delay adaptation method which helps to reduce the delay spread and improve 3dB channel bandwidth, hence increasing the SNR at the receiver. In terms of practical implementation, it should be noted that the diffraction limit has to be considered when considering commercially available spatial light modulators as the smallest pixel size that can be manufactured and operating wavelength to determine the maximum range of angles over which the beam can be steered [24]. This warrants further study.

5. Conclusions

The performance evaluation of the conventional CDS and non-adaptive LSMS can be significantly degraded by the transmitter/receiver mobility. In this chapter, the finite adaptive hologram using beam angle, power and delay adaptation techniques is introduced. All holograms are stored and pre-calculated in our adaptive system. A fast search algorithm based on the divide and conquer is reported. The fast algorithm reduced the time needed to generate hologram to select the best stored hologram in the system. The adaptive proposed system is combined with an angle diversity receiver. Nine beaches non-imaging angle diversity receiver was used to further improve the received optical signal in the presence of background noise and transmitter mobility. At 50 Mb/s, our simulation results show that the adaptive FAPA-Holograms system provides around 35 dB SNR gain over non-imaging diversity LSMS system. The proposed FAPA-Holograms system using beam angle and power adaptation methods is able to guide the spots nearer to the receiver location at each given transmitter-receiver location. The angles and powers associated with each hologram stored in the system are pre-calculated without adding any complexity at the transmitter to recomputed holograms. In order to further improve system performance and reduce the effect of multipath dispersion, beam delay adaptation method coupled was introduced when the system operated at high data rates. The proposed FDAPA-Holograms system was examined under eye safety regulations. A total transmit power of 80 mW was used. The SNR results of our 5 Gb/s FDAPA-Holograms system were around 13 dB, under the impact of mobility as well as background noise. A fast search algorithm based on the divide and conquer was proposed to reduce the time needed to generate a real hologram and select the best stored hologram in the system.

Acknowledgements

The authors extend their appreciation to the International Scientific Partnership Program ISPP at King Saud University for funding this research work through ISPP# 0093.

Author details

Mohammed T. Alresheedi[1]*, Ahmed Taha Hussein[2] and Jaafar M.H. Elmirghani[2]

*Address all correspondence to: malresheedi@ksu.edu.sa

1 Department of Electrical Engineering, King Saud University, Riyadh, Saudi Arabia

2 School of Electronic and Electrical Engineering, University of Leeds, Leeds, UK

References

[1] Alresheedi MT, Elmirghani JMH. Performance evaluation of 5 Gbit/s and 10 Gbit/s mobile optical wireless systems employing beam angle and power adaptation with diversity receivers. IEEE Journal on Selected Areas in Communications. 2011;29(6):1328-1340. DOI: 10.1109/JSAC.2011.110620

[2]   Alsaadi FE, Elmirghani JMH. High-speed spot diffusing mobile optical wireless system employing beam angle and power adaptation and imaging receivers. Journal of Lightwave Technology. 2010. **28**(16):2191-2206. DOI: 10.1109/JLT.2010.2042140

[3]   Alresheedi MT, Elmirghani JMH. 10 Gbit/s indoor optical wireless systems employing beam delay, angle and power adaptation methods with imaging detection. IEEE Journal of Lightwave Technology. 2012;**30**(12):1843-1856. DOI: 10.1109/JLT.2012.2190970

[4]   Alsaadi FE, Alhartomi MA, Elmirghani JMH. Fast and efficient adaptation algorithms for multi-gigabit wireless infrared systems. Journal of Lightwave Technology. 2013;**31**(23):3735-3751. DOI: 10.1109/JLT.2013.2286743

[5]   Hussein AT, Alresheedi MT, Elmirghani JMH. 20 Gbps mobile indoor visible light communication system employing beam steering and computer generated holograms. Journal of Lightwave Technology. 2015;**33**(24):5242-5260. DOI: 10.1109/JLT.2015.2495165

[6]   Wu L, Zhang Z, Liu H. Transmit beamforming for MIMO optical wireless communication systems. Wireless Personal Communications. 2014;**78**(1):615-628. DOI: 10.1007/s11277-014-1774-3

[7]   Kim S, Kim S. Wireless optical energy transmission using optical beamforming. Optical Engineering. 2013;2(4):205-210. DOI: 10.1117/1.OE.52.4.043205

[8]   Alresheedi MT, Elmirghani JMH. Hologram selection in realistic indoor optical wireless systems with angle diversity receivers. IEEE Journal on Optical and Networking (JOCN). 2015;**7**(8):797-813. DOI: 10.1364/JOCN.7.000797

[9]   Gfeller FR, Bapst UH. Wireless in-house data communication via diffuse infrared radiation. Proceedings of the IEEE. 1979;**67**(11):1474-1486.DOI: 10.1109/PROC.1979.11508

[10] Kahn JM, Barry JR. Wireless infrared communications. Proceedings of the IEEE. 1997;**85**(2):265-298

[11] AI-Ghamdi AG, Elmirghani JMH. Line strip spot-diffusing transmitter configuration for optical wireless systems influenced by background noise and multipath dispersion. IEEE Transactions on Communications. 2004;**52**(1):37-45. DOI: 10.1109/TCOMM.2003.822160

[12] Yun G, Kavehrad M. Spot diffusing and fly-eye receivers for indoor infrared wireless communications. In: Proceedings 1992 IEEE Conference. Selected Topics in Wireless Communications; Vancouver, BC, Canada; 1992. pp. 286-292 DOI: 10.1109/ICWC.1992.200761

[13] Fadlullah J, Kavehrad M. Indoor high-bandwidth optical wireless links for sensor networks. Journal of Lightwave Technology. 2010;**28**:3086-3094. DOI: 10.1109/JLT.2010.2076775

[14] Desurvire E. Erbium-Doped Fiber Amplifiers: Principles and Applications. New York; A Wiley-Interscience Publication; 1994, ISBN: 978-0-471-58977-8

[15] Moreira A, Valadas R, Duarte AO. Optical interference produced by artificial light. Wireless Networks. 1997;**3**(2):131-140. DOI: 10.1023/A:1019140814049

[16] Leskovar B. Optical receivers for wide band data transmission systems. IEEE Transactions on Nuclear Science. 1989;**36**(1):787-793. DOI: 10.1109/23.34550

[17] Alresheedi MT, Elmirghani JMH. High-speed indoor optical wireless links employing fast angle and power adaptive computer-generated holograms with imaging receivers. IEEE Transactions on Communications. 2016;**64**:1699-1710. DOI: 10.1109/TCOMM.2016.2519415

[18] Jivkova S, Kavehard M. Multispot diffusing configuration for wireless infrared access. IEEE Transactions on Communications. 2000;**48**(6):970-978. DOI: 10.1109/26.848558

[19] Seldowitz MA, Allebach JP, Sweeney DE. Synthesis of digital holograms by direct binary search. Applied Optics. 1987;**26**:2788-2798. DOI: org/10.1364/AO.26.002788

[20] Ramirez FA. Holography—Different Fields of Application. InTech. 2011. p. 158. ISBN 978-953-307-635-5. DOI: 10.5772/750

[21] Carnevali P, Coletti L, Patarnello S. Image processing by simulated annealing. IBM Journal of Research and Development. 1985;**29**(6):569-579

[22] Elmirghani JMH, Chan HH, Cryan RA. Sensitivity evaluation of optical wireless PPM systems utilizing PIN-BJT receivers. IEE Proceedings—Optoelectronics. 1996;**143**(6):355-359. 10.1049/ip-opt:19960880

[23] Available from: http://www.cablefree.co.uk/faqs/pin-photodiodes-faqs/. [Accessed 20 June 2015]

[24] McManamon PF et al. A review of phased array steering for narrow-band electrooptical systems. Proceedings of the IEEE. 2009;**97**(6):1078-1096. DOI: 10.1109/JPROC.2009.2017218

7

Selective Mode Excitation: A Technique for Advanced Fiber Systems

Dmitry V. Svistunov

Additional information is available at the end of the chapter

Abstract

Actual problems arising in development of fiber optical systems are increasing the information capacity and enhancing data security. Different encoding methods and data compression techniques have been developed to meet these requirements. The presented materials emphasize advantages of application of selective mode excitation in fiber systems that lead to both increasing the system information capacity and enhancing data security. Three-stage hierarchical scheme of data compression [where time division multiplexing (TDM) method is the content of the first stage, the second stage utilizes wavelength division multiplexing, and mode division multiplexing (MDM) is applied at the last stage] is discussed. Furthermore, it is highlighted that selective mode excitation is able to embarrass eavesdropping. It is shown that just application of the mentioned technique allows enhancing data security, while designing of special system architectures provides additional increase of data protection level. The examples of such system schemes are presented. Thus, application of selective mode excitation could improve the performances of fiber systems significantly, at least the ones such as short- and middle-haul communication lines and local area networks (LANs).

Keywords: selective mode excitation, information capacity, data compression, mode division multiplexing, data security

1. Introduction

The demands on increasing the information capacity of fiber systems and also on higher secrecy are declared the whole time since optical communication came to the practical stage of commercial systems, and intensive investigations have been performed in order to meet the mentioned requirements.

Regarding improvement of data security, conducted research resulted in a number of developed encoding algorithms and the specific methods as quantum cryptography that can embarrass decoding of eavesdropped information significantly.

As for information capacity, substantial progress has been achieved by employment of data compression methods. Various developed techniques can be combined in several groups by operating functions. The first group—time division multiplexing (TDM) method—had been developed for wire and radio communications and then was implemented also in optical fiber communication on appearance of fiber systems. That method bases on managing the launch of different input signal streams in turn to the same trunk line. In TDM method, time domains (frames) are defined, and each one is divided into several time slots which are filled with the data blocks of transmitted signal streams. While former applications of the method use managing the electric signals (ETDM), the latter application employs also extra techniques performing switching of optical signal streams (OTDM). Whereas a number of variants of TDM method have been developed, further works on TDM techniques and algorithms are still performed actively.

The next group—wavelength division multiplexing (WDM) method, was implemented later upon achieving appropriate performances and lower costs of required fiber system units. The method uses single optical fiber for parallel independent transmission of various light signals of different wavelengths. Performed standardization resulted in defining several wavelength windows and also wavelength channel spacing into the window. Maximal system information capacity can be reached under employment of dense WDM (DWDM) technique capable of providing minimal channel spacing defined by the standardized spectral grid for 12.5 and 25 GHz separations. Presently, WDM techniques have wide application in current fiber systems.

Evident way to get further increase of system information capacity is cooperative employment of different data compression techniques in advanced systems. Conducted research resulted in application of WDM method in combinations with ETDM or OTDM techniques mainly in single-mode fiber (SMF) systems and also in short-haul multimode systems. Developments of the latter systems have led to elaborating a promising approach to significant increasing the bandwidth-distant products of those systems: selective excitation of sole mode within multimode fiber that allows obtaining a regime of quasi-single-mode data transmission. The way to get further substantial rise of information capacity of those systems is employment of the specific data compression method named mode division multiplexing (MDM). Selective mode excitation is a crucial technique of that method, and the interest to research on that theme rises year-by-year.

Here, our subject is to consider possibility of complex employment of noted data compression methods. Furthermore, we shall show that application of selective mode excitation in the form of MDM technique allows enhancing data security due to just application of the method, while designing of special system architectures provides additional increase of data protection level.

2. Hierarchic scheme of data compression

The essence of MDM technique is the use of different fiber modes or mode groups of the same multimode fiber as independent information channels. High level of intermodal coupling in former commercial fibers delayed implementation of this method for a long time. However, achieved progress in fabrication technology reduced drastically the intermodal coupling (as for example, the average level of intermodal coupling coefficient as 0.007 km^{-1} for the case of neighbor-order modes had been reached already in the past decade [1]). Due to that progress, propagation of sole modes has been obtained in multimode commercial fibers, and further investigations concentrated just on realization of MDM technique. First of all, different selective mode couplers had been developed, and designing of experimental systems started. One group of developments was directed to application of MDM to coherent systems in order to reach long-haul trunk lines. However, these systems are complicated and high cost. On the other hand, a number of short- and middle-haul current fiber systems are based on multimode fibers. Bearing in mind, the possibility to reconstruct those systems by application of MDM technique; further, we consider low-cost direct detection systems that are quite appropriate for mentioned trunk lengths. By present, experimental MDM-based systems are developed demonstrating the available distances being sufficient for fiber systems such as metropolitan or toll lines and local area networks (LANs) (for example, [2, 3]). As feasibility of MDM systems is proven practically, the goal of the next stage is complex combination of this technique with other data compression methods.

Cascade scheme of data compression is shown in **Figure 1**. The scheme is capable of providing hierarchical three-level compression of data streams. Primary stage of data processing in the scheme bases on TDM techniques issuing the set of compressed data streams each intended for separate spectral channel. The second stage uses WDM method that joins a number of spectral channels in the same fiber. Application of MDM technique is the content of the third stage. At this stage, each compressed optical signal resulting from previous stages launches certain separate mode of multimode trunk fiber. The scheme has a symmetric structure, and reverse sequence of those stages at the receiver part provides recovering the initial data streams.

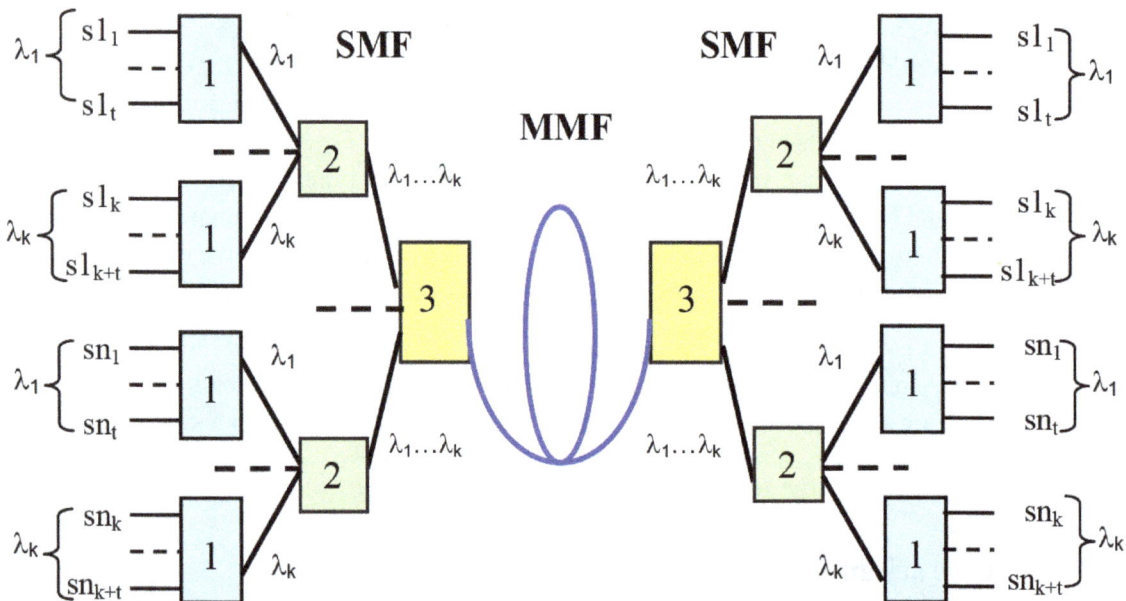

Figure 1. Hierarchical cascade scheme of data compression in multimode fiber systems. The blocks denote the devices of following types: 1 is TDM; 2 is WDM; 3 is MDM.

Multimode fiber (MMF) is utilized as a trunk fiber, while single-mode fibers (SMFs) required for optimal operating of mode multiplexers are employed in other optical interconnections into the scheme.

Every mode channel in multimode fiber can be considered as an analogue to separate single-mode fiber for which cooperative implementation of TDM and WDM techniques is already proven. So, the question concerning the proposed scheme is whether mutual compatibility of developed WDM and MDM operating units is limited restricting common application of the techniques.

Feasibility of the considered scheme depends on ability of mode multiplexers/demultiplexers to operate effectively in some spectral range. Those units are based on selective mode couplers which excite/detect independently fiber modes of different orders. A set of such elements basing on different operating principles have been developed (for example, [4–16]). In most cases, selective excitation of fiber modes is performed with external units that are built as waveguide (fiber or integrated optical) elements or bulk devices based on traditional optical elements. Optical matching of those units with the fibers is performed by traditional or GRIN lenses, and also butt joining is used for waveguide elements. Waveguide selective mode couplers require special consideration. Particularly, the attention should be paid to selective units whose operating principles lead to dependence of directions of intrinsic light propagation on the light wavelength. Waveguide mode multiplexer/demultiplexer described in Refs. [17, 18] is just the unit of this kind, and experimental samples of the unit were examined to determine the noted dependence.

Scheme of the considered unit is shown in **Figure 2**. Planar selective element operates as known input/output prism coupler and matches optically each mode of the multimode channel waveguide with the corresponding waveguide beam in single-mode planar region joined with the set of single-mode channel guides by horn transition structures. The angle α between the axis of certain planar beam and multimode channel guide depends on the ratio of mode indices of that beam and corresponding channel mode. As planar selective coupler is the key element of the unit, experimental samples of this element have been studied by measurement of directional diagrams of planar beams [19] in order to estimate whether those beams become superimposed if the channel waveguide is excited with light of certain spectral bandwidth.

Reliable accurate measurement of beam directivity diagram into planar waveguide could be difficult; therefore, extra prism input/output coupler having cylindrical base was applied, and directional diagrams of output beams were measured. **Figure 3** presents the results of examinations of the trial sample. Numerical angular values at the diagram axis correspond to goniometric readouts under arbitrary benchmark position. Obtained patterns are typical for all couplers of this kind. Every peak corresponds to the separate planar beam associated with the certain channel mode.

Study of the sample intended for operating at light wavelength as 1.3 μm and simulation of coupler excitation with light of whole standard O bandwidth showed that angular widths of three neighbor planar beams become as 22, 20, and 16 arcmin, while the angles between the

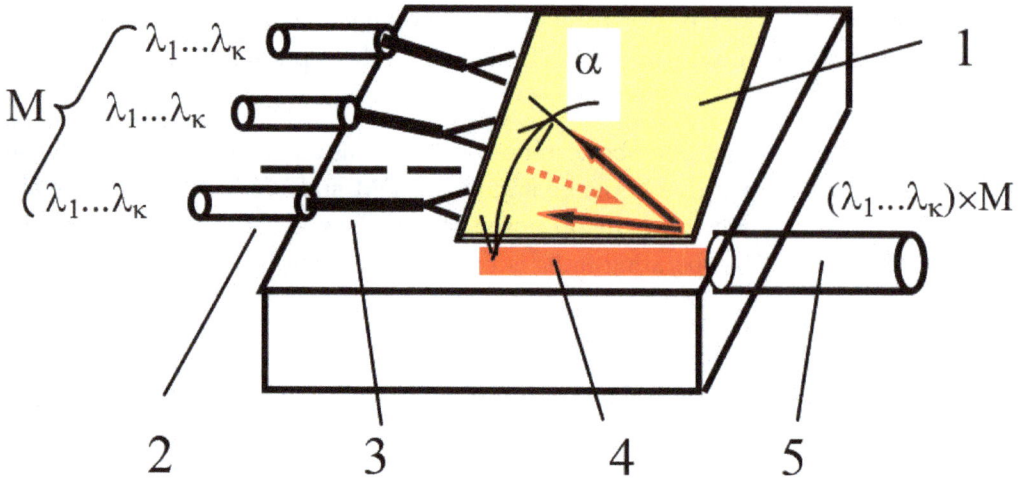

Figure 2. Scheme of waveguide unit for selective mode excitation/detection. 1 is single-mode planar section; 2 is SMF; 3 is single-mode channel waveguides with horn transitions; 4 is multimode channel waveguide; 5 is MMF.

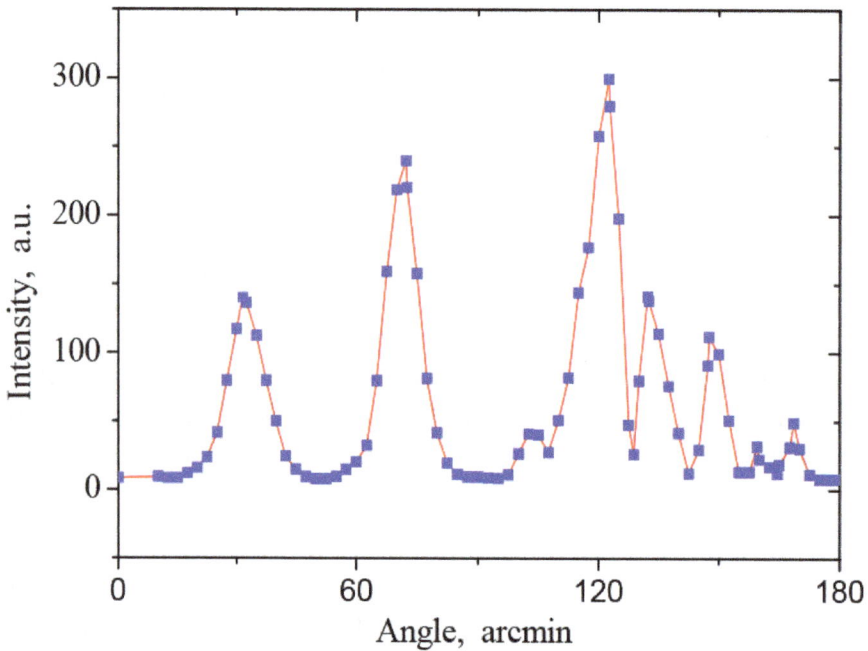

Figure 3. Angular distribution of light beams associated with the set of excited channel modes.

axes of adjacent beams are 64 and 33 arcmin [19]. These results indicate that the beams are spatially separated and can be directed to corresponding horn structures without appearance of significant crosstalk. So, the experimental results confirm that this mode multiplexer is capable of operating with optical signals compressed by WDM.

Considering that the mentioned mode multiplexer is assumed to be among the ones having perceptible spectral sensitivity, we can note compatibility of MDM and WDM units and conclude that the obtained results prove feasibility of cooperative application of MDM and WDM techniques. Thus, the mentioned hierarchic scheme of data compression can be realized.

3. Enhancement of data security

Nowadays, protection of transmitted data from eavesdropping is considered as a crucial problem, and intensive investigations are aimed to development of new approaches leading to effective solutions. One branch of developments is counteracting to eavesdropping under intrusion to the trunk fiber line. In order to understand the challenge, let us estimate what techniques could be applied for illegal data reading from the fiber bearing in mind that the essence of eavesdropping means that only minor part or optical power is spit off, otherwise alarm signals are issued by the control system blocks. Therefore, partial mode extraction to lateral direction is to be performed in some intrusion procedure, while the main part of optical power propagates further into the fiber. Some possible variants of the mentioned impact on the fiber are shown in **Figure 4**.

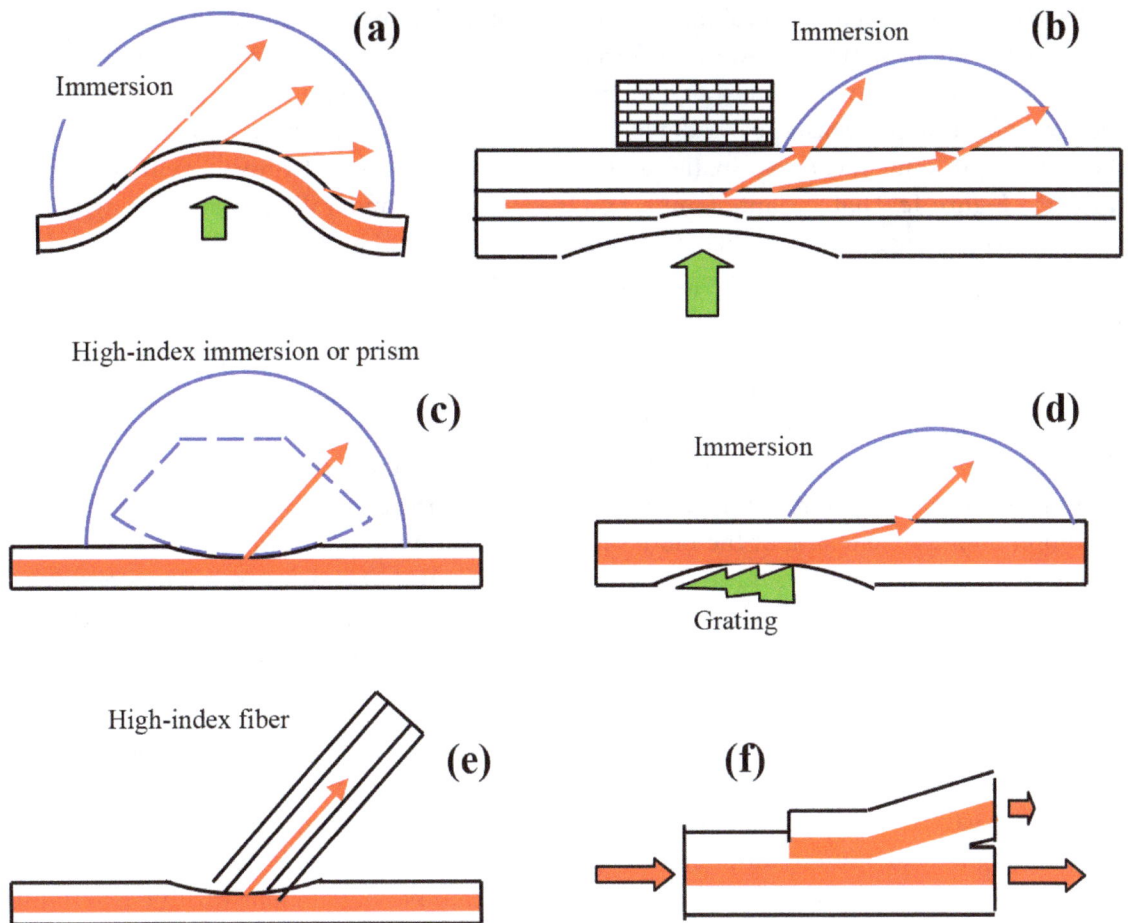

Figure 4. Variants of impact on the fiber resulting in partial extraction of the optical power from the fiber modes: fiber bending (a), local pressing (b), prism-like (c), and diffractive (d) couplers, higher-index fiber (e) and as all-fiber coupler (f).

Rather simple way to perform mentioned eavesdropping is to bend the fiber. Then, bent losses occur because the part of the tail of mode field cross distribution is cut off due to fiber bending and launches separate light beam (namely cladding mode) propagating into the fiber cladding. Those cladding modes form light beams in the adjacent medium due to scattering on cladding imperfections or refraction if the medium is higher index. Mode leakage increases as the bend radius decreases. As the leakage reasons remain the same along the bend, mode transformation occurs at every point of bent fiber, and the whole radiated beam is high-divergent as it is shown in **Figure 4a**. Similar technique uses microbendings (obtained with pressed special tools as it is performed in some pressure sensors) and/or deviations of the core radius caused by local impact to the fiber (see **Figure 4b**). In those cases, a part of fiber mode power is also split off being proportional to the pressure strength and forms cladding beams that can be outcoupled with immersion drop and registered.

The noted splitting techniques do not require preliminary treatment of fiber inner cladding, only outer protective jackets must be removed. The next variants exploit optical tunneling effect so they need affecting the cladding adjacent to fiber core. Appropriate gap between the applied tool and the fiber core can be obtained by partial remove of cladding layer, for example with etching by dropped chemical reagent or pressing under heating. Upon preparing the gap, higher index immersion is placed to the formed contact region or a bulk prism is pressed, and one obtain splitting tool operating as known longitudinal input/output prism couplers where the level of extracted optical power depends on combination of refractive indices and also on the gap length and thickness. Spatial extracted beams are formed directly with such tool drawn in **Figure 4c**. That variant of impact on the trunk fiber under illegal data reading seems more difficult in use but more dangerous with relation to the mentioned manner of data protection because longitudinal couplers can spatially separate output light beams associated with the modes of different orders. Similar tool can use external diffraction grating placed to the contact area with grating strokes normal to the fiber axis as shown in **Figure 4d**. The effective regime of that diffractive tool is realized when a long-period grating transfers the mode power to cladding modes. A higher index extra fiber can also replace the bulk prism as shown in **Figure 4e**. End face of the extra fiber is placed to the contact region, while its axis forms a certain angle with the trunk fiber. If inclination of the extra fiber provides meeting the condition of phase matching of the modes in both fibers, mode launching occurs in the extra fiber due to optical tunneling through the cladding gap. And of course, optical tunneling can be performed by exciting the extra fiber located along the trunk fiber as in all-fiber directional couplers (see **Figure 4f**). Treatment of fiber claddings is also necessary in this case in order to provide appropriate coupler structure.

It is shown in Ref. [20] that selective mode excitation has a valuable advantageous feature: being simply employed in the form of MDM technique, selective mode launching can resist seriously to eavesdropping performed with noted intrusion techniques. Indeed, fiber bending results in light irradiation in all points along the formed fiber curve. So, every fiber mode is associated with the certain highly divergent output beam having angular width equal to the central angle of the fiber bend arc. Evidently, these output beams superimpose when different fiber modes propagate simultaneously. Then, the information in the registered eavesdropped signal becomes mixed.

Similar situation occurs in cases of microbendings and core radius variations where every core mode excites a group of cladding modes of adjacent orders, and information carried by neighbor fiber modes becomes mixed just at this stage because the same cladding mode groups appear from different fiber modes. When these cladding mode groups are split off with placed immersion or fiber macrobending, or if scattering at cladding imperfections is registered, a degree of information mixing rises.

As for the scheme employing optical tunneling to high-index immersion drop or pressed bulk prism, consideration performed in Ref. [20] for commercial MMF showed that one can also expect a superposition of output space beams accompanied with information mixing if neighbor fiber mode groups are used for data transmission according to the MDM scheme. Similar assumption concerns the external grating tool because of likeness between performances of space beams formed by prism and grating directional couplers operating as longitudinal directional input/output units. Mixed information can also be expected in case of application of extra higher index fiber. Difference between mode spectra of trunk and mentioned fibers does not allow choosing the fiber inclination angle that could provide meeting the matching conditions for the set of modes simultaneously, and parasitic mode launching occurs in the extra fiber leading to cross-talk and mixing the data read from different mode channels.

The last noted splitting scheme is tunneling to the extra fiber along with the fiber length. In this case, particularly if both fibers are of the same type, simple application of MDM technique can fail in data protecting, and special system schemes should be built to counteract eavesdropping.

First of all, the fiber system must be capable of detecting the intrusion attempt and issue the alarm signal. This important feature is already realized in some fiber systems, for example, in the experimental system described in Ref. [21]. That system exploits selectively excited fundamental mode of graded-index MMF for high-bit-rate data transmission, and that mode is launched with external strip waveguide by on-axis butt joint. Another strip waveguide of that external chip excite several higher order fiber modes by off-axis butt joint, and these modes form together the monitor channel. Due to the character of mode field cross distributions, higher order modes are more sensitive to fiber bending and other variants of impact on the fiber than the fundamental mode. Therefore, monitor signal decreases immediately at the beginning of intrusion, and the control block issues the alarm signal timely. Because of the used mode coupler, only selectively excited lowest order fiber mode can be exploited for high-bit-rate data transmission in the built system. However, application of the mode coupler of another type could enable employing the MDM technique and providing parallel transmission of additional data streams and/or random signals, while the monitor channel still controls intrusion attempts. Besides the increase of the system information capacity that could allow achieving additional enhancement of data security in the system due to the reasons noted above. Data transmission part of the system of that kind is shown schematically in **Figure 5**.

Of course, the presented scheme can be reorganized to bidirectional transmission by adding the appropriate system blocks. The features to be remained in the scheme revisions are application of MDM technique for low-order modes and building the monitor channel for higher order modes.

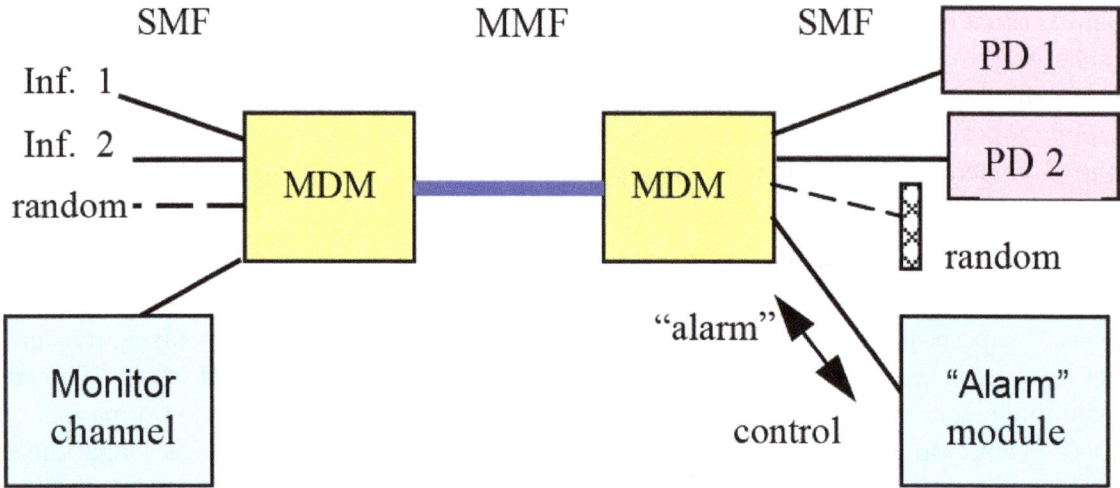

Figure 5. Scheme of transmission part of MDM system with monitor/alarm line. PD is photodetector.

Another possible system scheme could use a specific technique of data transmission. As in TDM technique, the data stream can be divided in time periods each containing data blocks. Each block of the same data period is directed individually to the certain optical delay line. The kit of delay lines provides time synchronization of blocks at the line outputs. Then, the optical signal from each line is transmitted over the certain mode of the trunk fiber. So, we obtain simultaneous transmission of originally sequential data blocks. At the end of trunk fiber, outcoupled mode signals pass via the output kit of delay lines performing reversal time shifts of data blocks and thus providing restoration of their original sequence. **Figure 6** shows the

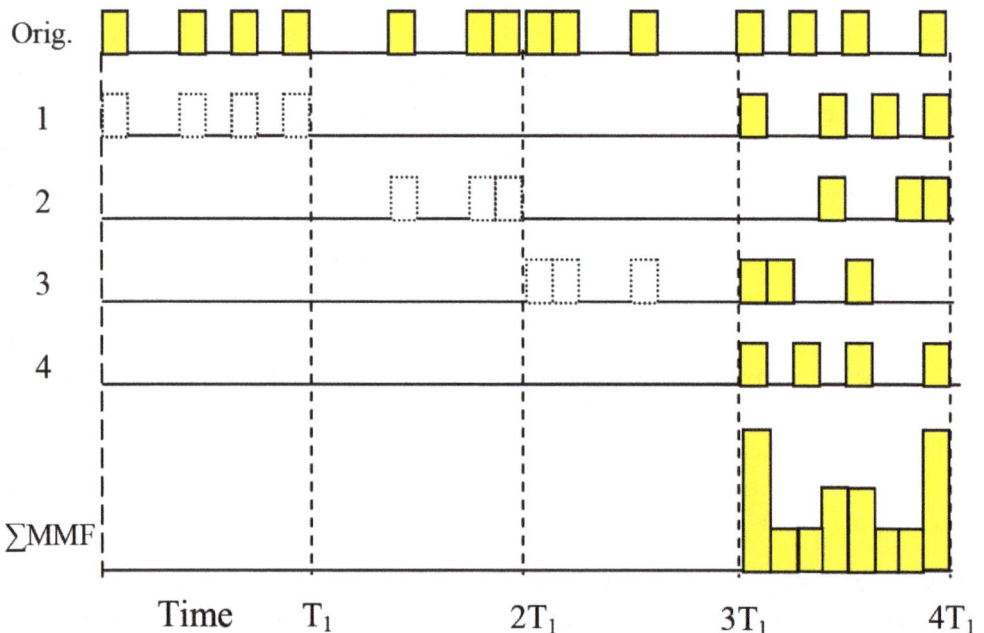

Figure 6. Time distribution of data bit blocks by the modes of the trunk fiber.

example of distribution of data stream by four trunk fiber modes. Here, one data time period is demonstrated as a bit stream in NRZ format.

In order to simplify the diagram, the period length is extremely shortened, and only 8 bits are included into each data block. T_1 denotes the time duration of one data block, while the numbers at the lines correspond to different operating trunk fiber modes. Four fiber modes having different principal mode numbers can be really excited independently (it follows, for example, from **Figure 3** where experimental characteristics of the selective mode coupler are presented). One can see that individually delayed data blocks come to the trunk fiber input simultaneously as they are shown within the time range $3T_1...4T_1$.

The signal denoted as $\sum$MMF represents the superposed data resulted from eavesdropping by fiber bending or similar nonselective technique. However, even if different fiber modes are distinguished in some intrusion procedure, decoding of the read data is massively impeded because the eavesdropper must determine right sequence of registered data blocks.

Figure 7 presents the scheme capable of performing the proposed technique for the noted case of four mode channels into the trunk fiber. The circles at SMF lines denote optical delay lines which can be represented by fiber loops of certain lengths. The extra line is reserved for random signal that could be transmitted over fifth selectively excited fiber mode filling the "empty" time windows $0...3T_1$ appeared in the trunk fiber.

The data distributor can be built as the integrated optical chip combining known wave-guide switches as it is shown in **Figure 8** together with the time diagram of tuning electric signals. For simplicity of the pattern, only one of every pair of switch electrodes is plotted.

Distributor provides division of data period to data blocks and individual directing each block to corresponding output channel. The input signal is the original data bit sequence. Every switch is activated with tuning electric impulse whose duration equals to the chosen duration of the data block. Consecutive turning on the switches performs directing the consecutive data blocks to different outputs of the distributor, and time distribution of those data blocks by device outputs is depicted in **Figure 6** as dotted bit images. According to the scheme shown in **Figure 7**, each output of the distributor is joined with the corresponding SMF delay line. Performing individual delaying, those lines synchronize data bit blocks at the input of the MDM unit as it is seen in **Figure 6** in the time range $3T_1 ... 4T_1$. The MDM unit matches each

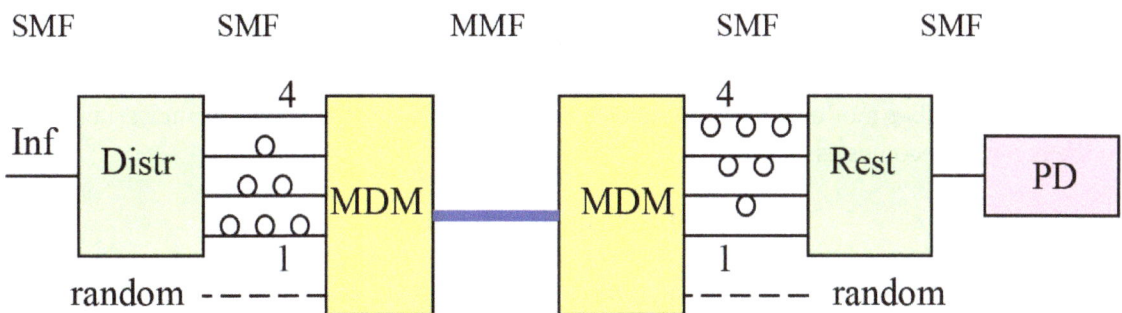

Figure 7. Scheme of transmission part of MDM system performing distribution of information data sequence by trunk modes.

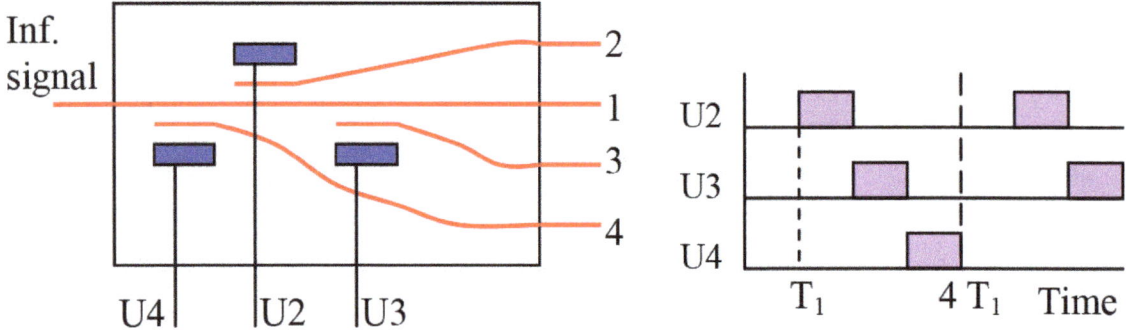

Figure 8. Scheme of data distributor/restorer together with the diagram of tuning voltage.

delay line with the certain trunk mode, and shown distribution of delayed data bit blocks represents time distribution of light pulses by different trunk fiber mode channels at the entire end of trunk line.

Let us evaluate the required lengths of SMF delay loops at the transmitter for the scheme with four mode information channels into the trunk fiber. The length of the certain SMF at the input of the MDM unit can be written as

$$L_i^{inp} = L_4^{inp} + (4 - i) \cdot \delta L_i \tag{1}$$

where i is the number of SMF (let us define that the signal from the first SMF is to be coupled to the trunk fiber of lower order); $\delta L_i = cT_1/N_{ms} = cK_b/fN_{ms}$ is the difference between the lengths of two neighbor SMFs; c is the light speed in vacuum; K_b is the amount of bits in the data block; N_{ms} is the mode index of SMF; f is the data stream frequency. L_4^{inp} corresponds to the SMF line whose signals are not delayed specially, and this length is minimal among four SMFs being defined by constructive reasons only. For $K_b = 8, f = 800$ MHz, $N_{ms} \approx 1.47$ (for $\lambda = 1.3$ μm), we obtain from Eq. (1) $\delta L_i \approx 2$ m. Although the used data block length is too small for practice, this rough evaluation resulted in the reasonable meaning of δL_i which could be still appropriate for block lengths of more than ten times longer. The frequency of tuning electrical signal applied to data distributor is $f_{el} = f/K_bK_{ic}$, where K_{ic} is the amount of mode informational channels. In the considered example $K_{ic} = 4$, and then $f_{el} = 25$ MHz that is also the reasonable value which will decrease inversely to the rise of data block length.

Regarding SMF loops in the receiver system part, corresponding SMF lengths should be in reversal relation in order to compensate data block delays performed in the transmitter and to recover original data sequence. Furthermore, here the differences in propagation times of different trunk fiber modes must be considered additionally. So, the length of the certain SMF in the receiver becomes as

$$L_i^{out} = L_1^{out} + (i - 1) \cdot \delta L_i + \delta L_i^{dmd} \tag{2}$$

where L_1^{out} is the length of shortest output SMF whose optical signal is not delayed specially in processing at the receiver. δL_i^{dmd} is the increment of SMF length defined by intermodal dispersion in the trunk MMF. That increment can be determined as

$$\delta L_i^{dmd} = c\delta t_i/N_{ms} \tag{3}$$

where δt_i is the differential delay time of the corresponding trunk mode. Depending on the types of employed trunk fibers, this delay time can vary in a wide range having significantly larger values in step-index fibers than the ones in graded-index fibers.

Maximal delay time caused by intermodal dispersion can be determined from optical pulse broadening when the whole mode spectrum is launched. Then, the specific delay related to the trunk fiber length is

$$\delta t_{sf}/L_{tr} = n_1\Delta/c \tag{4}$$

where $\Delta = (n_1 - n_2)/n_1$, L_{tr} is the fiber length; n_1 is the maximal index in the core cross-section (uniform value in a step-index fiber); n_2 is the cladding index [22]. Substituting to Eq. (4), the values $n_1 = 1.48$ and $\Delta = 0.01$ for the case of step-index fiber, we obtain $\delta t_{sf}/L_{tr} \approx 50$ ns/km that seems too big despite delay times for several low-order neighbor modes are evidently less. The first scheme with control/alarm line and independent data channels seems more appropriate for step-index fiber systems.

Unlike that fiber type, graded-index fibers demonstrate much less intermodal delay due to the nature of mode propagation via the fiber with radial refractive index cross distribution. The intermodal dispersion reaches minimal values when the index cross profile is close to a parabolic form. Optimization of profile decrement parameters for the chosen wavelength allows reaching the values of maximal delay less than 100 ps/km for $\Delta = 0.01$, but then, the real index profile must correspond precisely to the theoretical one. In practice, the value as about $\delta t_{gf}/L_{tr} \approx 1$ ns/km can be chosen in approximate estimations as a maximal delay time in commercial graded-index fibers. Now, let us evaluate a differential delay time in such fibers. Considering the performances of MMFs having parabolic-like index profile in circular cross-section, one finds equidistant dependence of mode propagation constant β in terms of β^2 on the principal number M (the value characterizing separate mode groups each containing the set of degenerate fiber modes; just those groups can be exploited as information channels), and group mode indices can also be assumed approximately equidistant. Total amount of mode groups is evaluated as

$$M_0 = kn_1a(\Delta/2)^{1/2} \tag{5}$$

where a is the core radius; $k = 2\pi/\lambda$; λ is the light wavelength in vacuum [23]. Equidistant mode indices mean equidistant distribution of mode group velocities that in turn lead to equal increments of delay time counted between adjacent mode groups. Therefore, for our case of four mode channels, we obtain following term that determines the specific delay time with respect to propagation of the fastest mode:

$$\delta t_i/L_{tr} \approx \delta t_{gf}(4 - i)/M_0L_{tr} \tag{6}$$

Then, the desired SMF length increment for the case of graded-index trunk fiber is

$$\delta L_i^{dmd} \approx c(i-1) \cdot \delta t_{gf}/M_0 N_{ms} \tag{7}$$

(the higher-order mode propagates faster and must be delayed in more degree). Substituting to Eqs. (5) and (7), the values $a = 25$ μm, $\Delta = 0.01$, $n_1 = 1.486$, $N_{ms} = 1.47$, $\lambda = 1.33$ μm, we obtain (for trunk line as $L_{tr} = 1$ km) the following set of SMF elongations δL_i^{dmd}: 0, 16.5, 33, and 49.5 mm placed here in priority according to the rise of the channel number. For longer trunk lines, one must multiply these values by the factor of the distance in km. Regarding the reasonable trunk distances (no more than 100 km), the SMF lengths required for recovering the original data sequence evidently dominate over these additional elongations in case of graded-index trunk fiber. Additional precise tuning of data block time shifting can be achieved if active fibers with longitudinal electrodes exploiting an electrooptical effect are employed as SMFs in the receiver. Appropriate levels of uniform voltage biases are to be determined in set-up procedures when installing the system.

The functions of restorer and distributor are reversal, and their constructions are symmetric. The restorer has four inputs (in our four-channel example) and one output. Input signals are the data blocks issued from delay lines at the receiver part where they have got individual time shifts that provide right time sequence of these blocks, whereas they propagate in parallel lines yet. These four fiber lines are joined with restorer inputs, and time distribution of the input blocks corresponds to location of dotted data bit images in **Figure 6**. Sequential turning on the restorer electrooptical switches by the tuning voltages (whose diagram is shown in **Figure 8**) enables transferring light pulses of all data blocks to one waveguide (one can imagine reversal light tracing in the scheme plotted in **Figure 8**). Then, the output restorer signal becomes as the original data bit sequence shown in **Figure 6**.

Simple filling the "empty" time windows by random signals in the fifth trunk mode can be replaced with different random bit streams (whose frequencies equal to the data stream frequency) each filling the certain local "empty" window in different trunk fiber mode. Then, the scheme transforms into the variant shown in **Figure 9**.

Every random sequence is divided to random bit blocks by the same manner as described above for the data sequence, and SMF loops are also employed to synchronize random blocks at the input of the MDM unit (those loops are plotted as corresponding circles at the scheme lines). That random blocks are directed from distributors to multiplexing connectors where the channel signals are formed containing each a continuous sequence of one data block and three different random blocks as it is depicted in **Figure 10**.

These signals are multiplexed by the MDM unit providing here filling each trunk mode channel with continuous bit sequences. That circumstance impedes significantly separation of random and data blocks in eavesdropped signal even if trunk modes are distinguished in intrusion procedure.

Upon passage the distance and executing a mode demultiplexing, bit sequences come to the extractor where random signals are omitted, while extracted data blocks are directed further to the restorer and processed there as described above for recovering the original data sequence. **Figure 11** demonstrates the schemes of the multiplexing connector and the extractor together with the diagram of tuning voltage. The devices are built on widespread elements—known 3dB-splitters and electrooptical switches.

Extractor includes four (for our example) switches each having one output joined with further system scheme line, while another output is empty. The signals that come to every extractor input are the sequences of data and random blocks shown in **Figure 10**. Corresponding blocks come to parallel extractor inputs simultaneously. Therefore, extractor switches are tuned with the same voltage signal (for simplicity, only one electrode of every pair is plotted at all switches). Activating the switches during the time range $T_1 \ldots T_4$ every data period, we direct

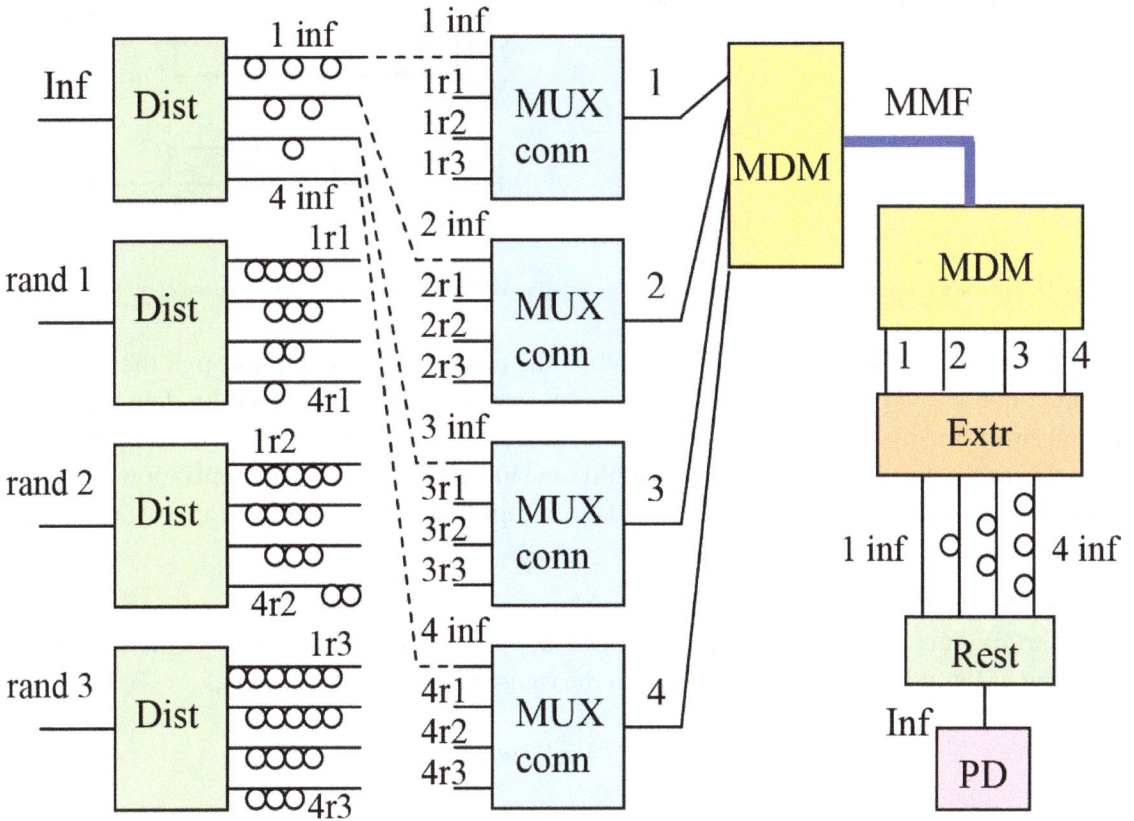

Figure 9. Scheme of transmission part of MDM system performing distribution of information data sequence by trunk modes and filling the empty time windows with random signals.

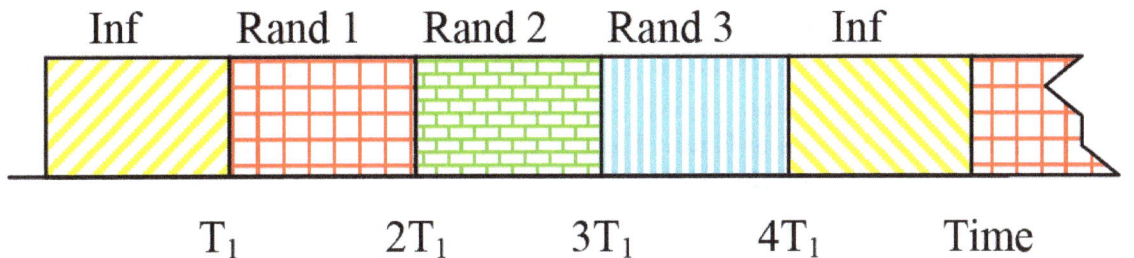

Figure 10. Sequence of data and random bit blocks in each mode of the trunk fiber.

Extractor

Multiplexing connector

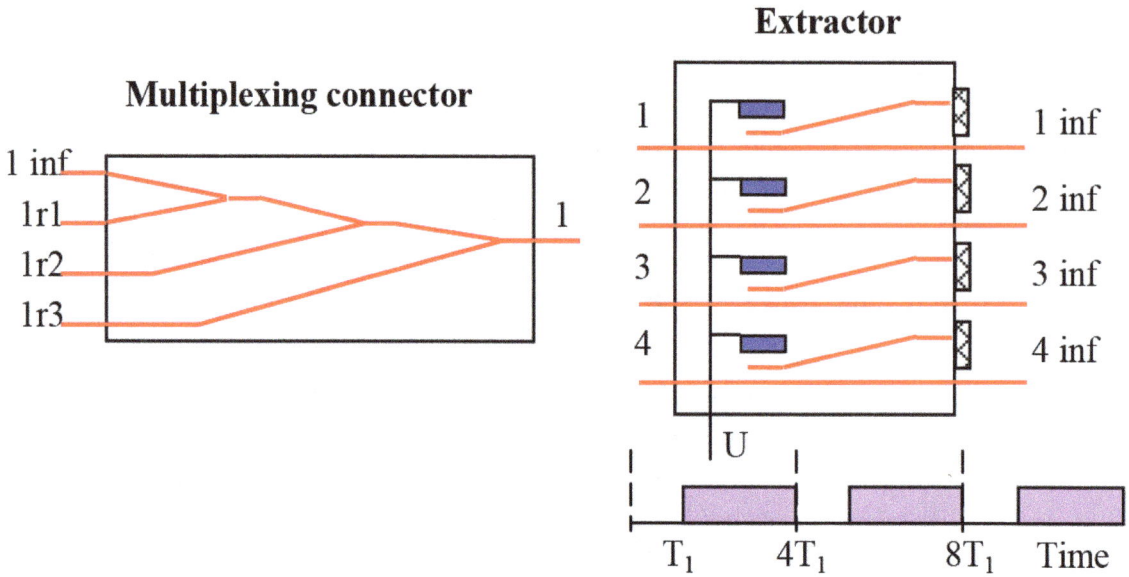

Figure 11. Schemes of the connector and the extractor depicted in **Figure 9** together with the diagram of tuning voltage.

the random bit blocks to empty outputs, while the data bit blocks pass through the extractor during times $0 \ldots T_1$. Then, the signal at the extractor output is represented by data bit blocks issued simultaneously from four parallel channels, and their distribution by those channels is the same as the one plotted in **Figure 6** within time limits $3T_1 \ldots 4T_1$. Synchronization of tuning voltages with data blocks can be executed with repeated timely service signals transmitted instead of data bit blocks.

To provide simultaneous coming of the data blocks to extractor inputs, differential mode delay in the trunk fiber must be considered preliminary. Therefore, in this scheme, the lengths of delaying SMFs at the input of MDM unit in the transmitter are determined as

$$l_i^{inp} = l_4^{inp} + (4 - i) \cdot \delta L_i + \delta L_i^{dmd} \tag{8}$$

where l_4^{inp} is the length of shortest SMF line, and other values are determined as above using Eqs. (1) and (7). The lengths of SMF delay lines that follow the extractor become as

$$l_i^{out} = l_1^{out} + (i - 1) \cdot \delta L_i \tag{9}$$

where l_1^{out} is the length of shortest SMF corresponding to the lowest order operating trunk mode. Elongation of delaying SMFs for random bit blocks is conducted by the same manner except that the required amount of SMF sections (each of δL_i length) for certain line can be determined from the scheme in **Figure 9** as a number of circles plotted at the corresponding line. The delayed signals are directed to the restorer where they are processed as described above. As the result, original data bit sequence becomes recovered.

Considering all above, one can conclude that application of selective mode excitation could really lead to substantial improvement of system data security.

4. Conclusion

The presented materials emphasize advantages of application of selective mode excitation that lead to both increasing the system information capacity and enhancing data security. Realizability of three-stage hierarchical scheme of data compression is shown, and also ability of the mentioned technique to embarrass eavesdropping is highlighted. Application of selective mode excitation in the form of MDM technique could improve significantly performances of fiber systems, at least the ones like short- and middle-haul communication lines and LANs.

Author details

Dmitry V. Svistunov

Address all correspondence to: svistunov@mail.ru

Peter-the-Great St.-Petersburg Polytechnic University, St.-Petersburg, Russia

References

[1] Tur M, Menashe D, Japha Y, Danziger Y. High-order mode based dispersion compensating modules using spatial mode conversion. Journal of Optical Fiber Communications Reports. 2007;**4**(2):110-172. DOI: 10.1007/s10297-007-0081-0

[2] Salsi M, Koebele C, Sperti D, Tran P, Mardoyan H, Brindel P, Bigo S, Boutin A, Verluise F, Sillard P, Bigot-Astruc M, Provost L, Charlet G. Mode-division multiplexing of 2×100 Gb/s channels using an LCOS based spatial modulator. Journal of Ligthwave Technology. 2012;**30**(4):618-623. DOI: 10.1109/JLT.2011.2178394

[3] Hu T, Li J, Ren F, Tang R, Yu J, Mo Q, Ke Y, Du C, Liu Z, He Y, Li Z, Chen Z. Demonstration of bidirectional PON based on mode division multiplexing. IEEE Photonics Technology Letters. 2016;**28**(11):1201-1204. DOI: 10.1109/LPT.2016.2530663

[4] Kawaguchi Y, Tsutsumi K. Mode multiplexing and demultiplexing devices using multimode interference couplers. Electronic Letters. 2002;**38**(25):1701-1702. DOI: 10.1049/el:20021154

[5] Lee BT, Shin SY. Mode-order converter in a multimode waveguide. Optical Letters. 2003;**28**(18):1660-1662. DOI: 10.1364/OL.28.001660

[6] Karpeev SV, Pavelyev VS, Khonina SN, Kazanskiy NL, Gavrilov AV, Eropolov VA. Fibre sensors based on transverse mode selection. Journal of Modern Optics. 2007;**54**(6):833-844. DOI: 10.1080/09500340601066125

[7]   Carpenter J, Thomsen BC, Wilkinson TD. Degenerate mode-group division multiplexing. Journal of Lightwave Technology. 2012;**30**(24):3946-3952. DOI: 10.1109/JLT.2012.2206562

[8]   Koonen AMJ, Chen H, Boom HPA, Raz O. Silicon photonic integrated mode multiplexer and demultiplexer. IEEE Photonics Technology Letters. 2012;**24**(21):1961-1964. DOI: 10.1109/LPT.2012.2219304

[9]   Love JD, Riesen N. Mode-selective couplers for few-mode optical fiber networks. Optical Letters. 2012;**37**(19):3990-3992. DOI: 10.1364/OL.37.003990

[10]   Li A, Chen X, Al Almin A, Shieh W. Fused fiber mode couplers for few-mode transmission. Photonics Technology Letters. 2012;**24**(21):1953-1956. DOI: 10.1109/LPT.2012.2218803

[11]   Vaissie L, Johnson EG. Selective mode excitation by non-axial evanescent coupling for bandwidth enhancement of multimode fiber links. Optical Engineering. 2002;**41**(8):1821-1828. DOI: 10.1117/1.1488607

[12]   Hoyningen-Huene J, Ryf R, Winzer P. LCoS-based mode shaper for few-mode fiber. Optics Express. 2013;**21**(15):18097-18110. DOI: 10.1364/OE.21.018097

[13]   Chen W, Wang W, Yang J. Mode multi/demultiplexer based on cascaded asymmetric Y junctions. Optics Express. 2013;**21**(21):25113-25119. DOI: 10.1364/OE.21.025113

[14]   Tsekrekos CP, Syvridis D. Symmetric few-mode fiber couplers as the key component for broadband mode multiplexing. Journal of Lightwave Technology. 2014;**32**(14):2461-2467. DOI: 10.1109/JLT.2014.2327479

[15]   Igarashi K, Souma D, Tsuritani T, Morita I. Selective mode multiplexer based on phase plates and Mach–Zehnder interferometer with image inversion function. Optics Express. 2015;**23**(1):183-194. DOI: 10.1364/OE.23.000183

[16]   Dong J, Chiang KS, Jin W. Compact three-dimensional polymer waveguide mode multiplexer. Journal of Lightwave Technology. 2015;**33**(22):4580-4588. DOI: 10.1109/JLT.2015.2478961

[17]   Svistunov DV. A planar prism for detection and selective excitation of modes in a multimode channel waveguide. Technical Physics Letters. 2004;**30**(4):332-334. DOI: 10.1134/1.1748616

[18]   Svistunov DV. Selective mode launching in a multimode channel waveguide by planar coupler. Journal of Optics A Pure Applied Optics. 2004;**6**(9):859-861. DOI: 10.1088/1464-4258/6/9/007

[19]   Svistunov DV. Evaluating the possibility of using mode compression in a cascade system for data compression in multimode fiber-optic communication lines. Journal of Optics Technology. 2012;**79**(1):29-35. DOI: 10.1364/JOT.79.000029

[20]   Svistunov DV. Cascade scheme of data compression in multimode fiber systems. In: Simmons M, editor. Network Coding and Data Compression: Theory, Applications and Challenges. New York: Nova Science; 2015. pp. 151-161. ISBN: 978-1-63483-185-7

[21] Asawa CK. Intrusion-alarmed fiber optic communication link using a planar waveguide bimodal launcher. Journal of Lightwave Technology. 2002;**20**(1):10-18. DOI: 10.1109/50.974813

[22] Lecoy P. Fiber-Optic Communications. London-Hoboken: ISTE-Wiley; 2008. p. 333. ISBN: 978-1-84821-049-3

[23] Olshansky R, Keck DB. Pulse broadening in graded-index optical fibers. Applied Optics. 1976;**15**(2):483-491. DOI: 10.1364/AO.15.000483

8

Power-Over-Fiber Applications for Telecommunications and for Electric Utilities

Joao Batista Rosolem

Additional information is available at the end of the chapter

Abstract

Beyond telecommunications, optical fibers can also transport optical energy to powering electric or electronic devices remotely. This technique is called power over fiber (PoF). Besides the advantages of optical fiber (immunity to electromagnetic interferences and electrical insulation), the employment of a PoF scheme can eliminate the energy supplied by metallic cable and batteries located at remote sites, improving the reliability and the security of the system. Smart grid is a green field where PoF can be applied. Experts see smart grid as the output to a new technological level seeks to incorporate extensively technologies for sensing, monitoring, information technology, and telecommunications for the best performance electrical network. On the other hand, in telecommunications, PoF can be used in applications, such as remote antennas and extenders for passive optical networks (PONs). PoF can make them virtually passives. We reviewed the PoF concept, its main elements, technologies, and applications focusing in access networks and in smart grid developments made by the author's research group.

Keywords: power over fiber, PoF, fiber powering, sensing, monitoring, photovoltaic converters, smart grid, fiber to the antennas

1. Introduction

Optical fiber (OF) technology has many decades of research and development mainly focused in telecommunication applications. Other classical applications for optical fibers include sensors, endoscopic imaging, and illumination.

Decades ago, in the 1970s, scientists of American Telephone and Telegraph (AT&T) had the idea to use the fiber to supply energy using optical fiber to make a sound alert in telephones, instead of the electric option. The concept of power over fiber (PoF) [1] was born this way. In other words, beyond their classical applications, optical fibers can also be used to transport optical energy to powering electric or electronic devices remotely. This is the concept of PoF.

Since that time in the 1970s, many applications have emerged for PoF mainly in two different areas: telecommunications and utilities.

Why PoF is interesting? The reasons depend on the applications but they are associated to the optical fiber characteristics.

For telecom applications, the key factors are (i) PoF eliminates the necessity of batteries, solar panels, and long cupper feeder wires in the remote sites, improving the reliability and the security of the system; (ii) PoF permits the reduction of the space and the installation cost in remote sites, which is very important to telecom companies.

For electrical utilities, there are at least four key factors: (i) PoF uses optical fiber, which is made of nonconductive material. This characteristic is important because in most applications in electrical world, the sensors are placed in high voltage. Any conductance in high-voltage elements can create current leaks; (ii) optical fiber is immune to electromagnetic interferences. The electrical world environment is polluted of electromagnetic interferences; therefore, optical fiber can transmit signals without quality degradation; (iii) the optical fiber eliminates the need to run conductive copper wire into a high-ground potential rise (GPR) zone. GPR arises when lightning strikes occur in substations and can cause severe interference problems in electronic equipment and systems. Considering that the sensor using PoF has a complete galvanic isolation to the ground potential, it is practically immune to the GPR effects; and (iv) there are many low-cost/low-power/high-efficiency electronic sensors available for transmission lines and substations monitoring that can be supplied by PoF without incurring in the described problems.

This chapter describes a revision of PoF, its technical principles, main elements, technologies, and the applications in telecom and in utilities, developed by the author's group.

2. Power over fiber

This section presents the technical principles of PoF including a short history, the main elements (high-power lasers, fibers, and photovoltaic converters (PVs)), and their main limitations (laser power, converter efficiency, and fiber-fuse effect), circuits, topologies and networks, application on multimode (MM), single mode (SMF) and special fibers, and finally the future perspectives.

2.1. Historical overview

2.1.1. Telecom

The concept of PoF was born together with the first optical fiber developments. Optical fiber with low attenuation (less than 20 dB/km) was developed in 1970, and the first fiber optic link was installed in Chicago in 1976, USA. To the best of author's knowledge, DeLoach et al. [1] published the first work about PoF in 1978. They proposed to use PoF to operate a sound alert of a telephone remotely. Continuing the previous experiment, in 1979, Miller and Lawry [2] implemented a two-way speech communication between an electrically powered station and an optically powered station, and in 1982, Miller et al. [3] demonstrated a bidirectional speech-television communication over a single optical fiber, with emergency optical powering of the remote station telephone. A great result of those researches was the development of a highly efficient photovoltaic converter based on GaAlAs.

In 1993, Banwell et al. investigated the issues of PoF application in fiber-in-the-loop (FITL) [4]. In this work, many technical and cost aspects have been considered in the analysis. Almost a decade after that, Miyakawa et al. [5–7] presented fiber optic power and signal transmission systems considering the application of DC (direct current) powering to information equipment such as personal computers.

Werthen and Cohen [8] in 2006 and 2008 [9] proposed the use of PoF for driving low-power switches or actuators to assuring path diversity in upstream data delivery in passive optical network (PON) architectures and for providing battery backup power in case of power outages.

In 2007, Wake et al. [10] studied the combination of radio over fiber and PoF, and in [11], they described optically powered radio-over-fiber remote units designed and constructed for distributed antenna system applications.

Sato and Matsuura [12] and Matsuura and Sato [13] demonstrated in 2013 the transmission of RF signal and high-power light using a double-clad fiber (DCF). A DCF has a single-mode core and a multimode inner clad. The RF signal and the optical powering signal travel, respectively, through core and in the inner clad of DCF.

In 2014, Penze et al. [14] proposed powering PON extenders using PoF. In this research, SOAs (semiconductor optical amplifier) were powered using PoF. The SOAs were used in a bidirectional amplification of signals of Gigabit PON (GPON) and 10 Gigabit PON (XGPON). In the same year, Ikeda [15] proposed a PoF instead of metal waveguides in order to protect microwave radio equipment from lightning.

Yan et al. [16], in 2015, developed a wireless sensor system based on PoF to realize a flexible distributed sensing over a middle distance, in environments of high voltage, strong magnetic field, flammable, and explosives. In the same year, Matsuura and Sato [17] demonstrated experimentally a bidirectional radio over fiber using a DCF for optically powered remote antenna units. The feasibility of the technique was demonstrated by bidirectional RoF transmission over a 100-m DCF optically feeding with 4.0 W, and in [18], Sato et al. demonstrated a bidirectional radio-over-fiber transmission using a double-clad fiber with 40-W optical power feeding. In the same year, Lee et al. [19] described the concept of cloud radio access network (CRAN) based on PON exploiting PoF, andSuto et al. [20] detailed the challenges of QoE-guaranteed and power-efficient network operation for CRAN based on PON exploiting PoF.

In 2016, Minamoto and Matsuura [21] and Matsuura and Minamoto [22] presented an optically controlled beam steering system using 60-W PoF in the fiber to feed remote antenna units, and in [23], Yoneyama et al. demonstrated a 1.3-μm dual-channel radio-over-fiber system using a 300-m double-clad fiber feed with 30-W optical power of a PoF system. In the same year, Umezawa et al. [24] reported a high-conversion gain in a high-speed optical receiver based on PoF supply and designed to work with radio over fiber at 100-GHz region. In [25, 26], the same authors proposed the use of a multicore-based radio and PoF in a transmission using a 100-GHz photo receiver.

2.1.2. Utilities and other industries

In 1980, Caspers and Neumann [27] first described a PoF method used to supply active electronic circuits at high potential. In the experiment, they used four infrared light-emitting diodes (LEDs) emitting at a wavelength of 940 nm which were connected in parallel and coupled to an uncladded glass rod of 40-mm diameter and a length of 1 m with an attenuation smaller than 200 dB/km at a wavelength of 850 nm. Four square silica solar cells of 20 mm length were connected in series in this experiment.

Ohte et al., in 1984 [28], demonstrated a transducer with optical-fiber data link to provide electrical isolation. A pulse-position–modulated optical signal was used in transmitted signal by optical fiber from a remote converter to mainframe. Wavelength division multiplexing (WDM) was used to transmit the modulated signal and the optical feed power in the same optical fiber.

In 1988, Trisno and Wobschall [29] described a method for improving the efficiency of the conversion of pulsed optical power to electrical power by photovoltaic. The system required a pulsed power source with a storage capacitor to hold energy for a time after the optical power is turned off. In the same year, Bjork et al. [30] demonstrated the ability to provide data communication a variety of off-the-shelf electronic transducers in a single fiber with proper attention to the synchronization issues and in [31] Lenz and Bjork presented a comparative study of four concepts for standardizing and multiplexing of fiber optic sensors for aircraft applications and compared them with PoF.

Kirkham and Johnston [32] presented in 1989 an optical frequency–modulated (FM) data link whose remote electronic sensors were optically powered. An application to current measurement is described in a high-voltage line by means of a linear coupler.

In 1991, Yamagata et al. [33] described a PoF system to measuring gas density of extra high-voltage substation gas-insulated switchgear. This system employed a method of matching the impedances of load and photodiode by optical electrical conversion of pulsed light with a photodiode and boosting the voltage with a transformer. In [34], Sai presented an optimization of this method. In the same year, Nieuwkoop et al. [35] described an alternative system to increase the voltage to the remote sensor using a low-cost coil.

Spillman and Crowne [36] presented in 1995 a throttle level angle (TLA)-sensing system applied to aircrafts utilizing a capacitance-based rotary position transducer which was powered and interrogated via light from a single MM optical fiber. The system used a unique GaAs device that served as both a power converter and optical data transmitter. In the same year, Tardy et al. [37] described a PoF system designed to measure current in high voltage;

Pember et al. [38] demonstrated a PoF network concept; and Dubaniewicz and Chilton [39] described a PoF system to gas monitoring in mines.

In 1996, Werthen et al. [40] presented an optically powered current transducer based on shunt technology using PoF and detailed a practical use of this technique. Wang et al. [41] and Zhijing [42] detailed the use of PoF to measure the pressure of liquids in 1998. In 2005, Turán et al. [43] review the basic properties of PoF, their key elements, and its industrial applications.

Bottger et al. [44], in 2008, presented an optically powered video camera link that allows acquiring and communicating a 100-Mb/s video stream over a distance of hundreds of meters. In [45], M. Roger demonstrated an optically powered sensor network with subscribers consuming less than 1-μW average power, and an optically powered high-speed video link transmitting data at a bit rate of 100 Mbit/s.

In 2010, Rosolem et al. [46] described the results of a PoF sensing system for monitoring partial discharges on hydro generators and de Nazaré and Werneck [47] proposed a monitoring system based on PoF to measure the temperature and current of transmission lines.

Audo et al. [48], in 2011, proposed PoF-based architecture to extend multidisciplinary-cabled networks or to create a dedicated submarine hydrophone or seismometer network, and in 2012, Lau et al. [49] reported a prototype optically remote-powered subsea video monitoring system that provides an alternative approach to powering subsea video cameras. In the same year, Tanaka and Kurokawa [50] presented a review of fiber optic sensor network that employs PoF. A hybrid sensor network with wireless sensors is also achievable by introducing wireless/optical interface nodes.

In 2014, Rosolem et al. [51] presented a PoF system to long-distance applications based on charge/discharge of super-capacitors. The system was installed in a transmission line tower, and it was connected to the substation using an optical ground wire cable (OPWG), and in [52], Silva et al. proposed a new MM fiber optic cable to transmit optical energy for long reach in PoF systems.

Rosolem et al. [53], in 2015, presented the results of a PoF system to monitor high-voltage switchgear using video cameras and free space optics (FSO), and in [54], Rosolem et al. proposed the use of PoF and FSO to measure current in high-voltage transmission lines.

In 2016, Zhang et al. [55] proposed a 15-kV silicon carbide (SiC) MOSFET gate drive using PoF and replaced the traditional design based on isolation transformer.

2.2. Technical principles of PoF

A generic PoF system is shown in **Figure 1**. In the left side, the control unit composed by the high-power optical source (HPOS) unit together with the optical reception unit (ORU) is shown, which receives signals from the remote sensor unit, shown on the right side of **Figure 1**. Two optical fibers connect the local control unit to the remote unit. The optical fibers may be standard SMF optical fibers or MM fibers. In the remote unit, a photovoltaic converter

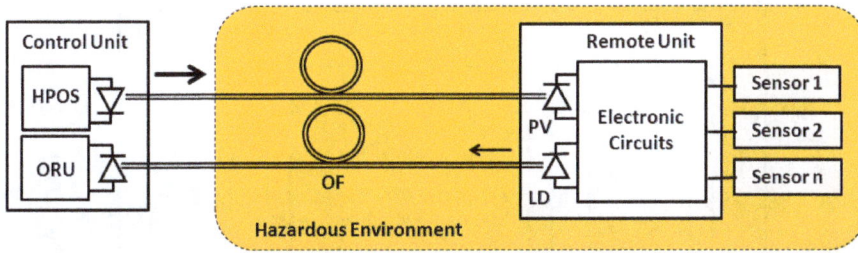

Figure 1. Generic PoF system diagram showing its main elements HPOS, OF, PV, LD, electronic circuit, and sensors.

detects the power transmitted by the HPOS. The electrical energy produced by the photovoltaic converter is used to power up a low-threshold laser (LD), electronic circuits, and sensors of the remote unit. The optical fibers and the remote unit can be installed in a hazardous environment, such as high-voltage substations, oil refineries, mines, oil tanks, water reservoirs, and sea depth.

The biggest challenge for a generic PoF system is to provide to a load the highest possible power, at the greatest possible distance, with the highest reliability. There are limits today to PoF applications and the limits are attributed to technological, physical, and cost aspects of PoF elements.

The amount of power that the PoF system can deliver is determined to a great extent by its components: laser, fiber, and photovoltaic cell. The following parameters can be used to evaluate the delivered electric power (P_{Load}) to an electronic load: the maximum transmitted optical power without causing damages in the optical fiber ($P_{MaxFiber}$); the optical power of the HPOS (P_{HPOS}); the optical power in the PV input (P_{In}); the total loss on the fiber (α_{Fiber}); the total loss of the optical connectors (α_{Conn}); the link distance (L); and the efficiency of the PV (η_{PV}). The power delivered to the extender can be expressed by

$$P_{Load} = P_{In} \cdot \eta_{PV} \tag{1}$$

$$P_{In} = P_{HPOS} \cdot \alpha_{Fiber} \cdot \alpha_{Conn} \tag{2}$$

$$\alpha_{Fiber} = 10^{(-L \cdot \alpha_{FiberdB}/10)} \tag{3}$$

$$\alpha_{Conn} = 10^{(-\alpha_{ConndB}/10)} \tag{4}$$

where $\alpha_{FiberdB}$ is the fiber attenuation in dB/km and α_{ConndB} is the total connector loss in dB. Notice that

$$P_{HPOS} \leq P_{MaxFiber} \tag{5}$$

In the following subsections, the main elements of PoF will be described, and **Figure 2** will be used as a reference for further discussion. **Figure 2** shows a qualitative graph of some parameters of the PoF elements, such as the loss of the optical fibers and the responsivity of PV cells. The main spectral bands for HPOS are also shown in **Figure 2**.

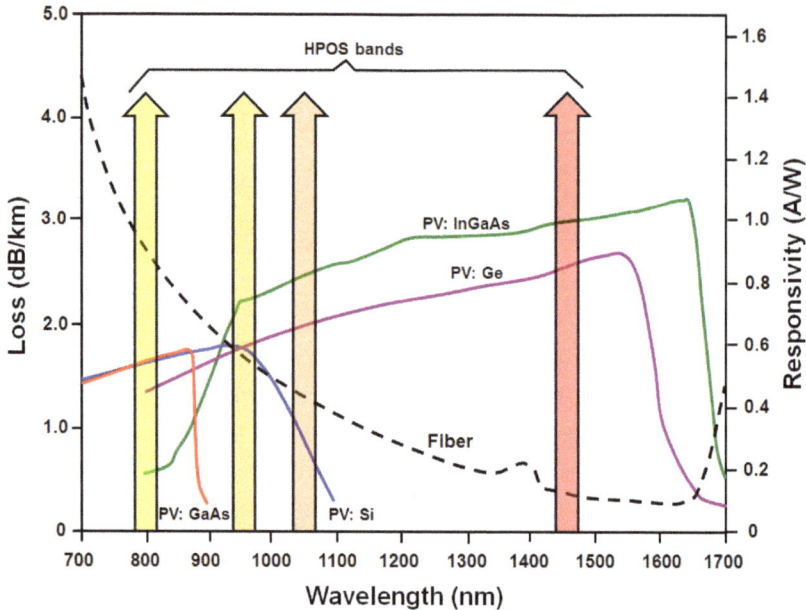

Figure 2. Qualitative graph of some PoF parameters.

2.3. Main elements of PoF

2.3.1. High-power optical source

Figure 2 shows the main spectral bands of HPOS. There are four main bands: 800, 950, 1050, and 1480 nm. The HPOSs for these bands have applications in medicine, pumping, thermal printing, and industrial applications. The power of these HPOSs can reach from 2 to 10 W for TO220 packaging depending on the wavelength band and from 6 to 650 W for laser modules. The output fiber core diameter for these HPOSs ranges from 50 to 600 μm depending on the HPOS output power. These fibers are MM types. For SMF fibers, the output power changes from 0.2 to 1.0 W.

Normally, an SMA (Sub-Miniature A) high-power type with metal ferrule is used in the HPOS output optical fiber. This metal ferrule is used for thermal dissipation in the HPOS connection to the fiber link. The high-power SMA connector utilizes air-gap-ferrule technology that eliminates the materials near the fiber end face that absorbs energy (e.g., epoxy). This absorption can damage the connector end face [56].

HPOS can also transmit information to the remote unit. The information signal is modulated in the continuous optical power generated by the HPOS. In general, this signal transmits commands to be executed in the remote unit, such as requesting the sensor parameters transmission, switching the sensor, and so on.

The most common modulation formats used to modulate the HPOS are on-off keying (OOK), pulse-width modulation (PWM), or frequency modulation. The extinction ratio of this signal can be high (100%) for few milliseconds like burst signals or low (less than 5%) during all the time of optical power sending. The extinction ratio is the ratio of the power used to transmit a logic level "1," to the power used to transmit a logic level "0." **Figure 3** shows an oscilloscope trace (voltage vs. time) of a telecommand signal transmitted over the optical supply power.

The main HPOS parameters required for a PoF project are maximum output power (P_{HPOS}), the operation wavelength (λ_{op}), the output fiber core diameter (D_{core}), the numerical aperture of the fiber (NA), and connector loss α_{ConndB}.

2.3.2. Optical fiber

The next element in the PoF system after the HPOS is the optical fiber, which is responsible to send the optical power from the HPOS to the PV. The fiber has two main parameters in the PoF system design: its attenuation and its limit of transmitted optical power.

The attenuation in modern optical fiber is basically caused by the intrinsic loss due to Rayleigh backscattering. Silica fibers are glasses that have materials with microscopic variations of density and refractive index. This effect gives rise to energy losses due to the scattered light. The loss due to the Rayleigh backscattering has high dependence with the operation wavelength ($1/\lambda_{op}^4$). The typical loss in the HPOS bands is 2.8 (800 nm), 1.8 (950 nm), 1.4 (1050 nm), and 0.25 dB (1480 nm).

The loss of optical fiber is not a big problem for the major PoF applications that has dozens of meters of fibers. It can be a problem for distances longer than 1 km. The major problem is regarding the maximum transmitted optical power. Next, the physical mechanism that limits the power in optical fiber will be detailed. After this limit, a degradation of the fiber core and the fiber coating occurs.

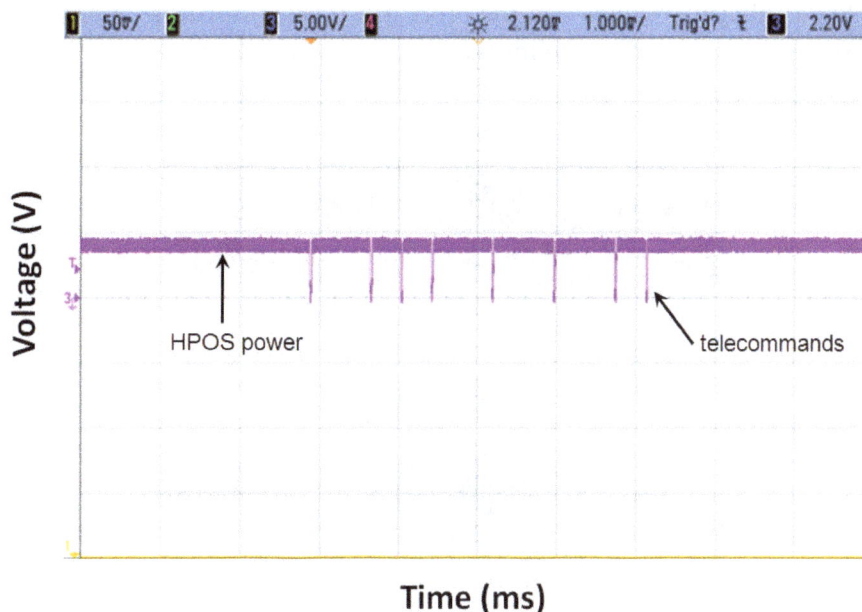

Figure 3. Oscilloscope trace showing a telecommand signal transmitted over the optical supply power.

Optical fibers when subjected to reduced diameter curvatures exhibit further attenuation of the optical signal. This attenuation may be significant when the bending diameter reaches a critical value. The attenuation of the optical signal in the bending zone of the fiber is related to the leakage of the fiber core region signal to the cladding region. The material of the outer-coating layer of the optical fiber absorbs the signal lost to the cladding region. When a high-power optical signal is propagated in the fiber, the energy absorbed by the coating in the zone of curvature is high, generating a local increase in temperature in the coating [56].

The heating of the coating of the optical fiber leads to its degradation, implying a consequent reduction of the useful life of the fiber. This degradation of the optical fiber is related to the reduction of the volume of the coating in the zone of curvature.

The second problem resulting from the propagation of high-power signals is the occurrence of the "fiber-fuse" effect on the optical fibers. This phenomenon was first observed in 1987 in a standard SMF fiber with optical intensity exceeding 5 MW/cm^2 at 1064 nm [57]. The fiber-fuse effect was initiated at a point of the fiber with a high-temperature value, then propagating toward the optical source with a velocity of about 1 m/s and emitting an optical signal in the spectral region of the visible. The propagation of this high-temperature zone is similar to the burning of a wick, which gave rise to the name of the "fiber-fuse" effect. After propagation of the fiber-fuse effect, the optical-fiber core presents a periodic bubble chain (see **Figure 4**) and is permanently destroyed, being unable to guide an optical signal.

Currently, the most accepted general explanation for the fiber-fuse effect relates to the ignition of this effect with the increase of fiber optic absorption at a point with a high-temperature value [56]. In turn, the increase in optical signal absorption is responsible for the catastrophic temperature increase in the fiber core, reaching values higher than the vaporization temperature of the silica. Through the thermal diffusion mechanism, this high-temperature zone is transmitted to neighboring regions and the process evolves toward the optical source. Thus, the "fiber-fuse" effect is related to the increase in fiber optic absorption and the thermal diffusion process.

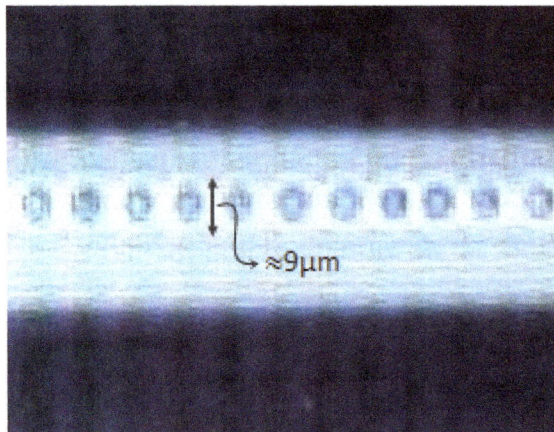

Figure 4. Schematic of a fiber-fuse effect and a photo of an optical fiber core presenting a periodic bubble chain due to fiber-fuse effect.

It should be noted that ignition and propagation of the fiber-fuse effect occurs only if the power of the optical signal in the fiber is maintained above a threshold value. According to [56], this threshold is proportional to the diameter of the fiber modal field (MDF) and dependent on the wavelength of the optical signal. A typical SMF fiber (MDF = 7.8 μm) presents a threshold of 1.5 W at 1467 nm. In other study [58], it was observed that an SMF fiber type 28F (MDF = 7.8 μm) presents a threshold of 1.0 W, and an MM fiber of 62.5-μm core (MDF = 12.0 μm) presents a threshold of 4.0 W at 1060 nm.

In order to increase the optical power in SMF fibers, many efforts have been made to transmit the high power by the cladding of the SMF fiber. In [21, 22], an optically controlled beam steering system using 60 W transmitted in the cladding of the fiber to feed remote antenna units is presented.

The propagation of signals with optical power exceeding the threshold required for ignition and propagation of the fiber-fuse effect is not sufficient to trigger it, and an ignition point characterized by having a high-temperature value is also required. This point of ignition occurs in contaminated and/or degraded connectors or in optical fibers subject to tight bending. As described previously, in optical fibers subjected to curvatures of reduced diameters, an additional attenuation of the optical signal occurs, which in combination with high-power signals generates a considerable localized heating.

Optical connectors can be easily contaminated with dust or organic particles that are common in outdoor environments. This additional attenuation in the connectors, contaminated and/or degraded, may be considerable and when associated with high-power signals also generate localized heating.

PoF is generally used in hazardous environment requiring rugged optical cables for this kind of environment. It is desirable that such cables containing fiber from 100- to 1000-μm core diameter be available on the market, since PoF requires fibers with large core diameter. The development of a rugged PoF optical cable using 100-μm fiber core for long-distance applications is described in [52].

The main OF parameters required for a PoF project are maximum transmitted optical power without causing damages ($P_{MaxFiber}$), fiber attenuation ($\alpha_{FiberdB}$) in dB/km at the operation wavelength (λ_{op}), the fiber core diameter (D_{core}), and the numerical aperture of the fiber (NA).

2.3.3. Photovoltaic converter

A photovoltaic power converter is the element where the light will be finally transformed in electricity to supply the control unit and the sensors. PV is one (or more) photodiode that operates in photovoltaic mode. When the supply optical power reaches the PV, electron-hole pairs (carriers) are generated creating a photocurrent. The total current is the photocurrent produced by the supply optical power minus the dark current, which is related to the reverse saturation current. This photocurrent (I_{ph}) is proportional to the incident supply optical power according to Eq. (6)

$$I_{ph} = P_{In} \cdot R(\lambda_{op})$$ (6)

$R(\lambda_{op})$ is the spectral responsivity of the cell, which is dependent on the HPOS operation wavelength (**Figure 2**).

The most important photovoltaic parameter is the conversion efficiency (η). It is defined as the maximum electric power produced by the PV ($P_{elecmax}$) divided by the incident light power under standard light conditions or as defined by Eq. (7):

$$\eta = P_{elecmax}/P_{Pin} = (V_{OC} \cdot I_{SC} \cdot FF)/(P_{Pin}) \tag{7}$$

where V_{oc} is the open-circuit voltage, I_{SC} is the short circuit current, and FF is the fill factor of PV.

Many PV types are actually made of micro-PV arrays [59, 60]. In this way, connecting the micro-PVs in parallel to the output voltage can reach an adequate level. **Figure 5** shows a schematic arrangement of a commercial PV array with four elements compared with 100-μm-core optical-fiber diameter. This type of fiber is normally used in fiber-packaged PVs.

The fill factor is the parameter that evaluates the junction quality and series resistance of the PV array. The fill factor is the ratio of the maximum electric output power to the product of VOC and ISC.

A comparative study of PV efficiency for the materials Si, GaAs, and Ge is founded in [61]. PV devices made of Si or GaAs proper for PoF uses can be founded in market. Unfortunately, these devices operate in the spectrum region where the fiber does not have the minimum attenuation (**Figure 2**). For operation in 1310 or 1550 nm, the PV arrays can be obtained easily connecting many InGaAs or Ge photodiodes in series or in parallel using an optical splitter. This is a topic for the next section.

The main PV parameters required for a PoF project are the maximum output electric power ($P_{elecmax}$), the maximum optical input power without damaging the PV (P_{Inmax}), the operation spectral band (λ_{op}), the efficiency (η), open-circuit voltage (V_{oc}), and the short-circuit current (I_{SC}).

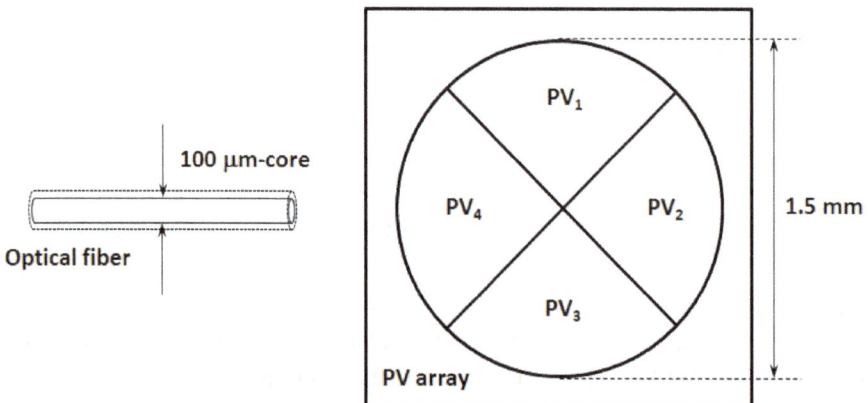

Figure 5. Schematic arrangement of a commercial PV array with four elements compared with 100-μm-core optical fiber diameter.

2.3.4. Basic remote unit circuits

The electronic circuits of unit control including LD driver and sensors need to work in a specific voltage (V_{cc}), and they consume an electrical current (I_{load}). To provide the required voltage and current, some basic circuits can be designed. The most basic circuits use just one PV and a DC-DC converter to stabilize the voltage, as shown in **Figure 6(a)**. In this case, the PV should be an array type to provide voltage higher than 3 V, although there are many integrated circuits that work below 3 V. If the PV chosen has low-output voltage (lower than 2V), the circuit shown in **Figure 6(b)** can be used. In this circuit, the DC-DC converter was substituted by a step-up converter, which elevates the PV output voltage to the voltage required by the circuit. Other possibility to elevate the voltage is using an optical splitter (with n output ports) connected to the optical fiber and the PVs (or photodetectors) connected in series (**Figure 6(c)**). Typically, splitters with $n = 4$–8 have been used, to provide V_{cc} ranging from 2.4 to 4.8 V. Each splitter output port is connected to one PV. This is an easy solution to use in spectral bands where there is no commercial PV array available, such as in 1310 and 1550 nm. **Figure 6(d)** shows a different version of **Figure 6(c)** circuit. In this case, the association of the PVs is made in parallel in order to increase the current available to the circuit. Certainly, the association of PVs can be made simultaneously in parallel and in series.

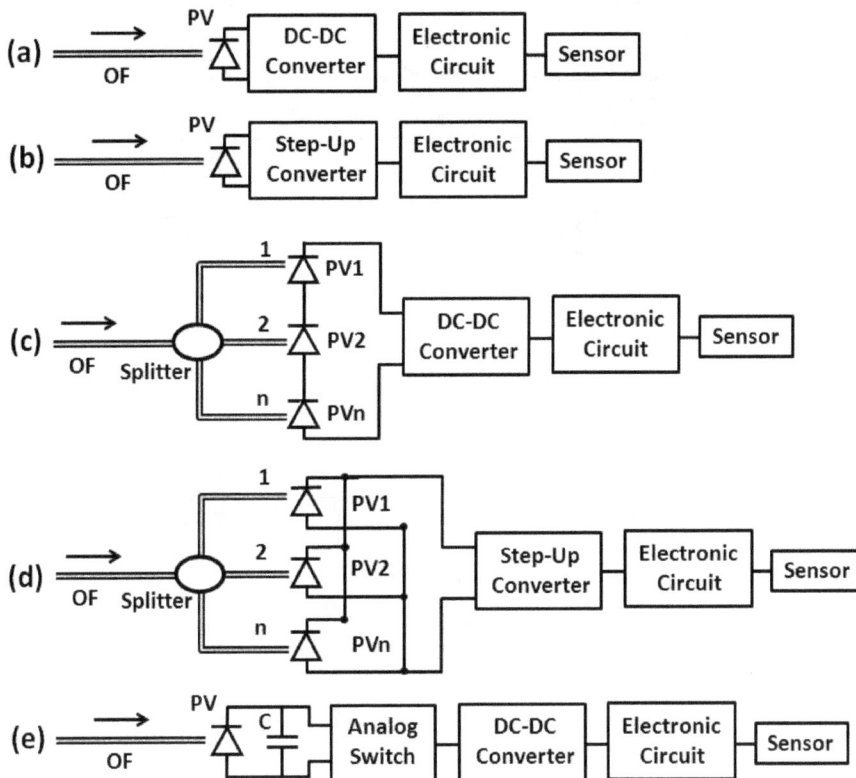

Figure 6. Some types of circuits used in the design of PoF remote units, (a) direct conversion, (b) conversion using a step-up converter, (c) PVs connected in series, (d) PVs connected in parallel, and (e) charging circuit using a capacitor.

Figure 6(e) shows a charging circuit type. This type of circuit is used to supply loads in long-distance PoF links where the available input optical power level in the PV is not enough to provide a necessary electric power. In this circuit, when the capacitor C connected in the PV output is charged in the maximum PV voltage output, an analog switch connects the capacitor to a DC-DC converter that stabilizes its output voltage for the load for a specific time period. This type of circuit should be used only if the sensors do not require measuring their parameters all the time.

2.4. PoF topologies

The connection between the control units and remote units in PoF systems is usually made in the same way of telecommunications links. **Figure 7(a)** shows a typical PoF link using two fibers. However, in some cases the quantity of fibers is a limiting factor in specific application. WDM can be used to multiply the supply optical power transmitted by the HPOS and the sensor signal transmitted by the LD in the remote unit. The WDM device must support the high power of the HPOS. WDM devices for high-power applications can support around 2 W of handling power. The wavelength pairs used in a WDM PoF system as the one shown in **Figure 7(b)** are as follows:

- λ_1 = 808–830 (only for MM fibers), 980, and 1480 nm;

- λ_2 = 1310 and 1550 nm.

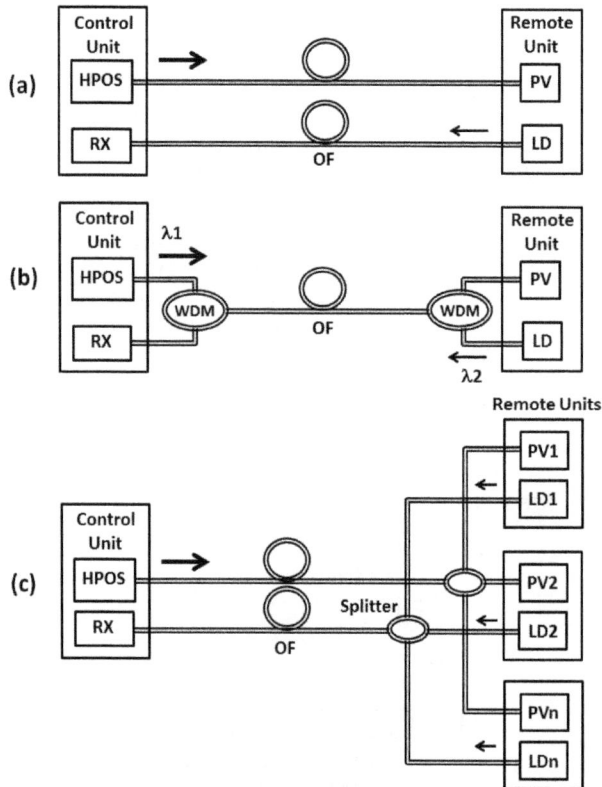

Figure 7. Illustrations of some topologies for PoF systems, (a) two fibers topology, (b) bidirectional WDM topology, and (c) tree topology.

PoF networks have also been reported. A typical tree topology (**Figure 7(c)**) permits the connection of a single control unit to many remote units. This type of network works in a similar way as a passive optical network used in telecommunications.

2.5. Example of PoF link calculations

Using Eqs. (1)–(5), it is possible to calculate the maximum PoF link for two cases, using SMF and MM fibers. **Table 1** shows the parameters used in this calculation.

Figure 8 shows a plot of the delivered power (P_{Load}) to an electronic load as a function of link distance. It can be observed that MM fiber option is better until 2.2 km. After this distance, the SMF option is better.

Parameter/fiber	SMF	MM
HPOS optical power	1.0 W	2.0 W
Wavelength	1480 nm	830 nm
Total connectors attenuation	1.0 dB	1.0 dB
Fiber attenuation	0.25 dB/km	3.0 dB/km
PV efficiency	0.2	0.4
PV material	InGaAs	GaAs

Table 1. Parameters used to calculate the maximum PoF link distance.

In utilities, the major applications of PoF occur in the substations, which the maximum internal distances are less than 0.2 km. For telecommunications, the distances in the major application are longer than 2 km.

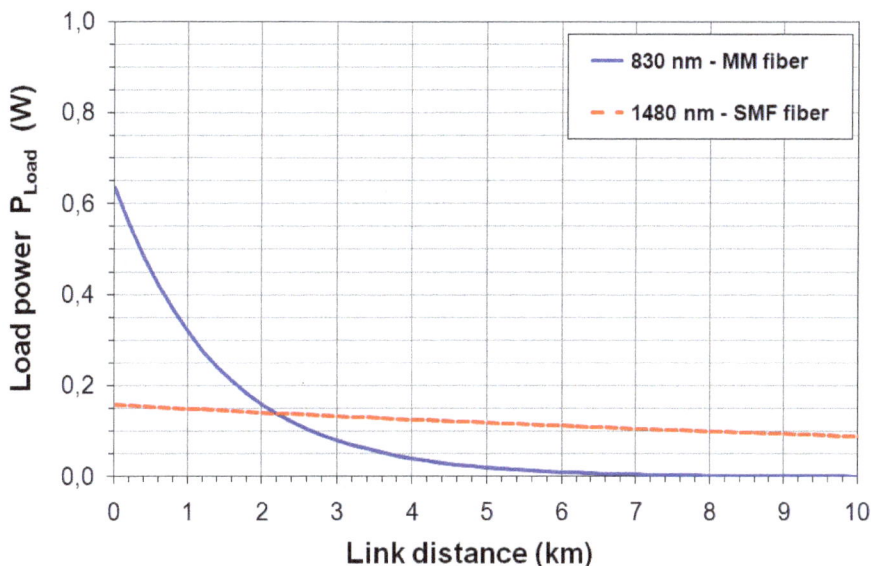

Figure 8. Plot of the delivered power to an electronic load as a function of link distance.

2.6. PoF applications developed at CPqD

This section presents some PoF systems and sensors developed by the author's group for telecommunications and for utilities applications. Most PoF systems for utilities have been tested in field trials.

2.6.1. PoF applications in telecommunications

In telecommunication, we developed a new approach to powering PON extenders using PoF [14]. PON extenders are elements that recover the downstream and upstream signals. These signals are highly attenuated by the splitters and by link length in any PON. However, the use of extenders in PON modifies its passive characteristic since extenders require electrical supply to work. Using PoF to supply, the extenders turns them virtually passives.

Using PoF to eliminate the batteries located at telecom remote sites, the reliability and the security of the system are improved. Furthermore, PoF can reduce the copper cables theft, used to power the extenders in many cases. We demonstrated the performance of a fiber-powered extender using semiconductor optical amplifiers in a 10-gigabit passive optical network system and a gigabit passive optical network system (G-PON) setup using a 1:32 splitter and 50-km reach. The extender was powered from a remote site placed in the access area using a 62.5-μm MM fiber at 830 nm. **Figure 9** shows a schematic of the PON using the fiber-powered extender and the photo of the extender.

2.6.2. PoF applications in utilities

In utilities, we developed applications for monitoring hydrogenerators, power transformers, switchgears, and for transmission lines.

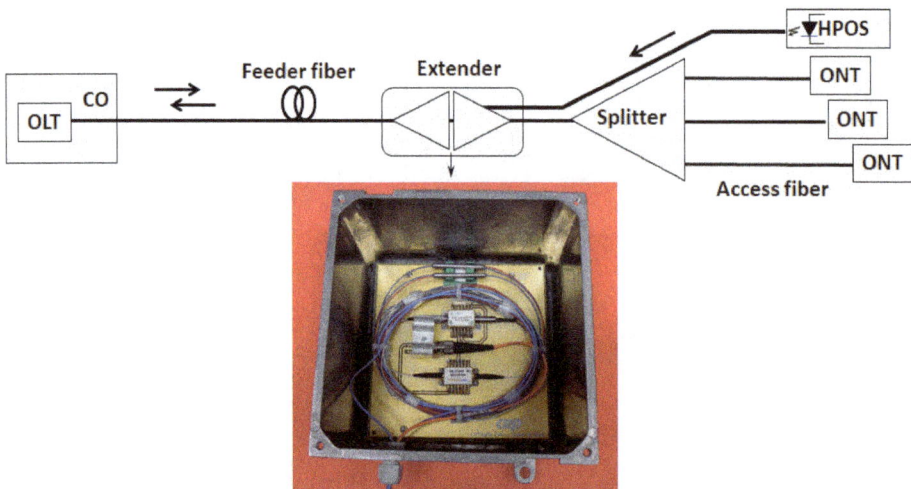

Figure 9. Schematic of a PON using extender powered by a PoF system and the extender photo.

Figure 10 shows the applications for partial discharges monitoring in hydrogenerators [46] and for 500-kV power transformers [62]. Partial discharges on high-voltage equipment insulation are a symptom of fragility of the dielectric capacity. The growing of partial discharges can cause serious consequences for the equipment and the electrical system. Partial discharges generate some physical effects, such as conducted and radiated electromagnetic pulses, light, acoustic noise, localized temperature variation, and chemical reactions. In the developed PoF-monitoring systems, we used a powered antenna to detect the radiated electromagnetic pulses of discharges.

For hydrogenerator monitoring (Figure 10(a)), the sensor is composed by one dipole meander antenna, one photovoltaic converter, and one semiconductor laser. Some passive components such as resistors, capacitors, and inductors are omitted in the schematic of sensor circuit to simplification. This system was installed in the Eletrobrás' Coaracy Nunes power plant, which is located in Amapá state, in the north of Brazil.

The sensor for monitoring the high-voltage power transformers (Figure 10(b)) is composed by one monopole antenna, one photovoltaic converter, one field-effect transistor (FET) amplifier, and one semiconductor laser. The FET was used in this sensor to increase the sensitivity of the sensor in this particular application. This system was installed in the Cemig's Neves substation, which is located in Minas Gerais state, in the southeast of Brazil.

Figure 10 shows the schematic circuit and sensor photo of antenna powered by fiber for partial discharges monitoring in (a) hydrogenerators and (b) high-voltage power transformers.

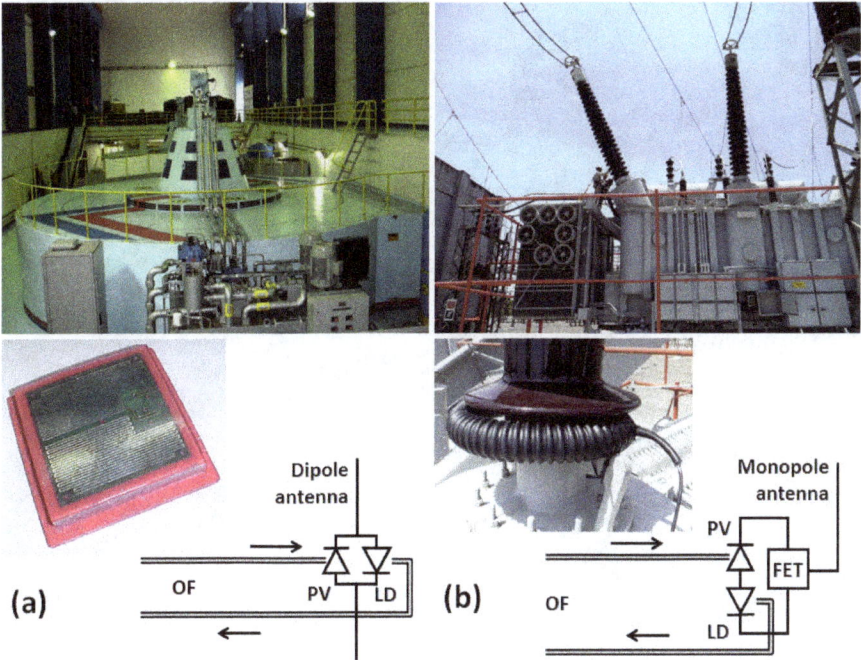

Figure 10. Schematic circuit and sensor photo of antenna powered by fiber for partial discharges monitoring in (a) hydrogenerators and (b) high-voltage power transformers.

Figure 11 shows some applications using more complex electronic circuits with PoF. **Figure 11(a)** shows a 138-kV switchgear-monitoring application using fiber-powered video cameras to inspect the quality contact of the switchgear [53]. In this application, the three phases of 138-kV switchgear were monitored using three sensors. The connection of the sensors to the control unit used the tree topology shown in **Figure 7(c)**. This system was installed in the Cemig's Bonsucesso substation, which is located in Minas Gerais state, in the southeast of Brazil.

Figure 11(b) shows a fiber-powered camera installed in a 138-kV transmission line tower [51]. The camera was used to monitor possible invasions in the security area of transmission line. The circuit of the camera works in a noncontinuous regime (circuit of **Figure 6(e)**) since the power transmission was done using an SMF fiber embedded in an optical ground wire cable. This system was installed in the Cemig's Bonsucesso/Gutierrez transmission line, which is located in Minas Gerais state, in the southeast of Brazil.

Remembering that for electrical utilities, there are key factors to use PoF. Optical fiber is made of nonconductive material. In high-voltage environment, any conductance can

Figure 11. PoF applications with fiber-powered video cameras, (a) 138-kV switchgear sensor shown in the circle (above) and a sensor image (below) and (b) 138-kV transmission line sensor shown in the circle (above) and a sensor image (below).

create current leaks. Optical fiber is immune to electromagnetic interferences. The electrical world environment is polluted of electromagnetic interferences; therefore, optical fiber can transmit signals without quality degradation. The optical fiber eliminates the need to run conductive copper wire into a GPR zone. GPR arises when lightning strikes occur in substations and can cause severe interference problems in electronic equipment and systems. PoF sensors have a complete galvanic isolation to the ground potential, then they are practically immune to the GPR effects and finally, as we show, there are many low-cost/low-power/high-efficiency electronic sensors available for transmission lines and substations monitoring that can be supplied by PoF increasing the monitoring capacity for utilities companies.

3. Conclusion and future perspectives

In conclusion, this chapter described a revision of PoF, its technical principles, main elements, technologies, and the applications focusing in telecom and in utilities, developed by the author's group.

The applications for PoF have evolved over the last years mainly in terms of power availability for the load in the remote units. Many publications describe incredible dozen watts power transmitted in the SMF fiber cladding. This evolution allows applications to become more complex using PoF, such as video cameras powering, antennas powering, or PoF networks.

The increase of supplier's options for PoF devices and fibers has also been occurring and it can reduce the cost of this interesting optical fiber technique.

Acknowledgements

The author wishes to thank his CPqD colleagues Claudio Floridia, Danilo C. Dini, Rivael S. Penze, Fabio R. Bassan, and João P. V. Fracarolli. Many PoF projects cited in this work were funded by ANEEL (Brazilian Electricity Regulatory Agency). CNPq (National Counsel of Technological and Scientific Development) sponsors the author under scholarship DT.

Author details

Joao Batista Rosolem

Address all correspondence to: rosolem@cpqd.com.br

CPqD—Research and Development Center in Telecommunications, Campinas, SP, Brazil

References

[1] DeLoach BC, Miller RC, Kaufman S. Sound alerter powered over an optical fiber. Bell System Technical Journal. 1978;**57**:3309–3316. DOI: 10.1002/j.1538-7305.1978.tb02205.

[2] Miller RC, Lawry RB. Optically powered speech communication over a fiber lightguide. Bell System Technical Journal 1979;**58**:1735–1741. DOI: 10.1002/j.1538-7305.1979.tb02280.x

[3] Miller RC, DeLoach BC, Stakelon TS, Lawry RB. Wideband, bidirectional lightguide communication with an optically powered audio channel. Bell System Technical Journal 1982;**61**:1538–7305. DOI: 10.1002/j.1538-7305.1982.tb04349.x

[4] Banwell TC, Estes RC, Reith LA, Shumate Jr PW, Vogel EM. Powering the fiber loop optically-a cost analysis. Journal of Lightwave Technology. 1993;**11**:481–494. DOI: 10.1109/50.219583

[5] Miyakawa H, Hyodo R, Tanaka Y, Kurokawa T. Photovoltaic cell characteristics for high-intensity laser light in fiber optic power transmission systems. In: Proceedings of Conference Record of the Twenty-Ninth IEEE Photovoltaic Specialists Conference; 19–24 May 2002; New Orleans, USA; **29**:1653–1655. DOI: 10.1109/PVSC.2002.1190934

[6] Miyakawa H, Herawaty E, Yoshimoto M, Tanaka Y, Kurokawa T. Power-over-optical local area network systems. In: Proceedings of 3rd World Conference on Photovoltaic Energy Conversion; 11–18 May 2003; Osaka, Japan; **3**:2466–2469.

[7] Miyakawa H, Tanaka Y, Kurokawa T. Design approaches to power-over-optical local-area-network systems. Applied Optics 2004;**43**:1379–1389. DOI: 10.1364/AO.43.001379

[8] Werthen JG, Cohen M. Photonic power: delivering power over fiber for optical networks. In: Proceedings of 2006 International Conference on Photonics in Switching; 16–18 October 2006; Heraklion, Crete; pp. 1–3. DOI: 10.1109/PS.2006.4350205

[9] Werthen JG. Powering next generation networks by laser light over fiber. In: Proceedings of 2008 Conference on Optical Fiber Communication/National Fiber Optic Engineers (OFC/NFOEC 2008); 24–28 February 2008; San Diego; USA, pp. 1–3. DOI: 10.1109/OFC.2008.4528749

[10] Wake D, Nkansah A, Gomes N. Optical powering of remote units for radio over fiber links. In: Proceedings of 2007 International Topical Meeting on Microwave Photonics; 3–5 Oct. 2007; VIC, Canada; pp. 29–32. DOI: 10.1109/MWP.2007.4378127

[11] Wake D, Nkansah A, Gomes NJ, Lethien C, Sion C, Vilcot J P. Optically powered remote units for radio-over-fiber systems. Journal of Lightwave Technology. 2008;**26**:2484–2491. DOI: 10.1109/JLT.2008.927171

[12] Sato J, Matsuura M. Radio-over-fiber transmission with optical power supply using a double-clad fiber. In: Proceedings of 18th OptoElectronics and Communications

Conference held Jointly with 2013 International Conference on Photonics in Switching (OECC/PS); 30 June–4 July 2013; Kyoto, Japan; pp. 1–2.

[13] Matsuura M, Sato J. Power-over-fiber using double-clad fibers for radio-over-fiber systems. In: Proceedings of 19th European Conference on Networks and Optical Communications (NOC); 4–6 June 2014; Milano, Italy; pp. 126–131. DOI: 10.1109/NOC.2014.6996840

[14] Penze RS, Rosolem JB, Duarte UR, Paiva GER, Filho RB. Fiber powered extender for XG-PON/G-PON applications. IEEE/OSA Journal of Optical Communications and Networking. 2014;6:250-258. DOI: 10.1364/JOCN.6.0002

[15] Ikeda K. Lightning protection of microwave radio equipment using radio on fiber and power over fiber experimental demonstration of communication quality. In: Proceedings of Microwave Photonics (MWP) and the 9th Asia-Pacific Microwave Photonics Conference (APMP) International Topical Meeting on; 20–23 October 2014; Sendai, Japan; pp. 185–188. DOI: 10.1109/MWP.2014.6994526

[16] Yan J, Wang J, Lu Y, Jiang J, Wan H. Novel wireless sensor system based on power-over-fiber technique. In: Proceedings of 14th International Conference on Optical Communications and Networks (ICOCN); 3–5 July 2015; Nanjing, China; pp. 1–3. DOI: 10.1109/ICOCN.2015.7203710

[17] Matsuura M, Sato J. Bidirectional radio-over-fiber systems using double-clad fibers for optically powered remote antenna units. IEEE Photonics Journal. 2015; 7:1–9. DOI: 10.1109/JPHOT.2014.2381669

[18] Sato J, Furugori H, Matsuura M. 40-watt power-over-fiber using a double-clad fiber for optically powered radio-over-fiber systems. In: Proceedings of Optical Fiber Communications Conference and Exhibition (OFC); 22–26 March 2015; Los Angeles, USA; pp. 1–3. DOI: 10.1364/OFC.2015.W3F.6

[19] Lee Y, Suto K, Nishiyama H, Kato N, Ujikawa H, Suzuki K. I. A novel network design and operation for reducing transmission power in cloud radio access network with power over fiber. In: Proceedings of IEEE/CIC International Conference on Communications in China (ICCC); 2–4 November; Shenzhen, China; pp. 1–5. DOI: 10.1109/ICCChina.2015.7448616

[20] Suto K, Miyanabe K, Nishiyama H, Kato N, Ujikawa H, Suzuki KI. QoE-guaranteed and power-efficient network operation for cloud radio access network with power over fiber. IEEE Transactions on Computational Social Systems;2:127–136. DOI: 10.1109/TCSS.2016.2518208

[21] Minamoto Y, Matsuura M. Optically controlled beam steering system with 60-W power-over-fiber feed for remote antenna units. In: Proceedings of Optical Fiber Communications Conference and Exhibition (OFC), Anaheim, USA; 20–24 March 2016; pp. 1–3.

[22] Matsuura M, Minamoto Y, Optically powered and controlled beam steering system for radio-over-fiber networks. Journal of Lightwave Technology. 2017; 35:979-988. DOI: 10.1109/JLT.2016.2631251

[23] Yoneyama A, Minamoto Y, Matsuura M. Power-over-fiber transmission using 1.3-μm dual-channel radio-over-fiber signals in a double-clad fiber. In: Proceedings of 21st OptoElectronics and Communications Conference (OECC) held jointly with 2016 International Conference on Photonics in Switching (PS); 3–7 July 2016; Niigata, Japan; pp. 1–3.

[24] Umezawa T, et al. 11-Gbps 16-QAM OFDM radio over fiber demonstration using 100 GHz high-efficiency photoreceiver based on photonic power supply. In: Proceedings of 21st OptoElectronics and Communications Conference (OECC) held jointly with 2016 International Conference on Photonics in Switching (PS); 3–7 July 2016; Niigata, Japan; pp. 1–33.

[25] Umezawa T, et al. Multi-core based 94-GHz radio and power over fiber transmission using 100-GHz analog photoreceiver. In: Proceedings of 42nd European Conference on Optical Communication (ECOC 2016), 18–22 Sept. 2016; Dusseldorf, Germany; pp. 1–3. DOI: 10.1109/JSTQE.2016.2611638

[26] Umezawa T, et al. 100-GHz fiber-fed optical-to-radio converter for radio- and power-over-fiber transmission. IEEE Journal of Selected Topics in Quantum Electronics. 2017; **23**:1–8. DOI: 10.1109/JSTQE.2016.2611638

[27] Caspers F, Neumann EG. Optical power supply for measuring or communication devices at high-voltage levels. IEEE Transactions on Instrumentation and Measurement. 1980; 29:73–74. DOI: 10.1109/TIM.1980.4314866

[28] Ohte A, Akiyama K, Ohno I. Optically-powered transducer with optical-fiber data link. In: Proceedings of SPIE 0478, Fiber Optic and Laser Sensors II; 17 September 1984; Arlington, USA. DOI: 10.1117/12.942655

[29] Trisno YS, Wobschall D. Optimization of an optically pulsed photocell array as a sensor power source. IEEE Transactions on Instrumentation and Measurement. 1988; **37**:142–144. DOI: 10.1109/19.2683

[30] Bjork P, Lenz J, Fujiwara K. Optically powered sensors. In: Proceedings of Optical Fiber Sensors, OSA Technical Digest Series; Mar 1988; New Orleans, USA. DOI: 10.1109/19.2683

[31] Lenz J, Bjork P, Optically powered sensors: a systems approach to fiber optic sensors. In: Proceedings of SPIE 0961, Industrial Optical Sensing, 29 November 1988; Dearborn, USA. DOI: 10.1117/12.947855

[32] Kirkham H, Johnston AR. Optically powered data link for power system applications. IEEE Transactions on Power Delivery. 1989; **4**:1997–2004. DOI: 10.1109/61.35623

[33] Yamagata Y, Kumagai T, Sai Y, Uchida Y, Imai K. A sensor powered by pulsed light (for gas density of GIS). In: Proceedings of Solid-State Sensors and Actuators, International Conference on Transducers '91, 24–27 June 1991; San Francisco, USA; pp. 824–827. DOI: 10.1109/SENSOR.1991.149011

[34] Sai Y. Optimization of optically powered sensors. In: Proceedings of International Conference on Industrial Electronics, Control and Instrumentation, IECON '91; 28 October–1 November 1991, Kobe, Japan, pp. 2439–2443. DOI: 10.1109/IECON.1991.238962

[35] Nieuwkoop E, Kapsenberg T, Steenvoorden GK, Bruinsma AJ. Optically powered sensor system using conventional electrical sensors. In: Proceedings of SPIE 1511, Fiber Optic Sensors: Engineering and Applications; 1 August 1991; The Hague, Netherlands. DOI:10.1117/12.45999

[36] Spillman Jr WB, Crowne DH. Optically powered and interrogated rotary position sensor for aircraft engine control applications. Optics Lasers Engineering. 1992; **16**:105–118. DOI:10.1016/0143-8166(92)90003-P

[37] Tardy A, Derossis A, Dupraz JP. A Current sensor remotely powered and monitored through an optical fiber link. Optical Fiber Technology; 1995; **1**:181–185. DOI: 10.1006/ofte.1995.1010

[38] Pember SJ, France CM, Jones BE. A multiplexed network of optically powered, addressed and interrogated hybrid resonant sensors. Sensors and Actuators A: Physical; 1995; **47**:474–477. DOI:10.1006/ofte.1995.1010

[39] Dubaniewicz Jr TH, Chilton JE. Fiber optic powered remote gas monitor. In: Proceedings of SPIE 2622, Optical Engineering Midwest '95; 18 August 1995; Chicago, USA. DOI:10.1117/12.216819

[40] Werthen JG, Andersson G, Weiss ST, Bjorklund HO. Current measurements using optical power. In: Proceedings of Transmission and Distribution Conference and Exposition; 15–20 Sept. 1998; Los Angeles, USA, pp. 213–218. DOI: 10.1109/TDC.1996.545937

[41] Wang L, Wang Y, Shi J, Zheng L. Optically powered hydrostatic tank gauging system with optical fiber link. In: Proceedings of SPIE 3555, Optical and Fiber Optic Sensor Systems; 13 August 1998; Beijing, China; DOI:10.1117/12.318165

[42] Zhijing Y. Optically powered oil tank multichannel detection system with optical fiber link. In: Proceedings of SPIE 3555, Optical and Fiber Optic Sensor Systems; 13 August 1998; Beijing, China. DOI:10.1117/12.318152

[43] Turán J, Ovseník L, Turán Jr J. Optically powered fiber optic sensors. Acta Electrotechnica et Informatica. 2005; **5**:1–7.

[44] Bottger G, et al. An optically powered video camera link. IEEE Photonics Technology Letters; 2008; **20**:39–41. DOI: 10.1109/LPT.2007.912695

[45] Röger M, Böttger G, Dreschmann M, Klamouris C, Huebner M, Bett AW, Becker J, Freude W, Leuthold J. Optically powered fiber networks. Optics Express. 2008; **16**:21821–21834. DOI:10.1364/OE.16.021821

[46] Rosolem JB, Floridia C, Sanz JPM. Field and laboratory demonstration of a fiber-optic/RF partial discharges monitoring system for hydrogenerators applications. IEEE Transactions on Energy Conversion. 2010; **25**:884–890. DOI: 10.1109/TEC.2010.2044507

[47] de Nazaré FVB, Werneck MM. Temperature and current monitoring system for transmission lines using power-over-fiber technology. In: Proceedings of IEEE Instrumentation & Measurement Technology Conference; 3–6 May 2010; Austin, USA; pp. 779–784. DOI: 10.1109/IMTC.2010.5488198

[48] Audo F, Guegan M, Quintard V, Perennou A, Bihan JL, et al. Quasi-all-optical network extension for submarine cabled observatories. Optical Engineering. 2011; 50:1-8. DOI: 10.1117/1.3560542

[49] Lau FK, Stewart B, McStay D. An optically remote powered subsea video monitoring system. In: Proceedings of SPIE 8372, Ocean Sensing and Monitoring IV, 837209; 12 June 2012; Baltimore, USA. DOI:10.1117/12.920834

[50] Tanaka Y, Kurokawa T. A fiber sensor network using fiber optic power supply. In: Proceedings of SPIE 8421, OFS2012 22nd International Conference on Optical Fiber Sensors; 84211M; 14 October 2012; Beijing, China. DOI:10.1117/12.981524

[51] Rosolem JB, Bassan FR, Pereira FR, Penze RS, Leonardi AA, et al. Fiber powered sensing system for a long reach single mode fiber link and non-continuous applications. In: Proceedings of SPIE 9157, 23rd International Conference on Optical Fibre Sensors, 9157AE; 02 June 2014; Santander, Spain. DOI:10.1117/12.2058672

[52] Silva JCV, Souza ELAS, Garcia V, Rosolem JB, Floridia C., Sanches MAB. Design of a multimode fiber optic cable to transmit optical energy for long reach in PoF systems. In: Proceedings of IWCS International Cable, Connectivity Symposium, 9–12 November 2014, Rhode Island, USA.

[53] Rosolem JB. Quantitative and qualitative monitoring system for switchgear with full electrical isolation using fiber-optic technology. IEEE Transactions on Power Delivery. 2015; **30**:449–1457 DOI: 10.1109/TPWRD.2014.2377122

[54] Rosolem JB, Bassan FR, Penze RS, Leonardi AA, Fracarolli JPV, Floridia C. Optical sensing in high voltage transmission lines using power over fiber and free space optics. Optical Fiber Technology. 2015; **26**:180–183. DOI: 10.1016/j.yofte.2015.09.003

[55] Zhang X, et al. A gate drive with power over fiber-based isolated power supply and comprehensive protection functions for 15-kV SiC MOSFET. IEEE Journal of Emerging and Selected Topics in Power Electronics. 2016; **4**:946–955. DOI: 10.1109/JESTPE.2016.2586107

[56] Seo K, Nishimura N, Shiino M, Yuguchi R, Sasaki H. Evaluation of high-power endurance in optical fiber links. Furukawa Review. 2003; **24**:17–22.

[57] Kashyap R, Blow KJ. Observation of catastrophic self-propelled self-focusing in optical fibers. Electronics Letters. 1988; **24**:47–49. DOI: 10.1049/el:19880032

[58] Davis DD, Mettler SC, DiGiovanni DJ. Experimental data on the fiber fuse", In: Proceedings of SPIE 2714, 27th Annual Boulder Damage Symposium: Laser-Induced Damage in Optical Materials; 1995; Boulder, USA; pp. 202–210. DOI:10.1117/12.240382

[59] Andreev V., et al. High current density GaAs and GaSb photovoltaic cells for laser power beaming. In: Proceedings of 3rd World Conference on Photovoltaic Energy Conversion; 11–18 May 2003; Osaka, Japan; pp. 761–764.

[60] Sohr S, Rieske R, Nieweglowski K, Wolter KJ. Laser power converters for optical power supply. In: Proceedings of 34th International Spring Seminar on Electronics Technology (ISSE); 11–15 May 2011; Tratanska Lomnica, Slovakia. DOI: 10.1109/ISSE.2011.6053563

[61] Allwood G, Wild G, Hinckley S. Power over fibre: material properties of homojunction photovoltaic micro-cells. In: Proceedings of Sixth IEEE International Symposium on Electronic Design, Test and Application; 17–19 Jan. 2011; Queenstown, New Zealand, pp. 78–82. DOI: 10.1109/DELTA.2011.66

[62] Rosolem JB, et al. A fiber optic powered sensor designed for partial discharges monitoring on high voltage bushings. In: Proceedings of SBMO/IEEE MTT-S International Microwave and Optoelectronics Conference (IMOC); 3–6 Nov. 2015; Porto de Galinhas, Brazil, pp. 1–5. DOI: 10.1109/IMOC.2015.7369072

9

Digital Signal Processing for Optical Communications and Networks

Tianhua Xu

Additional information is available at the end of the chapter

Abstract

The achievable information rates of optical communication networks have been widely increased over the past four decades with the introduction and development of optical amplifiers, coherent detection, advanced modulation formats, and digital signal processing techniques. These developments promoted the revolution of optical communication systems and the growth of Internet, towards the direction of high-capacity and long-distance transmissions. The performance of long-haul high-capacity optical fiber communication systems is significantly degraded by transmission impairments, such as chromatic dispersion, polarization mode dispersion, laser phase noise and Kerr fiber nonlinearities. With the entire capture of the amplitude and phase of the signals using coherent optical detection, the powerful compensation and effective mitigation of the transmission impairments can be implemented using the digital signal processing in electrical domain. This becomes one of the most promising techniques for next-generation optical communication networks to achieve a performance close to the Shannon capacity limit. This chapter will focus on the introduction and investigation of digital signal processing employed for channel impairments compensation based on the coherent detection of optical signals, to provide a roadmap for the design and implementation of real-time optical fiber communication systems.

Keywords: optical communications, optical networks, digital signal processing, coherent detection, chromatic dispersion, polarization mode dispersion, laser phase noise, fiber nonlinearities

1. Introduction

The performance of high-capacity optical communication systems can be significantly degraded by fiber attenuation, chromatic dispersion (CD), polarization mode dispersion (PMD), laser phase noise (PN), and Kerr nonlinearities [1–10]. Using coherent detection, the powerful com-pensation of transmission impairments can be implemented in electrical domain. With the full information of the received signals, the chromatic dispersion, the polarization mode dispersion, the carrier phase noise, and the fiber Kerr nonlinearities can be equalized and mitigated using digital signal processing (DSP) [11–22].

Due to the high sensitivity of the receiver, coherent optical transmission was investigated extensively in the eighties of last century [23, 24]. However, the development of coherent communication has been delayed for nearly 20 years after that period [25, 26]. Coherent optical detection re-attracted the research interests until 2005, since the advanced modulation formats, i.e., m-level phase shift keying (m-PSK) and m-level quadrature amplitude modulation (m-QAM), can be applied [27–30]. In addition, coherent optical detection allows the electrical mitigation of system impairments. With the two main merits, the reborn coherent detections brought us the enormous potential for higher transmission speed and spectral efficiency in current optical fiber communication systems [31, 32].

With an additional local oscillator (LO) source, the sensitivity of coherent receiver reached the limitation of the shot-noise. Furthermore, compared to the traditional intensity modula-tion direct detection system, the multilevel modulation formats can be applied using the phase modulations, which can include more information bits in one transmitted symbol than before.

Meanwhile, since the coherent demodulation is linear and all information of the received signals can be detected, signal processing approaches, i.e., tight spectral filtering, CD equali-zation, PMD compensation, laser PN estimation, and fiber nonlinearity compensation, can be implemented in electrical domain [33–40].

The typical block diagram of the coherent optical transmission system is shown in **Figure 1**. The transmitted optical signal is combined coherently with the continuous wave from the narrow-linewidth LO laser so that the detected optical intensity in the photodiode (PD) ends can be increased and the phase information of the optical signal can be obtained. The use of LO laser is to increase the receiver sensitivity of the detection of optical signals, and the perfor-mance of coherent transmission can even behave close to the Shannon limit [3, 12].

The development of the coherent transmission systems has stopped for more than 10 years due to the invention of Erbium-doped fiber amplifiers (EDFAs) [1, 2]. The coherent trans-mission techniques attracted the interests of investigation again around 2005, when a new stage of the coherent lightwave systems comes out by combining the digital signal processing techniques [41–46]. This type of coherent lightwave system is called as digital coherent communication system. In the digital coherent transmission systems, the electrical signals output from the photodiodes are sampled and transformed into the discrete signals using high-speed analogue-to-digital convertors (ADCs), which can be further processed by the DSP algorithms.

The phase locking and the polarization adjustment were the main obstacles in the traditional coherent lightwave systems, while they can be solved by the carrier phase estimation and the polarization equalization, respectively, in the digital coherent optical transmission systems [47–55]. Besides, the chromatic dispersion and the nonlinear effects can also be mitigated by using the digital signal processing techniques [56–62]. The typical structure of the DSP compensating modules in the digital coherent receiver is shown in **Figure 2**.

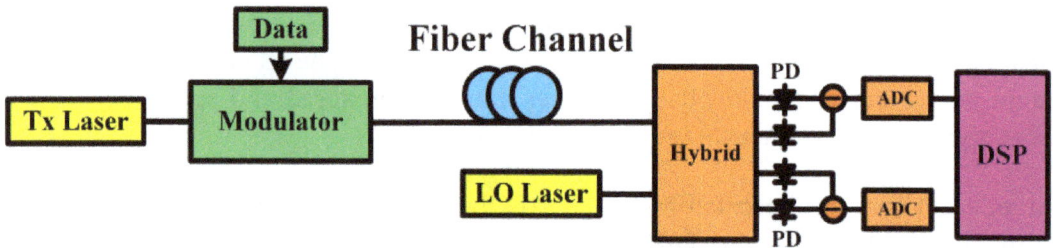

Figure 1. Schematic of coherent optical communication system with digital signal processing.

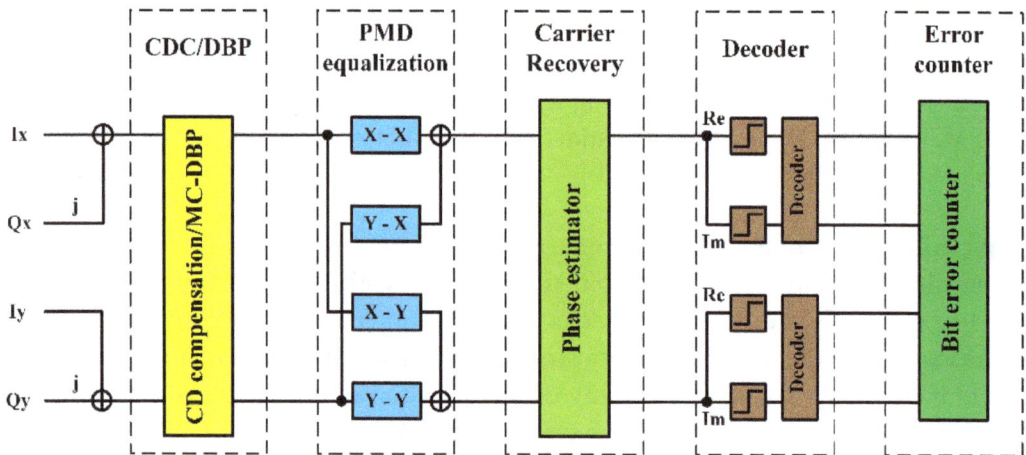

Figure 2. Block diagram of DSP in digital coherent receiver.

2. Digital signal processing for compensating transmission impairments

In this section, the chromatic dispersion compensation, polarization mode dispersion equalization, and carrier phase noise compensation are analyzed and discussed using corresponding DSP algorithms.

2.1. Chromatic dispersion compensation

Digital filters involving the time-domain least-mean-square (TD-LMS) adaptive filter, the static time-domain finite impulse response (STD-FIR) filter, and the frequency-domain equalizers (FDEs) are investigated for CD compensation. The characters of these filters are analyzed based on a 28-Gbaud dual-polarization quadrature phase shift keying (DP-QPSK) coherent transmission system using postcompensation of dispersion. It is noted that the STD-FIR filter and the FDEs can also be used for the dispersion predistorted coherent communication systems.

2.1.1. Time domain least-mean-square equalizer

The TD-LMS filter employs an iterative algorithm that incorporates successive corrections to weights vector in the negative direction of the gradient vector, which eventually leads to a minimum mean square error [34, 38, 63–65]. The transfer function of the TD-LMS digital filter can be described as follows:

$$y_{out}(n) = \vec{W}_{LMS}^{H}(n)\vec{x}_{in}(n) \tag{1}$$

$$\vec{W}_{LMS}(n+1) = \vec{W}_{LMS}(n) + \mu_{LMS}\vec{x}_{in}(n)e_{LMS}^{*}(n) \tag{2}$$

$$e_{LMS}(n) = d_{LMS}(n) - y_{out}(n) \tag{3}$$

where $\vec{x}_{in}(n)$ is the vector of received signals, $y_{out}(n)$ is the equalized output signal, n is the index of signal, $\vec{W}_{LMS}(n)$ is the vector of tap weights, H is the Hermitian transform operator, $d_{LMS}(n)$ is the desired symbol, $e_{LMS}(n)$ is the error between the desired symbol and the output signal, $*$ is the conjugation operator, and μ_{LMS} is the step size. To ensure the convergence of tap weights $\vec{W}_{LMS}(n)$, the step size μ_{LMS} has to meet the condition of $\mu_{LMS} < 1/U_{max}$, where U_{max} is the largest eigenvalue of the correlation matrix $R = \vec{x}_{in}(n)\vec{x}_{in}^{H}(n)$ [63]. The TD-LMS dispersion compensation filter can be applied in the "decision-directed" or the "sequence-training" mode [63].

The tap weights in TD-LMS adaptive equalizer for 20 km fiber CD compensation are shown in **Figure 3**. The convergence for 9 tap weights in the TD-LMS filter with step size equal to 0.1 is shown in **Figure 3(a)**, and it is found that the tap weights reach their convergence after ~5000 iterations. The distribution of the magnitudes of the converged tap weights is plotted in **Figure 3(b)**, and it is found that the central tap weights take more dominant roles than the high-order tap weights [34, 66].

2.1.2. Static time-domain finite impulse response filter

Compared with the iteratively updated TD-LMS filter, the tap weights in STD-FIR filter have a relatively simple specification [34, 67–69], the tap weight in STD-FIR filter is given by the following equations:

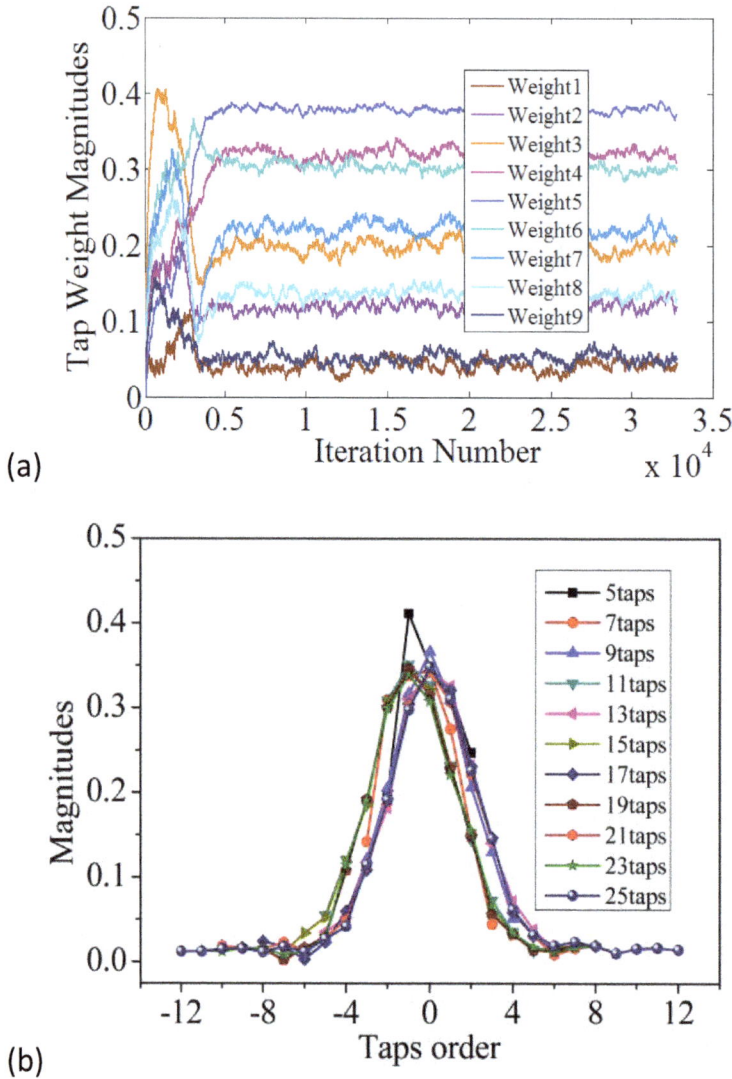

Figure 3. Taps weights of TD-LMS filter. (a) Tap weights magnitudes convergence and (b) converged tap weights magnitudes distribution.

$$a_k = \sqrt{\frac{jcT^2}{D\lambda^2 L}} \exp\left(-j\frac{\pi cT^2}{D\lambda^2 L}k^2\right) - \left\lfloor\frac{N}{2}\right\rfloor \leq k \leq \left\lfloor\frac{N}{2}\right\rfloor \tag{4}$$

$$N^A = 2 \times \left\lfloor\frac{|D|\lambda^2 L}{2cT^2}\right\rfloor + 1 \tag{5}$$

where D is the CD coefficient, λ is the carrier central wavelength, L is the length of fiber, T is the sampling period, N^A is the maximum number of taps, and $\lfloor x \rfloor$ means the nearest integer smaller than x.

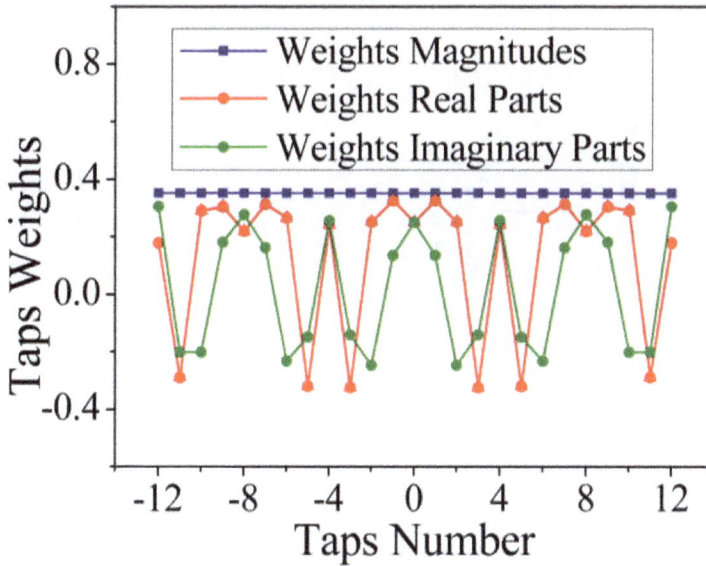

Figure 4. Tap weights of STD-FIR chromatic dispersion compensation filter.

For 20 km fiber with CD coefficient of $D = 16ps/(nm \cdot km)$, the distribution of the tap weights in the STD-FIR filter is shown in **Figure 4**.

2.1.3. Frequency domain equalizers

Since the complexity is very low for compensating large CD [34, 70], the most promising and popular chromatic dispersion compensation filters in coherent transmission systems are the frequency domain equalizers. The transfer function of the frequency domain equalizers is given by the following expression:

$$G_c(L, \omega) = \exp\left(\frac{-jD\lambda^2\omega^2L}{4\pi c}\right) \qquad (6)$$

where D is the chromatic dispersion coefficient, λ is the carrier central wavelength, ω is the angular frequency, L is the length of fiber, and c is the light speed in vacuum.

The frequency domain equalizers are generally implemented using the overlap-save (OLS) and the overlap-add (OLA) approaches based on the fast Fourier transform and the inverse fast Fourier transform (iFFT) convolution algorithms [71–73], as described in **Figure 5**.

2.2. Polarization mode dispersion equalization

Due to the random character of the polarization mode dispersion and the polarization rotation, the compensation of the PMD and the polarization rotation are generally realized by the adaptive algorithms such as the least-mean-square (LMS) and the constant modulus algorithm (CMA) filters.

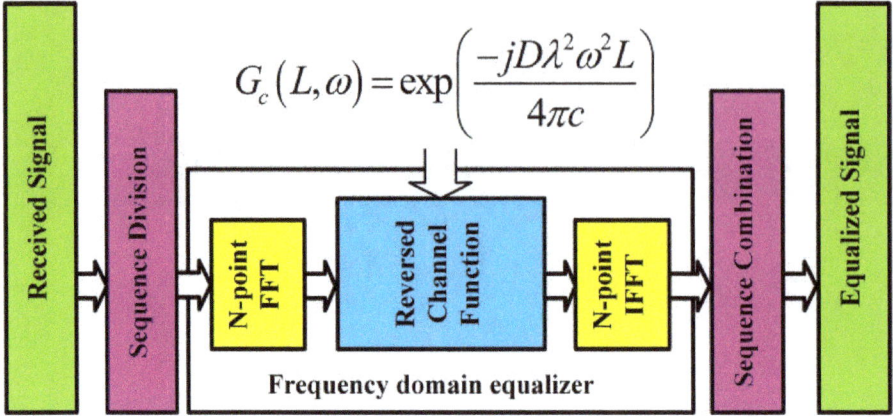

$$G_c(L,\omega) = \exp\left(\frac{-jD\lambda^2\omega^2 L}{4\pi c}\right)$$

Received Signal — Sequence Division — N-point FFT — Reversed Channel Function — N-point IFFT — Sequence Combination — Equalized Signal

Frequency domain equalizer

Figure 5. Schematic of frequency domain equalizer for chromatic dispersion compensation.

2.2.1. LMS adaptive PMD equalization

In the electrical domain, the impact of the PMD and the polarization fluctuation can be adaptively equalized using the decision-directed LMS (DD-LMS) filter [36, 63], of which the transfer function is given by:

$$\begin{bmatrix} x_{out}(n) \\ y_{out}(n) \end{bmatrix} = \begin{bmatrix} \vec{w}_{xx}^H(n) & \vec{w}_{xy}^H(n) \\ \vec{w}_{yx}^H(n) & \vec{w}_{yy}^H(n) \end{bmatrix} \cdot \begin{bmatrix} \vec{x}_{in}(n) \\ \vec{y}_{in}(n) \end{bmatrix} \tag{7}$$

$$\begin{cases} \vec{w}_{xx}(n+1) = \vec{w}_{xx}(n) + \mu_p \cdot \varepsilon_x(n) \cdot \vec{x}_{in}^*(n) \\ \vec{w}_{yx}(n+1) = \vec{w}_{yx}(n) + \mu_p \cdot \varepsilon_y(n) \cdot \vec{x}_{in}^*(n) \\ \vec{w}_{xy}(n+1) = \vec{w}_{xy}(n) + \mu_p \cdot \varepsilon_x(n) \cdot \vec{y}_{in}^*(n) \\ \vec{w}_{yy}(n+1) = \vec{w}_{yy}(n) + \mu_p \cdot \varepsilon_y(n) \cdot \vec{y}_{in}^*(n) \end{cases} \tag{8}$$

$$\begin{cases} \varepsilon_x(n) = d_x(n) - x_{out}(n) \\ \varepsilon_y(n) = d_y(n) - y_{out}(n) \end{cases} \tag{9}$$

where $\vec{x}_{in}(n)$ and $\vec{y}_{in}(n)$ are the vectors of the input signals, $x_{out}(n)$ and $y_{out}(n)$ are the equalized output signals, respectively, $\vec{w}_{xx}(n)$, $\vec{w}_{xy}(n)$, $\vec{w}_{yx}(n)$ and $\vec{w}_{yy}(n)$ are the complex tap weights vectors, $d_x(n)$ and $d_y(n)$ are the desired symbols, $\varepsilon_x(n)$ and $\varepsilon_y(n)$ are the estimation errors between the desired symbols and the output signals in the two polarizations, respectively, and μ_p is the step size in the DD-LMS algorithm.

2.2.2. CMA adaptive PMD equalization

The influence of the PMD and the polarization fluctuation can also be compensated employing the CMA adaptive filter [74, 75], of which the transfer function can be described as:

$$\begin{bmatrix} x_{out}(n) \\ y_{out}(n) \end{bmatrix} = \begin{bmatrix} \vec{v}_{xx}^{H}(n) & \vec{v}_{xy}^{H}(n) \\ \vec{v}_{yx}^{H}(n) & \vec{v}_{yy}^{H}(n) \end{bmatrix} \cdot \begin{bmatrix} \vec{x}_{in}(n) \\ \vec{y}_{in}(n) \end{bmatrix} \tag{10}$$

$$\begin{cases} \vec{v}_{xx}(n+1) = \vec{v}_{xx}(n) + \mu_q \cdot \eta_x(n) \cdot \vec{x}_{in}^{*}(n) \\ \vec{v}_{yx}(n+1) = \vec{v}_{yx}(n) + \mu_q \cdot \eta_y(n) \cdot \vec{x}_{in}^{*}(n) \\ \vec{v}_{xy}(n+1) = \vec{v}_{xy}(n) + \mu_q \cdot \eta_x(n) \cdot \vec{y}_{in}^{*}(n) \\ \vec{v}_{yy}(n+1) = \vec{v}_{yy}(n) + \mu_q \cdot \eta_y(n) \cdot \vec{y}_{in}^{*}(n) \end{cases} \tag{11}$$

$$\begin{cases} \eta_x(n) = 1 - |x_{out}(n)|^2 \\ \eta_y(n) = 1 - |y_{out}(n)|^2 \end{cases} \tag{12}$$

where $\vec{x}_{in}(n)$ and $\vec{y}_{in}(n)$ are the vectors of the input signals, $x_{out}(n)$ and $y_{out}(n)$ are the equalized output signals, respectively, $\vec{v}_{xx}(n)$, $\vec{v}_{xy}(n)$, $\vec{v}_{yx}(n)$, and $\vec{v}_{yy}(n)$ are the complex tap weights vectors, $\eta_x(n)$ and $\eta_y(n)$ are the estimation errors between the desired amplitude and the output signals in the two polarizations, respectively, and μ_q is the step size in the CMA algorithm.

It can be found that the CMA algorithm is based on the principle of minimizing the modulus variation of the output signal to update its weight vector.

2.3. Carrier phase estimation

In this section, the analyses on different carrier phase estimation algorithms, involving the one-tap normalized LMS, the differential phase estimation, the block-wise average (BWA), and the Viterbi-Viterbi (VV) methods in the coherent optical transmission systems, will be presented.

2.3.1. The normalized LMS carrier phase estimation

The one-tap normalized LMS filter can be employed effectively for carrier phase estimation [76–78], of which the tap weight is expressed as:

$$w_{NLMS}(n+1) = w_{NLMS}(n) + \frac{\mu_{NLMS}}{|x_{in}(n)|^2} x_{in}^{*}(n) e_{NLMS}(n) \tag{13}$$

$$e_{NLMS}(n) = d_{PE}(n) - w_{NLMS}(n) \cdot x_{in}(n) \tag{14}$$

where $w_{NLMS}(n)$ is the tap weight, $x_{in}(n)$ is the input signal, n is the symbol index, $d_{PE}(n)$ is the desired symbol, and $e_{NLMS}(n)$ is the carrier phase estimation error between the desired symbol and the output signal, and μ_{NLMS} is the step size in the one-tap normalized LMS filter.

It has been demonstrated that the one-tap normalized LMS carrier phase estimation behaves similar to the differential phase estimation [28, 53, 55, 76], of which the BER floor in the m-PSK coherent optical transmission systems can be approximately described by the following analytical expression:

$$BER_{floor}^{NLMS} \approx \frac{1}{\log_2 m} erfc\left(\frac{\pi}{m\sqrt{2}\sigma}\right) \tag{15}$$

where σ is the square root of the phase noise variance. The schematic of the one-tap normalized LMS carrier phase estimation is illustrated in **Figure 6**.

2.3.2. Differential carrier phase estimation

The differential signal demodulation can be also applied for carrier phase estimation in coherent transmission system [28, 53, 55], where the differentially encoded data can be recovered using the "delay and multiply" algorithm. Using differential carrier phase estimation, the encoded information can be recovered according to the phase difference between the two consecutive symbols, i.e., the decision variable $\Psi = x_n x_{n+1}^* \exp\{i\pi/m\}$, where x_n and x_{n+1} are the consecutive n-th and (n+1)-th received symbols. The BER floor of the differential carrier phase estimation can be evaluated using the principle of conditional probability. For the m-PSK coherent systems, the BER floor in differential phase estimation is expressed as the following equation [28, 53]:

$$BER_{floor}^{Differential} = \frac{1}{\log_2 m} erfc\left(\frac{\pi}{m\sqrt{2}\sigma}\right) \tag{16}$$

where σ is the square root of the phase noise variance. The schematic of the differential carrier phase estimation is described in **Figure 7**.

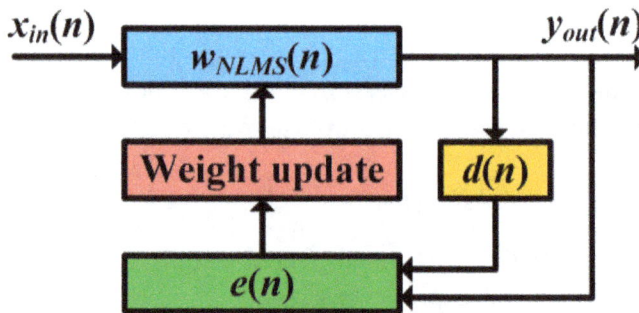

Figure 6. Schematic of one-tap normalized LMS carrier phase estimation.

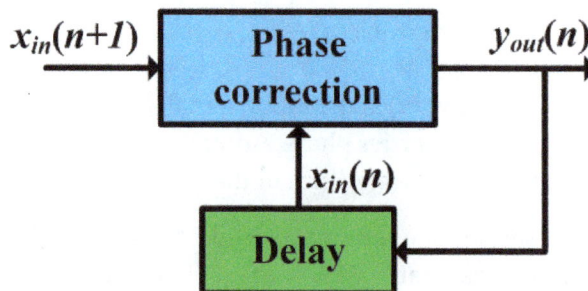

Figure 7. Schematic of differential carrier phase estimation.

2.3.3. The block-wise average carrier phase estimation

The block-wise average approach calculates the m-th power of the received symbols in each processing unit to remove the information of phase modulation, and the computed phase is summed and averaged over the entire process block, where the length of the process block is called block size. Then the averaged phase is divided by m, and the result leads to the phase estimate for the entire data block [79–81]. For the m-PSK coherent communication system, the estimated carrier phase in each process block using the block-wise average approach is given by the following expression:

$$\hat{\Phi}_{BWA}(n) = \frac{1}{m}\arg\left\{\sum_{k=1+(M-1)\cdot N_b}^{M\cdot N_b} x^m(k)\right\} \tag{17}$$

$$M = \left\lceil\frac{n}{N_b}\right\rceil \tag{18}$$

where N_b is the block size in the BWA approach, and $\lceil x \rceil$ means the nearest integer larger than x.

The performance of the block-wise average carrier phase estimation method in the m-PSK coherent optical communication system can be derived based on the Taylor expansion of the estimated carrier phase error, and the BER floor in the block-wise average carrier phase estimation can be described using the following expression [52, 53, 55, 79]:

$$BER_{floor}^{BWA} \approx \frac{1}{N_b \cdot \log_2 m} \cdot \sum_{k=1}^{N_b} erfc\left(\frac{\pi}{m\sqrt{2}\sigma_{BWA,k}}\right) \tag{19}$$

$$\sigma_{BWA,k}^2 = \frac{\sigma^2}{6N_b^2} \cdot \left[2(k-1)^3 + 3(k-1)^2 + 2(N_b-k)^3 + 3(N_b-k)^2 + N_b - 1\right] \tag{20}$$

where σ^2 represents the total phase noise variance in the coherent transmission system. The schematic of the block-wise average carrier phase estimation is shown in **Figure 8**.

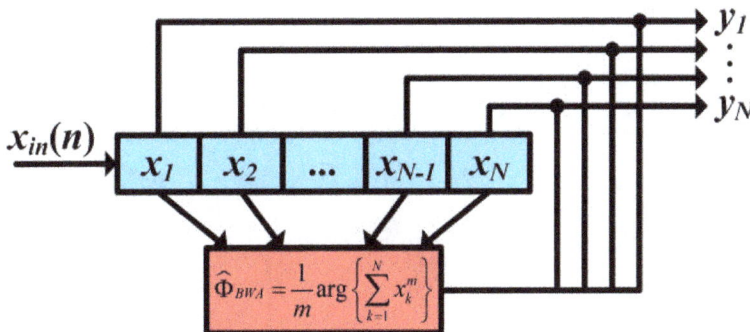

Figure 8. Schematic of block-wise average carrier phase estimation.

2.3.4. The Viterbi-Viterbi carrier phase estimation

The Viterbi-Viterbi carrier phase estimation approach also operates the symbols in each process block into the m-th power to remove the information of the phase modulation. The computed phase is also summed and averaged over the entire process block, where the length of the process block is also called block size. Then the averaged phase is divided by m as the estimated carrier phase. However, compared to the BWA approach, the estimated phase in the Viterbi-Viterbi carrier phase estimation approach is only applied in the phase recovery of the central symbol in each process block [55, 81–83]. The estimated carrier phase in the Viterbi-Viterbi approach in m-PSK optical communication systems is given by the following expression:

$$\hat{\Phi}_{VV}(n) = \frac{1}{m}\arg\left\{\sum_{k=-(N_v-1)/2}^{(N_v-1)/2} x^m(n+k)\right\}, \quad N_v = 1,3,5,7\ldots \tag{21}$$

where N_v is the block size in the Viterbi-Viterbi carrier phase estimation approach.

The performance of the Viterbi-Viterbi carrier phase estimation in the m-PSK coherent optical communication system can also be derived employing the Taylor expansion of the estimated carrier phase. The BER floor in the Viterbi-Viterbi carrier phase estimation for the m-PSK transmission system can be expressed as follows [52, 53, 55]:

$$BER_{floor}^{VV} \approx \frac{1}{\log_2 m}erfc\left(\frac{\pi}{m\sqrt{2}\sigma_{VV}}\right) \tag{22}$$

$$\sigma_{VV}^2 = \sigma^2 \cdot \frac{N_v^2 - 1}{12N_v} \tag{23}$$

where σ^2 represents the total phase noise variance in the coherent transmission system. The schematic of the Viterbi-Viterbi carrier phase estimation is illustrated in **Figure 9**.

According to Eqs. (20) and (23), it can be found that the phase estimate error in the Viterbi-Viterbi carrier phase estimation corresponds to the phase estimate error of the central symbol (the smallest error) in the block-wise average carrier phase estimation. Therefore, the Viterbi-

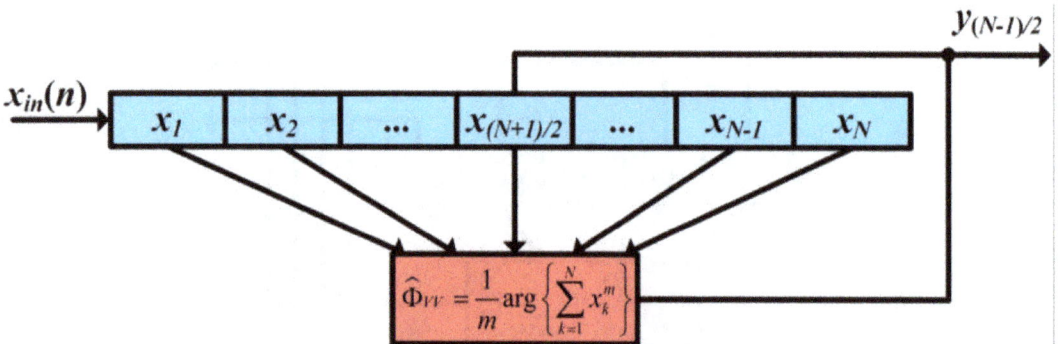

Figure 9. Schematic of Viterbi-Viterbi carrier phase estimation.

Viterbi approach will generally perform better than the block-wise average approach, in terms of the phase estimate error. However, it requires more computational complexity to update the process unit for the phase estimation of each symbol.

It is noted that the one-tap normalized LMS algorithm can also be employed for the m-QAM coherent transmission systems, while the block-wise average and the Viterbi-Viterbi methods cannot be easily used for the classical m-QAM coherent systems except the circular constellation m-QAM systems.

3. Conclusions

In this chapter, the digital signal processing techniques for compensating transmission impairments in optical communication systems including chromatic dispersion, polarization mode dispersion, and laser phase noise have been described and analyzed in detail. Chromatic dispersion can be compensated using the digital filters in both time domain and frequency domain. Polarization mode dispersion can be equalized adaptively using the least-mean-square method and the constant modulus algorithm. Phase noise from the laser sources can be estimated and compensated using the feed-forward and feed-back carrier phase recovery approaches.

Digital signal processing combined with coherent detection shows a very promising solution for long-haul high-capacity optical communication systems, which offers a great flexibility in the design, deployment, and operation of optical communication networks. Fiber nonlinearities, including self-phase modulation, cross-phase modulation, and four-wave mixing, can be mitigated using single-channel and multichannel digital back-propagation in the electrical domain, which will be discussed in future work.

Acknowledgements

This work is supported in part by UK Engineering and Physical Sciences Research Council (project UNLOC EP/J017582/1), in part by European Commission Research Council FP7-PEOPLE-2012-IAPP (project GRIFFON, No. 324391), in part by European Commission Research Council FP7-PEOPLE-2013-ITN (project ICONE, No. 608099), and in part by Swedish Research Council Vetenskapsradet (No. 0379801).

Author details

Tianhua Xu

Address all correspondence to: tianhua.xu@ucl.ac.uk

Department of Electronic and Electrical Engineering, University College London, London, United Kingdom

References

[1] Agrawal GP. Fiber-Optic Communication Systems. 4th ed. John Wiley & Sons, Inc, New Jersey, USA.; 2010. ISBN: 978-0-470-50511-3

[2] Kaminow I, Li T, Willner AE. Optical Fiber Telecommunications VB: System and Networks. 5th ed. Oxford: Academic Press; 2010. ISBN: 978-0-12-374172-1

[3] Semrau D. et al. Achievable information rates estimates in optically amplified transmission systems using nonlinearity compensation and probabilistic shaping. Optics Letters. 2017;**41**(21):121-124. DOI: 10.1364/OL.42.000121

[4] Li Y. et al. Dynamic dispersion compensation in a 40 Gb/s single-channeled optical fiber communication system, Acta Optica Sinica. 2007;**27**:1161-1165. ISSN: 0253-2239.2007.07.004

[5] IEEE, Xu T. et al. Overcoming fibre nonlinearities to enhance the achievable transmission rates in optical communication systems. In: Asia Communications and Photonics Conference, Workshop 3, Hong Kong; 2015. (Invited Talk)

[6] IEEE, Liga G. et al. Ultra-wideband nonlinearity compensation performance in the presence of PMD. In: European Conference on Optical Communication, Düsseldorf , Germany; 2016. 794-796. ISBN: 978-3-8007-4274-5

[7] KTH, Xu T. DSP Based Chromatic Dispersion Equalization and Carrier Phase Estimation in High Speed Coherent Optical Transmission Systems (KTH Ph.D. thesis), Stockholm, Sweden; 2012. ISBN 978-91-7501-346-6

[8] Sergeyev S. et al., All-optical polarisation control in fibre Raman amplifiers. International Scientific and Technical Conference on Quantum Electronics. 2013;**81**. ISBN: 978-985-553-157-0

[9] Jacobsen G. et al. Phase noise influence in coherent optical DnPSK systems with DSP based dispersion compensation. Journal of Optical Communications. 2014;**35**(1):57–61. DOI: 10.1515/joc-2013-0065

[10] Xu T. et al. Field trial over 820 km installed SSMF and its potential Terabit/s superchannel application with up to 57.5-Gbaud DP-QPSK transmission. Optics Communications. 2015;**353**:133-138. DOI: 10.1016/j.optcom.2015.05.029

[11] Ip E, Lau APT, Barros DJF, Kahn JM. Coherent detection in optical fiber systems. Optics Express. 2008;**16**(2):753-791. DOI: 10.1364/OE.16.000753

[12] Li G. Recent advances in coherent optical communication. Advances in Optics and Photonics. 2009;**1**(2):279-307. DOI: 10.1364/AOP.1.000279

[13] G. Jacobsen, et al., Phase noise influence in optical OFDM systems employing RF pilot tone for phase noise cancellation. Journal of Optical Communications. 2011;**32**(2):141-145. DOI: 10.1515/joc.2011.017

[14] IEEE, Maher R. et al. Digital pulse shaping to mitigate linear crosstalk in Nyquist spaced 16QAM WDM transmission systems. In: OptoElectronics and Communication Conference, MO2B2, Melbourne, Australia; 2014. ISBN: 978-1-922107-21-3

[15] IEEE, Xu T. et al. Phase noise mitigation in coherent transmission system using a pilot carrier. In: Asia Communications and Photonics Conference, Proceedings SPIE. Vol. 8309, 8309Z, Shanghai, China; 2011. DOI: 10.1117/12.904038

[16] IEEE, Liga G. et al. Digital back-propagation for high spectral-efficiency Terabit/s superchannels. In: Optical Fiber Communication Conference, San Francisco, California, USA; 2014, W2A.23. DOI: 10.1364/OFC.2014.W2A.23

[17] Savory SJ, Digital equalization in coherent optical transmission systems. In: Enabling Technologies for High Spectral-Efficiency Coherent Optical Communication Networks. John Wiley & Sons, Inc., New Jersey, USA; 2016. p. 311-332. DOI: 10.1002/9781119078289.ch8

[18] Xu T. et al. Digital adaptive carrier phase estimation in multi-level phase shift keying coherent optical communication systems. International Conference on Information Science and Control Engineering. 2016;1293-1297. DOI:10.1109/ICISCE.2016.276

[19] IEEE, Kikuchi K. Coherent transmission systems. In: European Conference on Optical Communication, Brussels, Belgium; 2008, Th.2.A.1. DOI: 10.1109/ECOC.2008.4729551

[20] IEEE, Shevchenko NA. et al. Achievable information rates estimation for 100-nm Raman-amplified optical transmission system. In: European Conference on Optical Communication, Düsseldorf , Germany; 2016. pp. 878-880. ISBN: 978-3-8007-4274-5

[21] IEEE, Maher R. et al. Linear and nonlinear impairment mitigation in a Nyquist spaced DP-16QAM WDM transmission system with full-field DBP. In: European Conference on Optical Communication, Cannes, France; 2014, P.5.10. DOI: 10.1109/ECOC.2014.6963971

[22] T. Xu, et al., Analytical estimation in differential optical transmission systems influenced by equalization enhanced phase noise. In: Progress in Electromagnetics Research Symposium; 2016. pp. 4844-4848. DOI: 10.1109/PIERS.2016.7735770

[23] Mynbaev DK, Scheiner LL. Fiber-Optic Communication Technology. Prentice Hall, New Jersey, USA; 2000. p. 1-68. ISBN: 978-0139620690

[24] Okoshi T, Kikuchi K. Coherent Optical Fiber Communications. Kluwer Academic Publishers, Tokyo, Japan; 1988. p. 1-32. ISBN: 978-90-277-2677-3

[25] Kikuchi K. Coherent optical communications - history, state-of-the-art technologies, and challenges for the future. In: Opto-Electronics and Communications Conference; 2008. pp. 1-4. DOI: 10.1109/OECCACOFT.2008.4610574

[26] Kikuchi K. History of coherent optical communications and challenges for the future. IEEE Summer Topical Meetings, TuC1.1; 2008. DOI: 10.1109/LEOSST.2008.4590512

[27] Xu T. et al. Modulation format dependence of digital nonlinearity compensation performance in optical fibre communication systems. Optics Express, 20147;25(4):3311-3326. DOI: 10.1364/OE.25.003311

[28] Xu T. et al. Analytical BER performance in differential n-PSK coherent transmission system influenced by equalization enhanced phase noise. Optics Communications. 2015;334:222-227. DOI: 10.1016/j.optcom.2014.07.094

[29] Bayvel P. et al. Maximising the optical network capacity. Philosophical Transactions of the Royal Society A. 2016;**374**(2062):20140440. DOI: 10.1098/rsta.2014.0440

[30] Jacobsen G. et al. Error-rate floors in differential n-level phase-shift-keying coherent receivers employing electronic dispersion equalization. Journal of Optical Communications. 2011;**32**(3):191-193. DOI: 10.1515/JOC.2011.031

[31] Kahn J, Ho KP. Spectral efficiency limits and modulation/detection techniques for DWDM systems. Journal of Selected Topics in Quantum Electronics. 2004;**10**(2):259-272. DOI: 10.1109/JSTQE.2004.826575

[32] IEEE, Griffin RA. et al. 10 Gb/s optical differential quadrature phase shift key (DQPSK) transmission using GaAs/AlGaAs integration. In: Optical Fiber Communication Conference, WX6, Anaheim, California, USA; 2002. DOI: 10.1109/OFC.2002.1036787

[33] Tsukamoto S. et al. Coherent demodulation of 40-Gbit/s polarization-multiplexed QPSK signals with 16-GHz spacing after 200-km transmission. In: Optical Fiber Communication Conference, PDP29, 2005. DOI: 10.1109/OFC.2005.193207

[34] Xu T. et al. Chromatic dispersion compensation in coherent transmission system using digital filters. Optics Express. 2010;**18**(15):16243-16257. DOI: 10.1364/OE.18.016243

[35] Jacobsen G. et al. Phase noise influence in coherent optical OFDM systems with RF pilot tone: IFFT multiplexing and FFT demodulation. Journal of Optical Communications. 2012;**33**(3):217-226. DOI: 10.1515/joc-2012-0038

[36] Ip E, Kahn JM. Digital equalization of chromatic dispersion and polarization mode dispersion. Journal of Lightwave Technology. 2007;**25**(8):2033-2043. DOI: 10.1109/JLT.2007.900889

[37] Maher R. et al. Spectrally shaped DP-16QAM super-channel transmission with multi-channel digital back propagation. Scientific Reports. 2015;**5**:08214. DOI: 10.1038/srep08214

[38] Xu T. et al. Normalized LMS digital filter for chromatic dispersion equalization in 112-Gbit/s PDM-QPSK coherent optical transmission system. Optics Communications. 2010;**283**:963-967. DOI: 10.1016/j.optcom.2009.11.011

[39] Jacobsen G. et al. Phase noise influence in long-range coherent optical OFDM systems with delay detection: IFFT multiplexing and FFT demodulation. Journal of Optical Communications. 2012;**33**(4):289-295. DOI: 10.1515/joc-2012-0047

[40] IEEE, Liga G, Czegledi CB, Xu T. PMD and wideband nonlinearity compensation: next bottleneck or fundamental limitation? In: European Conference on Optical Communication (ECOC), Workshop WS06, Düsseldorf, Germany; 2016. (Invited Talk)

[41] Jacobsen G. et al. Receiver implemented RF pilot tone phase noise mitigation in coherent optical nPSK and nQAM systems. Optics Express. 2011;**19** (15):14487-14494. DOI: 10.1364/OE.19.014487

[42] Xu T. et al. Analysis of chromatic dispersion compensation and carrier phase recovery in long-haul optical transmission system influenced by equalization enhanced phase noise. Optik, to appear, 2017.

[43] PIERS, Jacobsen G. et al. Capacity constraints for phase noise influenced coherent optical DnPSK systems. In: Progress in Electromagnetics Research Symposium, Guangzhou, China; 2014. 140319195903 (Invited Talk)

[44] Xu T. et al. Equalization enhanced phase noise in Nyquist-spaced superchannel transmission systems using multi-channel digital back-propagation. Scientific Reports. 2015;5:13990. DOI: 10.1038/srep13990

[45] SPIE, Yoshida T, Sugihara T, Uto K. DSP-based optical modulation technique for long-haul transmission. Next-Generation Optical Communication: Components, Sub-Systems, and Systems IV Proceedings SPIE; Vol. 9389, 93890K, San Francisco, California, USA; 2015. DOI: 10.1117/12.2078042

[46] IEEE, Xu T. et al. Quasi real-time 230-Gbit/s coherent transmission field trial over 820 km SSMF using 57.5-Gbaud dual-polarization QPSK. In: Asia Communications and Photonics Conference, Beijing, China; 2013. AF1F.3. DOI: 10.1364/ACPC.2013.AF1F.3

[47] Ip E, Kahn JM. Feedforward carrier recovery for coherent optical communications. Journal of Lightwave Technology. 2007;25(9):2675-2692. DOI: 10.1109/JLT.2007.902118

[48] Haunstein HF. et al. Principles for electronic equalization of polarization-mode dispersion. Journal of Lightwave Technology. 2004;22(4):1169-1182. DOI: 10.1109/JLT.2004.825333

[49] IEEE, Xu T. et al. Mitigation of EEPN in long-haul n-PSK coherent transmission system using modified optical pilot carrier. In: Asia Communications and Photonics Conference, Beijing, China; 2013. AF3E.1. DOI: 10.1364/ACPC.2013.AF3E.1

[50] Jacobsen G. et al. Study of EEPN mitigation using modified RF pilot and Viterbi-Viterbi based phase noise compensation. Optics Express. 2013;21(10):12351-12362. DOI: 10.1364/OE.21.012351

[51] Taylor MG. Phase estimation methods for optical coherent detection using digital signal processing. Journal of Lightwave Technology. 2009;17(7):901-914. DOI: 10.1109/JLT.2008.927778

[52] Xu T. et al. Analytical investigations on carrier phase recovery in dispersion-unmanaged n-PSK coherent optical communication systems. Photonics. 2016;3(4):51. DOI: 10.3390/photonics3040051

[53] Xu T. et al. Comparative study on carrier phase estimation methods in dispersion-unmanaged optical transmission systems. In: Advanced Information Technology, Electronic and Automation Control Conference, to appear, 2017.

[54] Fatadin I, Ives D, Savory SJ. Differential carrier phase recovery for QPSK optical coherent systems with integrated tunable lasers. Optics Express. 2013;21(8):10166-10171. DOI: 10.1364/OE.21.010166

[55] Xu T. et al. Carrier phase estimation methods in coherent transmission systems influenced by equalization enhanced phase noise. Optics Communications. 2013;293:54-60. DOI: 10.1016/j.optcom.2012.11.090

[56] Jacobsen G. et al. EEPN and CD study for coherent optical nPSK and nQAM systems with RF pilot based phase noise compensation. Optics Express. 2012;**20**(8):8862-8870. DOI: 10.1364/OE.20.008862

[57] IEEE, Ellis A. et al. The impact of phase conjugation on the nonlinear-Shannon limit: the difference between optical and electrical phase conjugation. In: IEEE Summer Topical Conference, Nassau, Bahamas; 2015. pp. 209-210. DOI: 10.1109/PHOSST.2015.7248271

[58] Jacobsen G. Influence of pre- and post-compensation of chromatic dispersion on equalization enhanced phase noise in coherent multilevel systems. Journal of Optical Communications. 2011;**32**(4):257-261. DOI: 10.1515/JOC.2011.053

[59] Liga G. et al. On the performance of multichannel digital backpropagation in high-capacity long-haul optical transmission. Optics Express. 2014;**22**(24):30053-30062. DOI: 10.1364/OE.22.030053

[60] IEEE, Xu T. et al. Receiver-based strategies for mitigating nonlinear distortion in high-speed optical communication systems. In: Tyrrhenian International Workshop on Digital Communications, IEEE Photonics in Switching, Florence, Italy; 2015. P2.2 (Invited Talk)

[61] IEEE, Killey RI. et al. Experimental characterisation of digital Nyquist pulse-shaped dual-polarisation 16QAM WDM transmission and comparison with the Gaussian noise model of nonlinear propagation. In: International Conference on Transparent Optical Networks, Graz, Austria; 2014. Tu.D1.3. DOI: 10.1109/ICTON.2014.6876439

[62] IEEE, Le S. et al. Optical and digital phase conjugation techniques for fiber nonlinearity compensation. In: OptoElectronics and Communication Conference, Shanghai, China; 2015. pp. 1-3. DOI: 10.1109/OECC.2015.7340113

[63] Haykin S. Adaptive Filter Theory. 5th ed. Prentice Hall, New Jersey, USA; 2013. p. 231-341. ISBN: 978-0132671453

[64] IEEE, Xu T. et al. Digital compensation of chromatic dispersion in 112-Gbit/s PDM-QPSK system. In: Asia Communications and Photonics Conference Proceedings; Vol. 7632, 763202, Shanghai, China; 2009. DOI: 10.1364/ACP.2009.TuE2

[65] SPIE, Xu T. et al. Variable-step-size LMS adaptive filter for digital chromatic dispersion compensation in PDM-QPSK coherent transmission system, In: International Conference on Optical Instruments and Technology, Shanghai, China; 2009. p. 7506, 75062I. DOI: 10.1117/12.837834

[66] www.diva-portal.org

[67] IEEE, Xu T. et al. Influence of digital dispersion equalization on phase noise enhancement in coherent optical system. In: Asia Communications and Photonics Conference, Guangzhou, China; 2012. AS1C.3. DOI: 10.1364/ACPC.2012.AS1C.3

[68] Savory SJ. Digital filters for coherent optical receivers. Optics Express. 2008;**16**(2):804-817. DOI: 10.1364/OE.16.000804

[69] IEEE, Xu T. et al. Digital chromatic dispersion compensation in coherent transmission system using a time-domain filter. In: Asia Communications and Photonics Conference, Shanghai, China; 2010. pp. 132-133. DOI: 10.1109/ACP.2010.5682798

[70] Kudo R. et al. Coherent optical single carrier transmission using overlap frequency domain equalization for long-haul optical systems. Journal of Lightwave Technology. 2009;**27**(16):3721-3728. DOI: 10.1109/JLT.2009.2024091

[71] Benvenuto N, Cherubini G. Algorithms for Communications Systems and Their Applications. New York: John Wiley & Sons, Inc.; 2004. ISBN: 978-0-470-84389-5

[72] IEEE, Lowe D, Huang X. Adaptive overlap-add equalization for MB-OFDM ultra-wideband. In: International Symposium on Communications and Information Technologies, Bangkok, Thailand; 2006. 644-648. DOI: 10.1109/ISCIT.2006.339826

[73] Xu T. et al. Frequency-domain chromatic dispersion equalization using overlap-add methods in coherent optical system. Journal of Optical Communications. 2011;**32**(2):131-135. DOI: 10.1515/joc.2011.022

[74] Kwon OW, Un CK, Lee JC. Performance of constant modulus adaptive digital filters for interference cancellation. Signal Processing. 1992;**26**(2):185-196. DOI: 10.1016/0165-1684(92)90129-K

[75] Benesty J, Duhamel P. Fast constant modulus adaptive algorithm. IEE Radar and Signal Processing. 1991;**138**(4):379-387. DOI: 10.1049/ip-f-2.1991.0049

[76] Xu T. et al. Analytical estimation of phase noise influence in coherent transmission system with digital dispersion equalization. Optics Express. 2011;**19**(8):7756-7768. DOI: 10.1364/OE.19.007756

[77] Mori Y. et al. Unrepeated 200-km transmission of 40-Gbit/s 16-QAM signals using digital coherent receiver. Optics Express. 2009;**17**(3):1435-1441. DOI: 10.1364/OE.17.001435

[78] SPIE, Xu T. et al. Close-form expression of one-tap normalized LMS carrier phase recovery in optical communication systems. International Conference on Wireless and Optical Communications; Vol. 9902, 990203, Beijing, China; 2016. DOI: 10.1117/12.2261932

[79] Goldfarb G, Li G. BER estimation of QPSK homodyne detection with carrier phase estimation using digital signal. Optics Express. 2006;**14**(18):8043-8053. DOI: 10.1364/OE.14.008043

[80] Ly-Gagnon DS. et al. Coherent detection of optical quadrature phase-shift keying signals with carrier phase estimation., Journal of Lightwave Technology. 2006;**24**(1):12-21. DOI: 10.1109/JLT.2005.860477

[81] www.mdpi.com

[82] Viterbi AJ, Viterbi AM. Nonlinear estimation of PSK-modulated carrier phase with appli-
      cation to burst digital transmission. IEEE Transactions on Information Theory. 1983;**29**
      (4):543-551. DOI: 10.1109/TIT.1983.1056713

[83] IEEE, Xu T. et al. Analysis of carrier phase extraction methods in 112-Gbit/s NRZ-PDM-
      QPSK coherent transmission system. In: Asia Communications and Photonics Confer-
      ence, Guangzhou, China; 2012, AS1C.2. DOI: 10.1364/ACPC.2012.AS1C.2

10

OFDM Systems for Optical Communication with Intensity Modulation and Direct Detection

Jian Dang, Liang Wu and Zaichen Zhang

Additional information is available at the end of the chapter

Abstract

Intensity modulation and direct detection (IM/DD) is a cost-effective optical communication strategy which finds wide applications in fiber communication, free-space optical communication, and indoor visible light communication. In IM/DD, orthogonal frequency division multiplexing (OFDM), originally employed in radio frequency communication, is considered as a strong candidate solution to combat with channel distortions. In this research, we investigate various potential OFDM forms that are suitable for IM/DD channel. We will elaborate the design principles of different OFDM transmitters and investigate different types of receivers including the proposed iterative receiver. In addition, we will analyze the spectral efficiency and decoding complexities of different OFDM systems to give a whole picture of their performance. Finally, simulation results are given to assess the detection performance of different receivers.

Keywords: OFDM, optical communication, IM/DD, modulation, detection

1. Introduction

Optical communication is an important part of modern communication techniques due to the excessive bandwidth of the light spectrum. Theoretically, optical communication has much higher system throughput than its radio frequency (RF) communication counterpart. Therefore, it finds many applications and facilitates our lives. Some typical optical communication scenarios include optical fiber communication, free-space optical communication, and visible light communication. In those communication scenarios, intensity modulation and direct detection (IM/DD) is a cost-effective communication scheme compared to coherent ones. In IM/DD, the intensity, or power, of the light beam from a laser or a light-emitting diode (LED) is modulated by the information bits and no phase information is needed. Due to this nature, no local oscillator is required for IM/DD communication, which greatly eases the cost of the hardware.

In IM/DD channel, there are still some non-ideal factors that may deteriorate the quality of communication. One key factor is the multipath effect. This effect is caused by several mechanisms. First, in wireless communications, the light could be reflected at multiple locations and by many times by the surroundings before arriving at the receiver side. Second, the modulation bandwidth of LED is limited, typically below 100 MHz. When the bandwidth of signal exceeds the modulation bandwidth of LED, multipath effect occurs. Third, in fiber communication, light components of different wavelength propagate through different paths, which also cause multipath effect. Therefore, effective means of mitigating the multipath effect are necessary in IM/DD optical communications.

In RF communication, orthogonal frequency division multiplexing (OFDM) is a powerful multi-carrier modulation scheme to combat the multipath effect. Compared to the single-carrier modulation schemes, OFDM avoids the usage of a complicated high-order time-domain equalizer. Instead, it employs frequency domain equalizer that only has a single tap. This greatly simplifies the equalization task and can perfectly resolves the multipath effects without any residual errors at high signal-to-noise ratio (SNR) region. Thus, introducing OFDM to IM/DD optical communication is a natural choice. However, different from RF communication, IM/DD requires that the transmitted signal must be real and positive, which imposes strict constraint on the modulation scheme and the original OFDM transceiver must be modified carefully to satisfy the new scenario. In addition, different applications may have diverse emphasis such as spectral efficiency, power efficiency, detection capability, as well as computational complexity.

Within such perspectives, the purpose of this chapter was to analyze the potential forms of OFDM that are suitable for IM/DD transmission as well as various receiver designs in optical communication. We first study the concepts and basic modulation schemes of OFDM systems in IM/DD optical communication. They can be generally classified into three categories: direct-current-biased optical OFDM (DCO-OFDM), non-DC-biased optical OFDM, and hybrid optical OFDM. We will elaborate the system models and explain the validity of some fancy designs in those systems through analysis. Second, we investigate the preliminary receivers of those OFDM systems. Besides, we will propose a new receiver that is capable of improving the detection performance based on the inherent signal structures of the specific transmitted signal. Third, the spectral efficiencies and computational complexities of different systems and receivers are analyzed and compared. Finally, the bit error rate (BER) performance of different systems is compared through computer simulations to give the reader a whole picture of different candidate OFDM systems in IM/DD optical communication.

2. OFDM principles

This section gives a brief introduction on the principles of OFDM in radio frequency communication which serves as a basis for further reading. The baseband diagram of OFDM is shown in **Figure 1**. At the transmitter side, coded information bits are first mapped to symbols through digital modulation such as pulse amplitude modulation (PAM), quadrature amplitude modulation (QAM), and phase shift keying (PSK). Typically, complex-valued QAM modulation is used

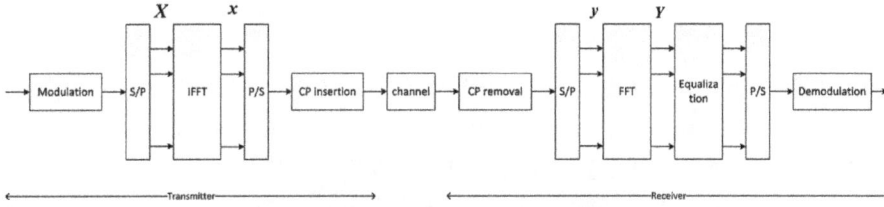

Figure 1. The "IFFT" module at the receiver side should be "FFT" module.

in OFDM. Then, the modulated symbols are divided into multiple groups and each group consists of N-modulated symbols, defined as $X = [X(0), X(1), ..., X(N-1)]^T$, where the group index is omitted here for simplicity. In OFDM, each modulated symbol $X(k)$ is loaded on a *subcarrier* with center frequency $\frac{2\pi}{N}k$ and there are N subcarriers in total. All the symbols are transmitted on their subcarriers simultaneously. Mathematically, this is equivalent to transform the vector X by an N-point inverse fast Fourier transform (IFFT) module, resulting in a new vector $x = [x(0), x(1), ..., x(N-1)]^T$, that is,

$$x(n) = \frac{1}{\sqrt{N}} \sum_{k=0}^{N-1} X(k) e^{j\frac{2\pi}{N}kn}. \tag{1}$$

In OFDM, X is generally considered as *frequency* domain signal and x is viewed as *time*-domain signal. In addition, x is typically called an *OFDM symbol*, which is different with the *modulated symbol* aforementioned. Finally, x is appended at its head with a cyclic prefix (CP), which is just the copy of the last few samples of x. In general, the length of CP is no smaller than the length of transmission channel to avoid inter-symbol interference (ISI) between adjacent OFDM symbols.

Assuming the impulse response of the multipath channel is denoted by $h = [h(0), h(1), ..., h(L-1)]^T$; then, the received signal at the receiver after channel transmission is given by

$$r(n) = h(n) * x_c(n) + z(n), \tag{2}$$

where $x_c(n)$ is the CP-appended version of the time-domain transmitted signal, $z(n)$ is the additive white Gaussian noise (AWGN) with zero mean, and * denotes linear convolution. The receiver first removes the CP parts of received signal, which results in a new vector y of length N, which can be rewritten as

$$y(n) = h(n) \otimes x(n) + z(n), \tag{3}$$

where $\otimes$ denotes cyclic convolution. We can see that due to the insertion and removal of CP, the linear convolution is now transformed to a cyclic one, which would be beneficial for equalization, as will be shown in the following text.

From signal processing theory, cyclic convolution in time domain is equivalent to product in frequency domain. Based on this fact, by defining $Y(k)$, $H(k)$, and $Z(k)$ as the N-point fast Fourier transform of $y(n)$, $h(n)$, and $z(n)$, respectively, one has

$$Y(k) = H(k)X(k) + Z(k), k = 0, 1, ..., N - 1. \tag{4}$$

As we can see, each frequency-domain symbol $X(k)$ is transmitted as if in a *flat* channel of response $H(k)$ and different symbols transmit in different subchannels (subcarriers) without interfering with each other. This greatly simplifies the equalization task. For example, both zero forcing (ZF) and minimum mean square error (MMSE) equalization can be performed to recover $X(k)$ which only involves single-tap equalizer per subcarrier:

$$\widehat{X}(k) = \begin{cases} \dfrac{Y(k)}{H(k)}, & ZF \\ \dfrac{H^*(k)Y(k)}{|H(k)|^2 + \sigma_n^2}, & MMSE \end{cases} \tag{5}$$

where σ_n^2 is the variance of noise. The processing in Eqs. (4) and (5) can be realized by performing FFT and per-subcarrier equalization, as shown in **Figure 1**.

3. Optical OFDM systems for IM/DD channel

3.1. Preliminaries

In IM/DD channel, there is a key difference with RF channel: the transmitted signal must be real and positive. This results from the fact that the intensity of light must be a real and positive quantity. Therefore, the structure for OFDM shown in **Figure 1** cannot be directly used in IM/DD optical channel. Necessary changes must be made instead. A common approach is to generate a real time-domain signal first. This can be realized by imposing *Hermitian symmetry* on the frequency domain signal **X**, which is defined as follows:

$$X(N - k) = X^*(k), k = 1, ..., N - 1, X(0) = X(N/2) = 0. \tag{6}$$

It can be easily shown that the IFFT of **X** having property Eq. (6) is a pure *real*-valued signal *x*. Based on this real signal, one can further generate a positive signal to drive the optical source by various means. Those resultant new OFDM systems are typically referred to as *optical OFDM* systems.

3.2. DC-biased optical OFDM

The most straightforward approach to generate a positive signal from a real signal is to impose a proper DC bias. In optical communication, the DC bias is typically chosen such that the mean value of the positive signal just lies on the center point of the linear range of the optical source. This system is called direct current-biased optical OFDM, or DCO-OFDM, whose transmitter is shown in **Figure 2**.

The clipping module shown in **Figure 2** is necessary. Since $x(n)$ is Gaussian distributed, it is possible that $x(n)$ plus a DC bias is still out of the linear range of the optical source. For example, if an LED accepts driving current within the range of $[a, b]$, where $0 \le a < b$, then the

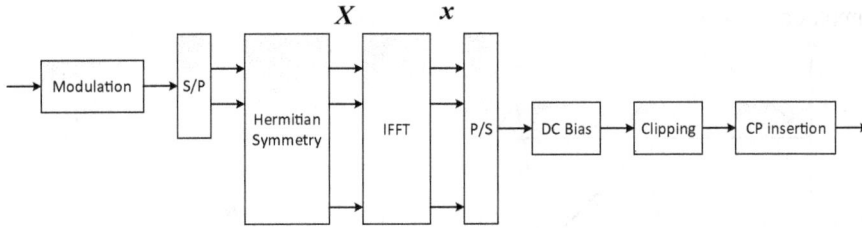

Figure 2. Diagram of DCO-OFDM transmitter.

clipping is needed to confine $x(n)$ plus a DC bias into this range. Otherwise, the LED could not be illumined due to under-driving or even be damaged due to over-driving.

3.3. Non-DC-biased optical OFDM

Besides DCO-OFDM, there are many forms of optical OFDM systems that are not relying on DC bias. The most famous ones are introduced in this subsection.

3.3.1. ACO-OFDM

Asymmetrically clipped optical OFDM (ACO-OFDM) is the most famous non-DC-biased optical OFDM system and has been extensively studied in literature [1]. The basic idea of ACO-OFDM is to generate an asymmetrically structured time-domain signal such that direct clipping at zero (without adding DC bias) is allowed without any information loss. To do so, the frequency-domain input symbol X has a special structure besides satisfying Hermitian symmetry. Specifically, the *odd* components of X contain useful information U but the *even* components of X are set to zeros. This is shown in **Figure 3**.

After IFFT, the time-domain signal x has an asymmetrical structure:

$$x(n) = -x\left(n + \frac{N}{2}\right), n = 0, 1, ..., N/2. \tag{7}$$

As shown in **Figure 4**, signal x can be directly clipped at zero without adding any DC bias, yet the information is kept after clipping thanks to the asymmetrical structure.

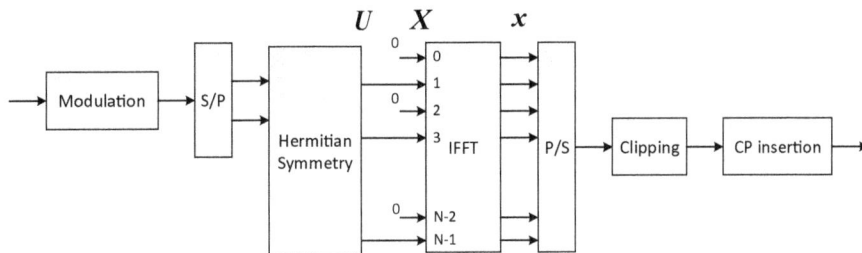

Figure 3. Diagram of ACO-OFDM transmitter.

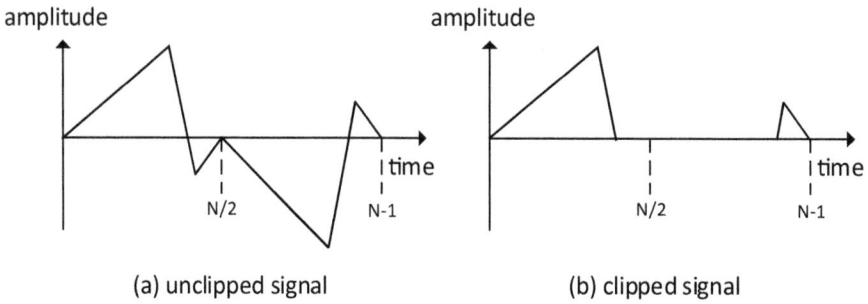

(a) unclipped signal   (b) clipped signal

Figure 4. Asymmetrical structure before and after clipping at zero in ACO-OFDM.

3.3.2. PAM-DMT

Pulse amplitude modulation-discrete multi-tone (PAM-DMT) is another non-DC-biased optical OFDM system [2]. It is similar to ACO-OFDM, in that direct clipping is used. However, the difference is that in PAM-DMT, only the imaginary part of subcarrier input X carries useful information U while the real part is set to zero, as shown in **Figure 5**. Note that PAM modulation should be used in PAM-DMT rather than QAM in DCO-OFDM and ACO-OFDM.

It can be easily shown that the resultant time-domain signal x also has an asymmetric structure but is slightly different with that of ACO-OFDM, which is shown in Eq. (8) and **Figure 6**:

$$x(n) = -x(N-n), n = 1, 2, ..., \frac{N}{2}, x(0) = x\left(\frac{N}{2}\right) = 0. \tag{8}$$

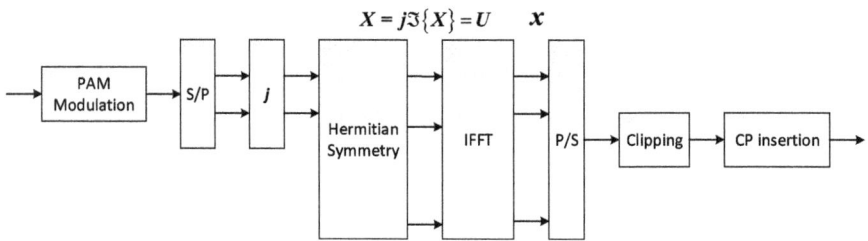

Figure 5. Diagram of PAM-DMT transmitter.

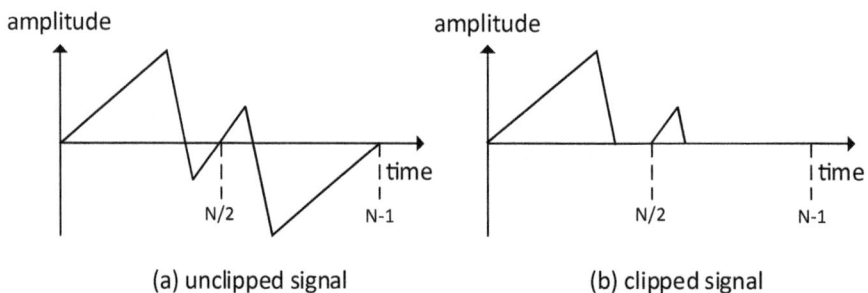

(a) unclipped signal   (b) clipped signal

Figure 6. Asymmetrical structure before and after clipping at zero in PAM-DMT.

3.3.3. Flip-OFDM

Both ACO-OFDM and PAM-DMT rely on specially designed signal structures on the frequency- and time-domain signals. In contrast, flip-OFDM employs a simpler way such that a general frequency-domain signal X without any fancy structure is accepted [3]. Instead, the real time-domain signal x, without any symmetry, is split into two parts: the first part only contains the samples of positive ones in x, the negative ones are set to zeros; the second part only contains the samples of negative ones, but with flipped signs, and leaves the positive ones as zeros. This is shown in **Figure 7**.

Mathematically, the first and second parts of the flipped signal are given by

$$x_1(n) = \frac{x(n) + |x(n)|}{2}, x_2(n) = \frac{-x(n) + |x(n)|}{2}, n = 0, 1, ..., N - 1. \tag{9}$$

After flip processing, the two signal parts are appended with CP, respectively, and are transmitted on channel consecutively. In some literature, flip-OFDM is also referred to as unipolar OFDM (U-OFDM).

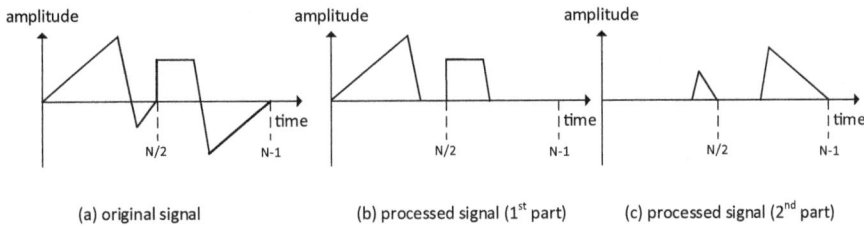

(a) original signal   (b) processed signal (1st part)   (c) processed signal (2nd part)

Figure 7. Signal structure before and after flipping processing in flip-OFDM.

3.4. Duality of non-DC-biased optical OFDM systems

This section gives a brief introduction on the duality of the non-DC-biased optical OFDM systems. Based on this duality, many receiver design methods could be easily extended from one system to other systems.

For ACO-OFDM, the transmitted signal is given by

$$x_c = (x + |x|)/2, \tag{10}$$

and x can be written as $x = [x_1; -x_1]^T$, where x_1 is the first half of x. Therefore, the first and second halves of x_c can be written as

$$x_{c,1}(n) = \frac{x_1(n) + |x_1(n)|}{2}, x_{c,2}(n) = \frac{-x_1(n) + |x_1(n)|}{2}, \tag{11}$$

which is exactly the same as the model in Eq. (9) except that the size is changed from N to $N/2$.

For PAM-DMT, the transmitted signal is also given by Eq. (10). However, the unclipped signal x is slightly different, that is, $x = [x_1; x_2]^T$, where $x_2 = -Jx_1$ with J being a matrix whose anti-diagonal elements are 1s and other elements are all 0s. Now, the first and second halves of the transmitted signal are given by

$$x_{c,1} = \frac{x_1 + |x_1|}{2}, x_{c,2} = \frac{-Jx_1 + J|x_1|}{2}. \tag{12}$$

However, $JJ = I$; therefore, by defining $x_{c,3} = Jx_{c,2}$, we have

$$x_{c,1} = \frac{x_1 + |x_1|}{2}, x_{c,3} = \frac{-x_1 + |x_1|}{2}. \tag{13}$$

Now, Eq. (13) is exactly the same as Eqs. (9) and (11).

Therefore, we can see that ACO-OFDM, PAM-DMT, and flip-OFDM essentially share the same signal structure and there is a duality between them. Based on this fact, the receivers designed for one system can be readily extended to other systems with simple substitution of variables.

3.5. Hybrid systems

Beside the basic forms of DC and non-DC-biased optical OFDM systems, there also exist some hybrid ones where multiple basic systems are superimposed in a specially designed fashion. In general, hybrid systems can be further classified into three categories.

3.5.1. Hybrid optical OFDM based on DCO-OFDM and a non-DC-biased one

A representative for this kind of system is ADO-OFDM, which combines DCO-OFDM and ACO-OFDM in a special way [4]. Specifically, in ACO-OFDM, the useful data are only loaded on the odd subcarriers, as illustrated in **Figure 3**, the even subcarriers are forced to be zero. After clipping in time domain, the clipping noise only falls onto even subcarriers and the odd subcarriers are not affected by the clipping noise. At the receiver side, the data could be recovered by using only the odd subcarriers. With the recovered data, one can further perfectly reconstruct the clipping noise on even subcarriers. Therefore, in ACO-OFDM, the even subcarriers can be exploited to load more data, which is the basic idea of ADO-OFDM.

In ADO-OFDM, the odd subcarriers are performed exactly the same as the ACO-OFDM. For the even subcarriers, a modified DCO-OFDM signal is generated, in which only the even subcarriers are used. Then, the signals generated from ACO-OFDM and DCO-OFDM are added together to obtain the ADO-OFDM signal. At the receiver side, ACO-OFDM signal, which is on odd subcarriers, are first detected. Then, the clipping noise on even subcarriers is estimated and subtracted. After that, the even subcarriers contain only DCO-OFDM signal, which is finally decoded.

3.5.2. Hybrid optical OFDM based on two different non-DC-biased ones

HACO-OFDM, or hybrid ACO-OFDM, combines ACO-OFDM and PAM-DMT in one system [5]. The basic idea is similar to that of ADO-OFDM, that is, the odd subcarriers are used for ACO-OFDM transmission while the even subcarriers are used for PAM-DMT transmission.

At the receiver side, interference cancellation is used for even subcarriers before decoding PAM-DMT signal. An alternative form for HACO-OFDM is also proposed [6].

3.5.3. Hybrid optical OFDM based on a same non-DC-biased one

Another form of hybrid optical OFDM is to superimpose multiple blocks of signals from a same non-DC-biased OFDM system. For example, enhanced unipolar OFDM (eU-OFDM) involves multiple blocks of signals from flip-OFDM [7]. **Figure 8** shows its time-domain structure with three layers [8], where the symbols $x_{i,j}^{+}$ and $x_{i,j}^{-}$ denote, respectively, the positive and flipped negative parts of the original j-th bipolar signal $x_{i,j}$ from layer-i. Each layer is just a repetition of flip-OFDM time-domain signal. We can see that for the first layer, four normal flip-OFDM symbols are used. For the second layer, two normal flip-OFDM symbols are repeated two times. For the third layer, one normal flip-OFDM symbol is repeated four times. Then, all the time symbols from three layers are added for transmission.

The receiver detection is very simple. The first layer is firstly decoded using normal subtraction. The second and third layers do not interfere in this procedure due to perfect self-cancellation. After the first layer is decoded, its impact is subtracted from the received signal. Then, the second layer is decoded subsequently. Then, the second layer signal is subtracted from the received signal and the third layer is decoded finally.

Besides eU-OFDM, the overlapping of ACO-OFDM or PAM-DMT is also proposed in literatures [9–11]. They all share similar idea with eU-OFDM and the receiver is based on layer-by-layer decoding.

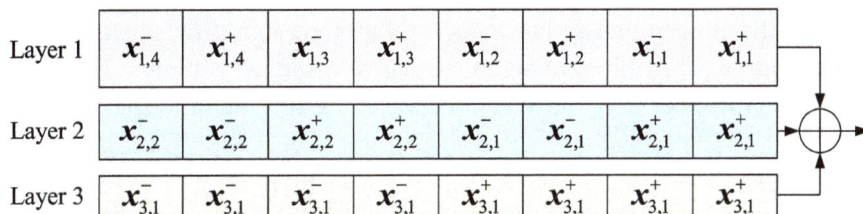

Layer 1	$x_{1,4}^{-}$	$x_{1,4}^{+}$	$x_{1,3}^{-}$	$x_{1,3}^{+}$	$x_{1,2}^{-}$	$x_{1,2}^{+}$	$x_{1,1}^{-}$	$x_{1,1}^{+}$
Layer 2	$x_{2,2}^{-}$	$x_{2,2}^{-}$	$x_{2,2}^{+}$	$x_{2,2}^{+}$	$x_{2,1}^{-}$	$x_{2,1}^{-}$	$x_{2,1}^{+}$	$x_{2,1}^{+}$
Layer 3	$x_{3,1}^{-}$	$x_{3,1}^{-}$	$x_{3,1}^{-}$	$x_{3,1}^{-}$	$x_{3,1}^{+}$	$x_{3,1}^{+}$	$x_{3,1}^{+}$	$x_{3,1}^{+}$

Figure 8. Illustration of a three-layer eU-OFDM time-domain signal components.

4. Receiver design for optical OFDM systems

In this section, we investigate the receiver design for optical OFDM systems. For DCO-OFDM, the receiver is straightforward. For non-DC-biased optical OFDM systems, there exist multiple candidate receivers which will be detailed later. For hybrid systems, since they are constructed mainly based on non-DC-biased ones, the receivers designed for non-DC-biased systems are also applicable for hybrid systems. Moreover, as the duality between non-DC-biased systems, in this chapter we focus on ACO-OFDM. We review the basic receiver design, diversity combining receiver design, and propose an iterative receiver design. All the formulations are based on an AWGN channel model but the results can be readily extended to multipath channels.

4.1. Basic receiver

The basic receiver for ACO-OFDM is very simple. The received signal after CP removal is given by

$$y(n) = x_c(n) + z(n). \tag{14}$$

In frequency domain, one has

$$Y(k) = X_c(k) + Z(k). \tag{15}$$

As proved by Ref. [1], the odd subcarriers of $X_c(k)$ is related to $X(k)$ by

$$X_c(k) = \frac{1}{2}X(k), \text{ for odd } k. \tag{16}$$

Therefore, the data $\mathbf{U}$, which is only on odd subcarriers of $\mathbf{X}$, can be recovered from $Y(k)$ by

$$\widehat{U}(k) = \begin{cases} 2Y(2k+1), & ZF \\ \dfrac{2Y(2k+1)}{1 + \sigma_n^2}, & MMSE \end{cases} \quad k = 0, 1, \ldots, \frac{N}{2} - 1 \tag{17}$$

The receiver diagram is shown in **Figure 9**.

4.2. Diversity combining receiver

The basic receiver only utilizes the odd subcarriers for signal detection. The even subcarriers, bearing pure clipping noise, are simply discarded. Therefore, half of the received power is wasted. However, the clipping noise has a special inherent signal structure that is dependent on the unclipped signal. This inherent signal structure could be exploited for better detection performance. This is the basic idea of diversity combining receiver and the iterative receiver.

To unveil the relationship between the clipping noise and the unclipped signal, we rewrite the clipped signal as

$$x_c(n) = \frac{1}{2}[x(n) + |x(n)|], \forall n. \tag{18}$$

Based on Eqs. (16) and (18), one can see that the clipping noise falls only onto the even subcarriers and has a special form as

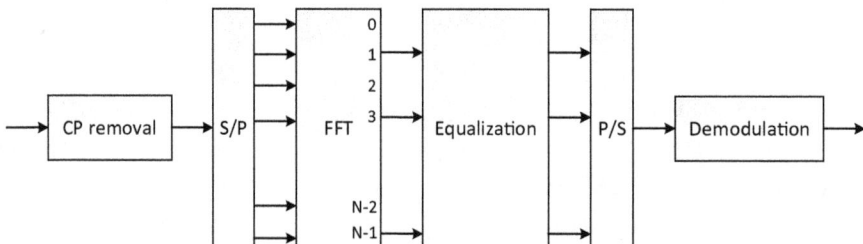

Figure 9. Diagram of the basic receiver for ACO-OFDM.

$$X_c(k) = \frac{1}{2} FFT\{|x(n)|\}, \text{ for even } k, \tag{19}$$

which says that the clipping noise on the even subcarriers is just the FFT of $|x(n)|$. Thus, we can generate two signals based on $Y(k)$: the first one is $\mathbf{Y_0} = [0 \quad Y(1) \quad 0 \quad Y(3)...0 \quad Y(N-1)]^T$, and the second one is $\mathbf{Y_e} = [Y(0) \quad 0 \quad Y(2) \quad 0...Y(N-2) \quad 0]^T$. Denoting their time-domain signal by $\mathbf{y_0}$ and $\mathbf{y_e}$, respectively, we have

$$y_0 = \frac{1}{2}x + z_0, \tag{20}$$

$$y_e = \frac{1}{2}|x| + z_e. \tag{21}$$

Eq. (21) is obtained from Eqs. (16) and (19). A new signal $y_c(n)$ is generated based on Eqs. (20) and (21):

$$y_c(n) = \begin{cases} y_e(n), \text{ if } & y_0(n) \geq 0, \\ -y_e(n), \text{ if } & y_0(n) < 0. \end{cases} \tag{22}$$

Now, we get two branches of signals that are related to $x(n)$. Thus, diversity combining technique could be used to enhance the detection performance. The diversity combining is performed by

$$r(n) = ay_0(n) + (1-a)y_c(n), \forall n. \tag{23}$$

The combining coefficient a is usually a bit larger than 0.5 since $y_c(n)$ is not as accurate as $y_0(n)$ [12]. Based on $r(n)$, the data could be estimated just as in the basic receiver. The whole procedure of diversity combining receiver is shown in **Figure 10**.

A pairwise-ML receiver based on noise cancellation has been proposed in [13]. It has been proved that this receiver is in fact a special case of diversity combining receiver with $a = 0.5$ [14].

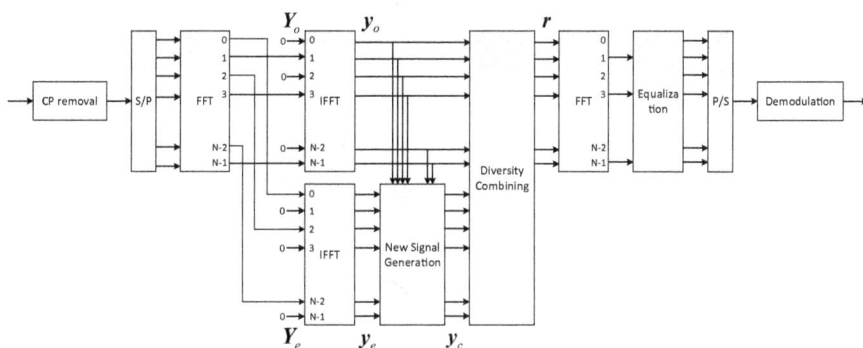

Figure 10. Diagram of diversity combining receiver for ACO-OFDM.

4.3. Proposed iterative receiver

Although the diversity combining receiver exploits the signal on even subcarriers, it is not performed in an optimal way, resulting in possible performance loss compared to optimal joint detection. Here, we propose an iterative receiver that has a better way to exploit the signal on even subcarriers [14]. The basic idea is to re-estimate the modulated data in a complete mathematical model at each iteration. At the very first iteration, the basic receiver is used for initialization. The details are given as follows.

Define an N by $N/2$ matrix P_0 whose odd rows form an identity matrix and even rows are all zeros. Similarly, define another N by $N/2$ matrix P_e whose even rows form an identity matrix and odd rows are all zeros. Then, we have

$$X = P_0 U, U = P_0{}^T X = P_0{}^T P_0 U. \tag{24}$$

In addition, based on Eqs. (16) and (19), we have

$$P_0{}^T X_c = \frac{1}{2} U, \tag{25}$$

$$P_e{}^T X_c = \frac{1}{2} W|x| = \frac{1}{2} WSx = \frac{1}{2} WSW^H X = \frac{1}{2} WSW^H P_0 U, \tag{26}$$

where W is the FFT matrix and S is a diagonal matrix whose entries on the main diagonal are the signs of x. Then, Eq. (15) could be decomposed to

$$Y_{odd} \overset{\Delta}{=} P_0{}^T Y = P_0{}^T X_c + P_0{}^T Z = \frac{1}{2} U + Z_{odd}, \tag{27}$$

$$Y_{even} \overset{\Delta}{=} P_e{}^T Y = P_e{}^T X_c + P_e{}^T Z = \frac{1}{2} P_e{}^T WSW^H P_0 U + Z_{even}. \tag{28}$$

Collecting Eqs. (27) and (28) together, we have

$$R = QU + V, \tag{29}$$

where

$$R = \begin{bmatrix} Y_{odd} \\ Y_{even} \end{bmatrix}, Q = \frac{1}{2} \begin{bmatrix} I \\ P_e{}^T WSW^H P_0 U \end{bmatrix}, V = \begin{bmatrix} Z_{odd} \\ Z_{even} \end{bmatrix}, \tag{30}$$

where I denotes the identity matrix of proper size. Eq. (29) is a complete signal model of the received signal with respect to the information symbol. Therefore, based on Eq. (29), we can readily get the estimation of U by

$$\widehat{u} = \begin{cases} (Q^H Q)^{-1} Q^H R, & ZF \\ (Q^H Q + \sigma_n{}^2 I)^{-1} Q^H R, & MMSE \end{cases} \tag{31}$$

Note that Q is in fact a function of U due to the component S. However, at each iteration, we assume Q is known by substituting $\widehat{U}$ from previous iteration to get Q. At the first iteration, the basic receiver is used to get the initial estimate of $\widehat{U}$.

5. Performance comparison

5.1. Spectral efficiency

In this section, we give a comparison on the spectral efficiencies of different optical OFDM systems.

For DCO-OFDM, each OFDM symbol only contains $N/2$ information-bearing complex-modulated symbols. Assuming the modulation order is M, then the spectral efficiency of DCO-OFDM is given by

$$\eta_{DCO-OFDM} = \frac{1}{2}\log_2 M \text{ bits/s/Hz.} \tag{32}$$

For all the non-DC-biased optical OFDM systems, as redundancy (zeros) is used in either frequency domain (ACO-OFDM and PAM-DMT) or time expansion is used (flip-OFDM), the spectral efficiencies are only half of DCO-OFDM:

$$\eta_{ACO-OFDM} = \eta_{PAM-DMT} = \eta_{Flip-OFDM} = \frac{1}{4}\log_2 M \text{ bits/s/Hz.} \tag{33}$$

For hybrid systems, things are a little bit complex. There is no general expression but one has to analyze the specific system.

For ADO-OFDM, in addition to a conventional ACO-OFDM transmission on odd subcarriers, a half-rate DCO-OFDM is used on even subcarriers. Therefore, its spectral efficiency is

$$\eta_{ADO-OFDM} = \eta_{ACO-OFDM} + \frac{1}{2}\eta_{DCO-OFDM} = \frac{1}{2}\log_2 M \text{ bits/s/Hz.} \tag{34}$$

For HACO-OFDM, a similar expression could be obtained:

$$\eta_{HACO-OFDM} = \eta_{ACO-OFDM} + \frac{1}{2}\eta_{PAM-DMT} = \frac{3}{8}\log_2 M \text{ bits/s/Hz.} \tag{35}$$

For eU-OFDM, the spectral efficiency depends on the number of layers. For an L-layer system, the spectral efficiency is given by

$$\eta_{eU-OFDM}(L) = \sum_{l=1}^{L}\left(\frac{1}{2}\right)^{l-1}\eta_{Flip-OFDM} = 2\left(1 - \frac{1}{2^{L-1}}\right)\eta_{Flip-OFDM}. \tag{36}$$

When L approaches infinity, we have the upper bound of spectral efficiency:

$$\eta_{eU-OFDM} = 2\eta_{Flip-OFDM} = \frac{1}{2}\log_2 M \text{ bits/s/Hz.} \tag{37}$$

There is a tradeoff between the spectral efficiency and decoding complexity with respect to L: a larger L means the spectral efficiency is closer to its upper bound but the decoding complexity, mainly from signal cancellation for decoded layers, will increase linearly with L. In practice, a fairly small L is desired to achieve a balance between the spectral efficiency and decoding complexity, say, for example, $L = 5$ is a good choice. In fact, when $L = 5$, we have

$$\eta_{eU-OFDM}(5) = 2\left(1 - \frac{1}{2^{5-1}}\right)\eta_{Flip-OFDM} = 0.94\eta_{eU-OFDM'} \tag{38}$$

which shows that the spectral efficiency is very close to the upper bound.

On summarizing, we can see that DCO-OFDM has the highest spectral efficiency. However, its power efficiency is not very good due to the non-information-bearing DC. On the contrary, the non-DC-biased optical OFDM systems have better power efficiency due to the elimination of DC offset but their spectral efficiency is only half of that of DCO-OFDM. The hybrid systems, especially ADO-OFDM and eU-OFDM, have better balance between the spectral efficiency and power efficiency. With those facts, one can choose a proper implementation form in practice under specific communication requirement and constraint.

5.2. Receiver complexity

The computational complexity of different receivers is analyzed here using the order notation. For the basic receiver, the main computation burden is the FFT and equalization, which have complexities of $O(N\log_2 N)$ and $O(N)$, respectively. In total, it is just $O(N\log_2 N)$. For the diversity combining receiver, it involves finite number of FFT/IFFT and a final equalization. Therefore, although it is more complex than the basic receiver, there is no difference when considering the order notation, that is, it is still $O(N\log_2 N)$. For the proposed iterative receiver, its main computation burden is the matrix inversion of Eq. (31) and this operation should be repeated at each iteration. Thus, the total complexity is in the order of $O(TN^3)$, where T is the total number of iterations. As we can see, this receiver is the most complicated one among the three receivers. However, as will be shown later, its performance is the best and can be far better than the other two. Thus, it is acceptable considering the performance gains. In addition, with the rapid development of modern signal-processing hardware, the computation burden will not be a limiting factor for these small-scale computations.

6. Simulations

In this section, we compare the average uncoded BER performance of different receivers in VLC channels through simulations. The channels are generated using the method in [15] with the following configurations: an empty room of size $8 \times 6 \times 4$ m with reflection coefficients 0.8, 0.8, and 0.3 for the ceiling, the walls, and the floor, respectively; LEDs are used as the optical source and they are attached 0.1-m below the ceiling. The photodetectors (PDs) are 1 m above the floor with an 80° of field of view (FOV). Both line-of-sight (LOS) and nonline-of-sight (NLOS) channels are tested (the LEDs point straight downward and upward, respectively). Multiple LEDs and PDs are used to enhance the performance and robustness of the communication link, resulting in a multiple-input multiple-output (MIMO) channel model. The receiver design methods described in Section 4 can be easily extended to this channel model by using vector notation. For each MIMO channel realization, the positions of the LEDs and the photodetectors are randomly drawn from their corresponding plains and the channels are normalized to have power $N_R N_T$, where N_R and N_T denote the number of PDs and LEDs, respectively. The ill-conditioned channels are rejected for fair comparison. The number of subcarriers is $N = 64$. In the legend of figures, "conventional" denotes the basic receiver, "pairwise ML" denotes the receiver in [13], which is a special case of the diversity combining receiver. "Lower bound" denotes the ideal curve of the proposed receiver with perfect estimation of matrix Q. In all receivers, MMSE equalization is used.

Figure 11 shows a sample view of the impulse response of the LOS and NLOS channels with a sampling rate of 300 MHz. It can be seen that the LOS channel is more like a delta function but NLOS channel has relatively longer delay spread, which means it can be viewed as a multipath channel.

First, we compare the BER performance in single-input single-output (SISO) channel. **Figure 12** shows the performance with modulation order $M = 64$. It can be seen that in LOS channel, the

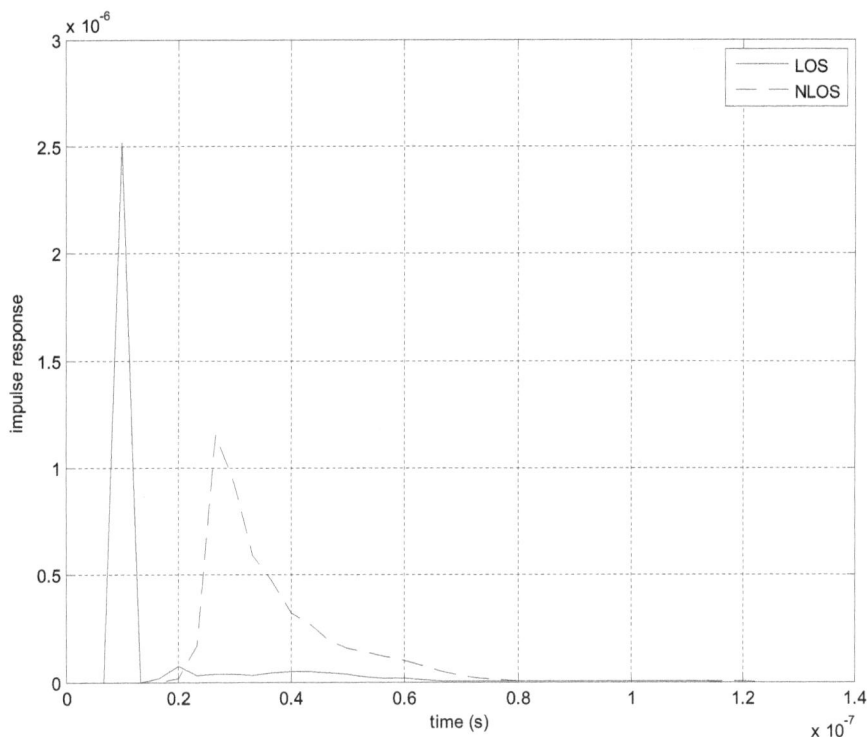

Figure 11. A sample view of LOS and NLOS channels with 300-MHz sampling rate.

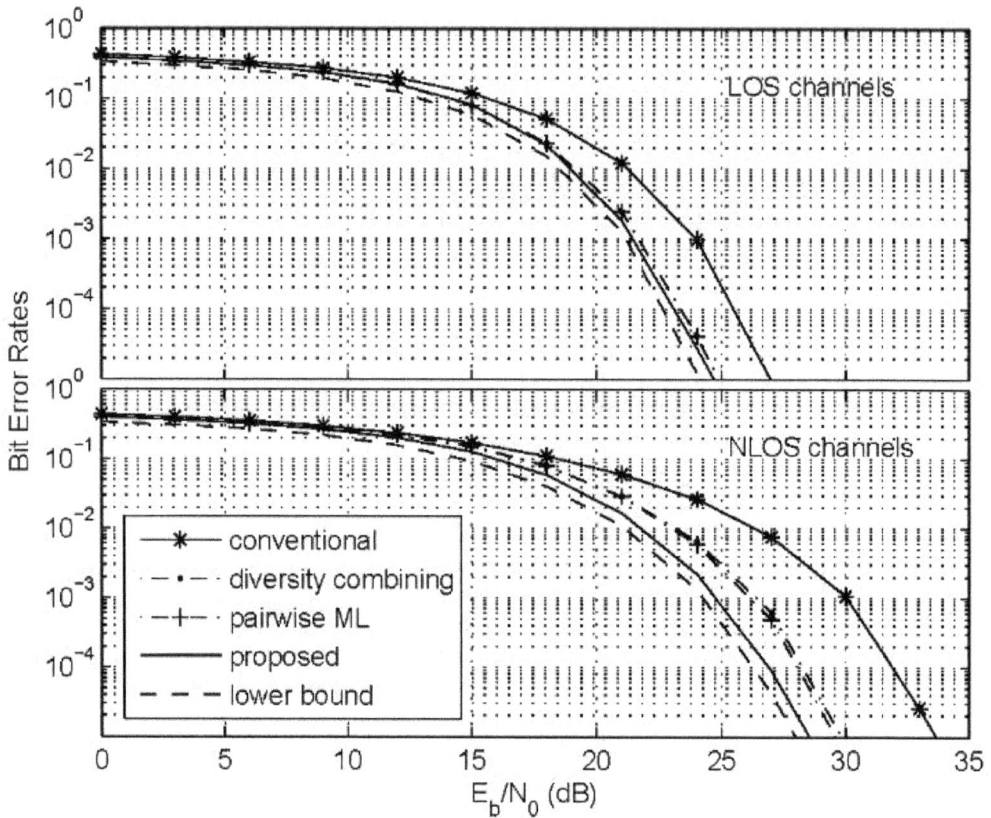

Figure 12. BER comparison of SISO ACO-OFDM using modulation sizes of $M = 64$.

diversity combining receiver and the proposed iterative receiver have similar performance and are much better than the basic receiver. In addition, their performance gap to the lower bound is limited. However, in NLOS channel, things are different: the proposed iterative receiver has the best performance. Compared to the basic receiver, its performance gain is more than 10 dB at high SNR range, which is significant. Even compared to the diversity combining receiver, 1-dB gain could be observed.

Now, we turn to MIMO channels. **Figure 13** shows the BER performance of a 4×4 MIMO ACO-OFDM using modulation sizes of $M = 16$. It can be seen that the performance of different receivers has similar behavior as the SISO case. However, compared to the SISO case, the performance gain of the proposed receiver is even larger: compared to the diversity combining receiver, it is more than 6 dB at high SNR regime; compared to the basic receiver, it is much more than 10 dB. Nonetheless, there is still a fairly large performance gap between the proposed receiver and its lower bound, which indicates that more advanced signal processing at the receiver side is desired in the future.

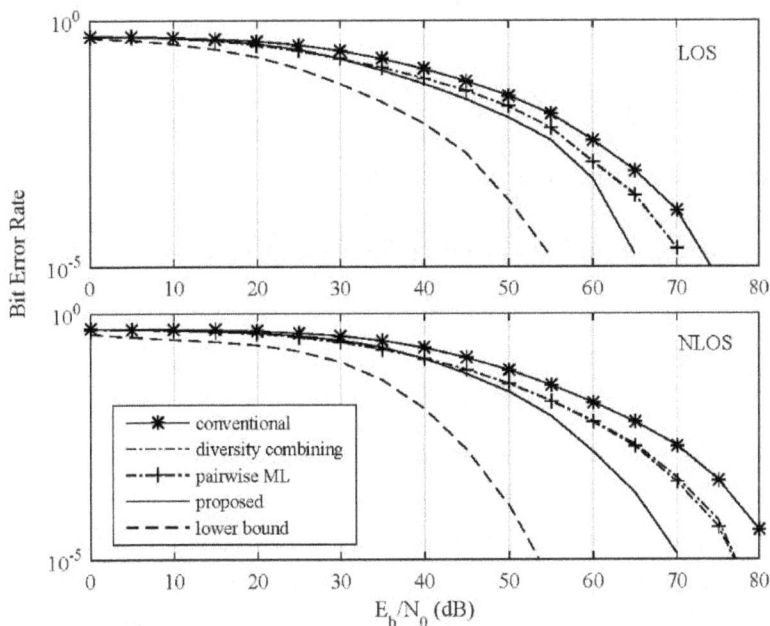

Figure 13. BER comparison of 4×4 MIMO ACO-OFDM using modulation sizes of $M = 16$.

7. Conclusions

In this research, we have investigated various forms of optical OFDM systems that are suitable for IM/DD optical channel. Different receivers are described with a proposed iterative one. Spectral efficiencies, computational complexities, as well as BER performance in LOS and NLOS channels of different systems and receivers are given. It is found that DCO-OFDM is more spectrally efficient than the non-DC-biased systems. The hybrid systems achieve a better tradeoff between the spectral efficiency and power efficiency. The proposed iterative receiver has the highest complexity but is far superior than other receivers, especially the basic receiver. Those results reveal the potential of OFDM systems in IM/DD channels for optical communication.

Acknowledgements

This work is supported by Southeast University 3-Category Academic Programs Project ("Optical Wireless Communication" and "Software-Defined Radio") and Top-notch Academic Programs Project of Jiangsu Higher Education Institutions (no. PPZY2015A035).

Author details

Jian Dang*, Liang Wu and Zaichen Zhang

*Address all correspondence to: newwanda@seu.edu.cn

School of Information Science and Engineering, Southeast University, Nanjing, P.R. China

References

[1] J. Armstrong, A. J. Lowery. Mint: Power efficient optical OFDM. Electronics Letters. 2006; **42**(6): 370–372. DOI: 10.1049/el:20063636.

[2] S. C. J. Lee, S. Randel, F. Breyer, A. M. J. Koonen. Mint: PAM-DMT for intensity-modulated and direct-detection optical communication systems. IEEE Photonics Technology Letters. 2009;**21**(23):1749–1751. DOI: 10.1109/LPT.2009.2032663.

[3] N. Fernando, Y. Hong, E. Viterbo. Mint: Flip-OFDM for unipolar communication systems. IEEE Transactions on Communications. 2012;**60**(12):3726–3733. DOI: 10.1109/TCOMM.2012.082712.110812.

[4] S. D. Dissanayake, K. Panta, J. Armstrong. A novel technique to simultaneously transmit ACO-OFDM and DCO-OFDM in IM/DD systems. In: IEEE GLOBECOM Workshops (GC Wkshps); 5–9 December 2011; Houston, Texas. IEEE; 2012. p. 782–786. DOI: 10.1109/GLOCOMW.2011.6162561.

[5] B. Ranjha, M. Kavehrad. Mint: Hybrid asymmetrically clipped OFDM-based IM/DD optical wireless system. Journal of Optical Communications and Networking. 2014;**6**(4):387–396. DOI: 10.1364/JOCN.6.000387.

[6] Q. Wang, Z. Wang, L. Dai. Mint: Asymmetrical hybrid optical OFDM for visible light communications with dimming control. IEEE Photonics Technology Letters. 2015;**27**(9):974–977. DOI: 10.1109/LPT.2015.2404972.

[7] D. Tsonev, S. Videv, H. Haas. Mint: Unlocking spectral efficiency in intensity modulation and direct detection systems. IEEE Journal on Selected Areas in Communications. 2015;**33**(9):1758–1770. DOI: 10.1109/JSAC.2015.2432530.

[8] J. Dang, Z. Zhang, L. Wu. Mint: Improving the power efficiency of enhanced unipolar OFDM for optical wireless communication. Electronics Letters. 2015;**51**(21):1681–1683. DOI: 10.1049/el.2015.2024.

[9] A. J. Lowery. Mint: Enhanced asymmetrically-clipped optical OFDM. Mathematics. 2015;**40**(1):36–40. DOI: 10.1364/OE.24.003950.

[10] M. S. Islim, D. Tsonev, H. Haas. On the superposition modulation for OFDM-based optical wireless communication. In: IEEE Global Conference on Signal and Information

Processing (Global SIP); 14–16 December 2015; Orlando, Florida, USA. IEEE; 2016. p. 1022–1026. DOI: 10.1109/GlobalSIP.2015.7418352.

[11] M. S. Islim, H. Haas. Mint: Augmenting the spectral efficiency of enhanced PAM-DMT-based optical wireless communications. Optics Express. 2016;24(11):11932–11949. DOI: 10.1364/OE.24.011932.

[12] L. Chen, B. Krongold, J. Evans. Diversity combining for asymmetrically clipped optical OFDM in IM/DD channels. In: IEEE Global Telecommunications Conference; 30 November–4 December 2009; Honolulu, Hawaii, USA. IEEE; 2010. p. 1–6. DOI: 10.1109/GLOCOM.2009.5425293.

[13] K. Asadzadeh, A. Dabbo, S. Hranilovic. Receiver design for asymmetrically clipped optical OFDM. In: IEEE GLOBECOM Workshops (GC Wkshps); 5–9 December 2011; Houston, Texas, USA. IEEE; 2012. p. 777–781. DOI: 10.1109/GLOCOMW.2011.6162559.

[14] J. Dang, Z. Zhang, L. Wu. Mint: A novel receiver for ACO-OFDM in visible light communication. IEEE Communications Letters. 2013;17(12):2320–2323. DOI: 10.1109/LCOMM.2013.111113.132223.

[15] J. B. Carruthers, P. Kannan. Mint: Iterative site-based modeling for wireless infrared channels. IEEE Transactions on Antennas Propagation. 2002;50(5):759–765. DOI: 10.1109/TAP.2002.1011244.

Permissions

All chapters in this book were first published in OFWC, by InTech Open; hereby published with permission under the Creative Commons Attribution License or equivalent. Every chapter published in this book has been scrutinized by our experts. Their significance has been extensively debated. The topics covered herein carry significant findings which will fuel the growth of the discipline. They may even be implemented as practical applications or may be referred to as a beginning point for another development.

The contributors of this book come from diverse backgrounds, making this book a truly international effort. This book will bring forth new frontiers with its revolutionizing research information and detailed analysis of the nascent developments around the world.

We would like to thank all the contributing authors for lending their expertise to make the book truly unique. They have played a crucial role in the development of this book. Without their invaluable contributions this book wouldn't have been possible. They have made vital efforts to compile up to date information on the varied aspects of this subject to make this book a valuable addition to the collection of many professionals and students.

This book was conceptualized with the vision of imparting up-to-date information and advanced data in this field. To ensure the same, a matchless editorial board was set up. Every individual on the board went through rigorous rounds of assessment to prove their worth. After which they invested a large part of their time researching and compiling the most relevant data for our readers.

The editorial board has been involved in producing this book since its inception. They have spent rigorous hours researching and exploring the diverse topics which have resulted in the successful publishing of this book. They have passed on their knowledge of decades through this book. To expedite this challenging task, the publisher supported the team at every step. A small team of assistant editors was also appointed to further simplify the editing procedure and attain best results for the readers.

Apart from the editorial board, the designing team has also invested a significant amount of their time in understanding the subject and creating the most relevant covers. They scrutinized every image to scout for the most suitable representation of the subject and create an appropriate cover for the book.

The publishing team has been an ardent support to the editorial, designing and production team. Their endless efforts to recruit the best for this project, has resulted in the accomplishment of this book. They are a veteran in the field of academics and their pool of knowledge is as vast as their experience in printing. Their expertise and guidance has proved useful at every step. Their uncompromising quality standards have made this book an exceptional effort. Their encouragement from time to time has been an inspiration for everyone.

The publisher and the editorial board hope that this book will prove to be a valuable piece of knowledge for researchers, students, practitioners and scholars across the globe.

List of Contributors

Isiaka Alimi, Ali Shahpari, Artur Sousa, Ricardo Ferreira, Paulo Monteiro and António Teixeira
Instituto de Telecomunicações, Department of Electronics, Telecommunications and Informatics, Universidade de Aveiro, Aveiro, Portugal

Waqas Ahmed Imtiaz, Affaq Qamar and Haider Ali
Abasyn University, Peshawar, Pakistan

Sevia M. Idrus
Sarhad University of Science and Information Technology, Peshawar, Pakistan

Javed Iqbal
Faculty of Electrical Engineering, Universiti Teknologi Malaysia, Malaysia

Waleed Tariq Sethi and Mohamed Himdi
Institute of Electronics and Telecommunications of Rennes University (IETR), University of Rennes, France
Institute of Electronics and Telecommunications of Rennes University (IETR), University of Rennes, France

Hamsakutty Vettikalladi
Electrical Engineering Department, King Saud University, Riyadh, Saudi Arabia

Habib Fathallah
Institute of Electronics and Telecommunications of Rennes University (IETR), University of Rennes, France

Bodhisattwa Gangopadhyay
Coriant Portugal, Amadora, Portugal

João Pedro
Coriant Portugal, Amadora, Portugal
Instituto de Telecomunicações, Lisboa, Portugal

Stefan Spälter
Coriant GmbH, Munich, Germany

Vjaceslavs Bobrovs, Sergejs Olonkins, Sandis Spolitis, Jurgis Porins and Girts Ivanovs
Riga Technical University, Institute of Telecommunications, Latvia

Mohammed T. Alresheedi
Department of Electrical Engineering, King Saud University, Riyadh, Saudi Arabia

Ahmed Taha Hussein and Jaafar M.H. Elmirghani
Department of Electrical Engineering, King Saud University, Riyadh, Saudi Arabia

Dmitry V. Svistunov
Peter-the-Great St.-Petersburg Polytechnic University, St.-Petersburg, Russia

Joao Batista Rosolem
CPqD—Research and Development Center in Telecommunications, Campinas, SP, Brazil

Tianhua Xu
Department of Electronic and Electrical Engineering, University College London, London, United Kingdom

Jian Dang, Liang Wu and Zaichen Zhang
School of Information Science and Engineering, Southeast University, Nanjing, P.R. China

Index

www.ingramcontent.com/pod-product-compliance
Lightning Source LLC
Chambersburg PA
CBHW061937190326
41458CB00009B/2764